Low Carbon Transition

Low-carbon transition is a shift from an economy that depends heavily on fossil fuels to a sustainable, low-carbon energy economy. This book analyzes the role of renewables in driving the low-carbon transition in agriculture, explores the circular bio-based economy, examines policies and strategies designed to facilitate low-carbon transition in agriculture, and greenhouse gas emission trends in the European Union agriculture sector. It provides new knowledge and understanding about the impact of low-carbon energy transition in the agriculture sector, emphasizes the key role of renewable energy in a wide range of agricultural activities, and offers alternative sustainable solutions to current practices.

Features

- Discusses a novel approach on low-carbon transition that is not considered by the majority of studies.
- Emphasizes the urgent need to minimize the carbon and environmental footprint of the EU agriculture and food industries through low-carbon energy transition.
- Provides a theoretical background of sustainable agriculture and explains the decarbonization path of agriculture.
- Investigates the role of renewables, new technologies, business models, and practices in agriculture while assessing their socioeconomic and environmental effects.
- Presents a case study on the applications of low-carbon transition policies in selected EU member states and analyses in detail various implications.

This book is suitable for senior undergraduate and graduate students, professionals in agriculture, researchers, and policymakers interested in sustainable agriculture and renewable energy usage and their economics.

Low Carbon Transition
Sustainable Agriculture in the European Union

Dalia Streimikiene, Indre Siksnelyte-Butkiene, and Tomas Balezentis

CRC Press is an imprint of the
Taylor & Francis Group, an **informa** business

Designed cover image: © Stone36, Shutterstock

First edition published 2024
by CRC Press
2385 NW Executive Center Drive, Suite 320, Boca Raton FL 33431

and by CRC Press
4 Park Square, Milton Park, Abingdon, Oxon, OX14 4RN

CRC Press is an imprint of Taylor & Francis Group, LLC

© 2024 Dalia Streimikiene, Indre Siksnelyte-Butkiene, and Tomas Balezentis

Reasonable efforts have been made to publish reliable data and information, but the author and publisher cannot assume responsibility for the validity of all materials or the consequences of their use. The authors and publishers have attempted to trace the copyright holders of all material reproduced in this publication and apologize to copyright holders if permission to publish in this form has not been obtained. If any copyright material has not been acknowledged please write and let us know so we may rectify in any future reprint.

Except as permitted under U.S. Copyright Law, no part of this book may be reprinted, reproduced, transmitted, or utilized in any form by any electronic, mechanical, or other means, now known or hereafter invented, including photocopying, microfilming, and recording, or in any information storage or retrieval system, without written permission from the publishers.

For permission to photocopy or use material electronically from this work, access www.copyright.com or contact the Copyright Clearance Center, Inc. (CCC), 222 Rosewood Drive, Danvers, MA 01923, 978-750-8400. For works that are not available on CCC please contact mpkbookspermissions@tandf.co.uk

Trademark notice: Product or corporate names may be trademarks or registered trademarks and are used only for identification and explanation without intent to infringe.

Library of Congress Cataloging-in-Publication Data
Names: Streimikiene, Dalia, author. | Siksnelyte-Butkiene, Indre, author. | Balezentis, Tomas, author.
Title: Low carbon transition : sustainable agriculture in the European Union / Dalia Streimikiene, Indre Siksnelyte-Butkiene, and Tomas Balezentis.
Description: First edition. | Boca Raton, FL : CRC Press, 2024. | Includes bibliographical references.
Identifiers: LCCN 2023055056 (print) | LCCN 2023055057 (ebook) | ISBN 9781032607900 (hardback) | ISBN 9781032607924 (paperback) | ISBN 9781003460589 (ebook)
Subjects: LCSH: Sustainable agriculture—European Union countries. | Energy transition—European Union countries.
Classification: LCC S452 .S77 2024 (print) | LCC S452 (ebook) | DDC 630.2/086094—dc23/eng/20240310
LC record available at https://lccn.loc.gov/2023055056
LC ebook record available at https://lccn.loc.gov/2023055057

ISBN: 978-1-032-60790-0 (hbk)
ISBN: 978-1-032-60792-4 (pbk)
ISBN: 978-1-003-46058-9 (ebk)

DOI: 10.1201/9781003460589

Typeset in Times
by codeMantra

Contents

About the Authors viii
Preface ix

Introduction 1

Chapter 1 Low-Carbon Transition 3

Dalia Streimikiene

    1.1   International Political Support to Climate Change
           Mitigation and Low-Carbon Transition 3
    1.2   Barriers of Low-Carbon Transition and Policies
           to Promote Low-Carbon Transition with Emphasis
           on Agriculture 16
    1.3   Green Deal Policy Areas with Emphasis on
           Agriculture 29
    1.4   Cooperation for the Transition to a Carbon-Neutral
           Society among Various Actors and Sectors
           with Emphasis for Agriculture 42
    1.5   Conclusions and Recommendations 58
    References 60

Chapter 2 GHG Emission and Energy Use in the EU Agriculture 77

Tomas Balezentis

    2.1   Introduction 77
    2.2   GHG Emission in the EU and Its Agriculture 78
    2.3   Energy Consumption and GHG Emission
           in the EU Agriculture 81
    2.4   Use of Renewables in EU Agriculture 87
    2.5   Conclusions 94
    References 96

Chapter 3 Sustainable Agriculture 97

Dalia Streimikiene

    3.1   Agriculture and Sustainable Development 97
    3.2   Sustainable Agriculture Development Concepts 101
           3.2.1   Sustainable Agriculture Dimensions 101

		3.2.2	Obstacles Preventing Sustainable Agriculture Development 108
		3.2.3	Challenges of Agricultural Sustainability 110
	3.3	Climate Smart Agriculture 113
		3.3.1	Climate-Smart Agriculture: Principles and Objectives 115
		3.3.2	Concept and Agroecological Characteristics of Traditional Agriculture 115
		3.3.3	Alternative Practices for Climate Change Mitigation in Traditional Agriculture 116
		3.3.4	Traditional Agriculture: A Sustainable Adaptation Strategy to Climate Change 117
	3.4	Sustainability Indicators for Agriculture 118
		3.4.1	Soil Quality 120
		3.4.2	Water Conservation 121
		3.4.3	Preservation of Biodiversity 122
		3.4.4	Use of Agrochemicals 123
		3.4.5	Energy Conservation and Energy Efficiency 125
		3.4.6	Footprint of Carbon 126
		3.4.7	Economic Viability 127
		3.4.8	Social Prosperity 128
		3.4.9	Climate Change Resilience 129
		3.4.10	Certification and Standards 131
	3.5	MCDM-based Agricultural Sustainability Measurements 133
		3.5.1	Agricultural Sustainability Indicators Frameworks 133
		3.5.2	Multi-Criteria Decision Aiding Models for Sustainability Assessment in Agriculture 136
	3.6	Conclusions 138
	References 138

Chapter 4 Use of Renewables in Agriculture 153

Dalia Streimikiene

4.1	Renewable Energy Production in Agriculture 153
4.2	Renewable Energy Usage in Agriculture 157
4.3	Renewable Energy Sources and Circular Economy in Agriculture 163
4.4	Renewables – The Primary Drivers of Agriculture's Low-Carbon Transition 174
4.5	Conclusions 181
References 181

Contents vii

Chapter 5 The Role of Agricultural Business in Low-Carbon
Transition 193

Dalia Streimikiene

5.1 Green Growth and Agricultural Business 193
5.2 Business in Low-Carbon Energy Transition 207
    5.2.1 Definition of Low-Carbon Energy 207
5.3 Green Agricultural Initiatives 219
    5.3.1 The Importance of Green Agricultural
          Initiatives 223
5.4 Corporate Social Responsibility in Agriculture 236
    5.4.1 Definition of CSR in Agriculture 236
5.5 Conclusions 245
References 246

Chapter 6 Assessment of Low-Carbon Transition Policy
in Agriculture 261

Indre Siksnelyte-Butkiene

6.1 Low-Carbon Energy Transition Challenges
    to Agriculture 261
6.2 Policies to Promote Low-Carbon Energy
    Transition in Agriculture 265
6.3 Multi-Criteria Decision Making (MCDM)
    Instruments for Sustainable Agriculture
    Decision Making 273
    6.3.1 Application of MCDM Methods for
          Sustainable Agriculture Decision Making 273
    6.3.2 Comparison of the Most Popular MCDM
          Techniques 284
    6.3.3 Guidelines for Criteria Selection and the
          Whole Process of the Agriculture Sustainability
          Assessment 288
6.4 The Assessment of Success of Low-Carbon
    Transition in Agriculture: A Case Study
    of the EU Countries 292
    6.4.1 Selection of Indicators 293
    6.4.2 General Region Overview 294
    6.4.3 Evaluation Tool 298
    6.4.4 Data for Comparative Assessment 299
    6.4.5 Results 299
6.5 Conclusions 303
References 303

Index 313

About the Authors

Prof. Dr. Dalia Streimikiene is a Full Professor at Vilnius University, Kaunas faculty and Leading Researcher at Lithuanian Centre for Social Sciences, Institute of Economics and Rural Development. Her main areas of research are sustainable development, sustainability assessment, and corporate social responsibility. She is an author of more than 200 papers in international journals referred at WoS. H index 40. She is invited Editor for several special issues of *Sustainability* journal and Editor-in-Chief of the international journal *Transformations in Business and Economics* referred at WoS.

Dr. Indre Siksnelyte-Butkiene is a Senior Researcher at Vilnius University, Kaunas Faculty and Senior Researcher at Lithuanian Centre for Social Sciences, Institute of Economics and Rural Development. She has been conducting scientific research for more than ten years in the field of sustainable energy development. The main areas of her research are sustainable energy development, sustainability assessment, multi-criteria decision-making, and energy policy. She is co-author of two monographs and more than 40 scientific articles.

Prof. Dr. Tomas Balezentis is a Full Professor at Vilnius University and Leading Researcher at Lithuanian Centre for Social Sciences, Institute of Economics and Rural Development. He holds PhD degrees from Vilnius University and the University of Copenhagen. He has received the Lithuanian National Scientific Prize. His research focuses on efficiency and productivity analysis, agricultural economics, and energy economics. He has published in such outlets as *Energy Economics, Energy Policy, China Economic Review*, and the *European Journal of Operational Research* among others. He is the author of more than 200 papers in international journals referred at WoS. H index 39.

Preface

The European Green Deal sets target for European Union to become climate-neutral society by 2050. It is not possible to achieve this target without new actions to reduce greenhouse gas (GHG) emissions in all sectors including agriculture. Recent global crises linked to Covid-19 and Russian–Ukrainian war create a lot of new challenges for low-carbon transition in agriculture sector. The Farm to Fork Strategy included in European Green Deal setting the strategic goals or a fair, healthy and environmentally friendly food system. In addition, agriculture and food systems including various processes of manufacturing, processing, and transportation of food are significant sources of GHG emissions including other negative environmental effects. It is necessary to stress that agriculture sector was responsible for 10% (2020) of total GHG emissions in EU excluding GHG emissions from fuel combustion in agriculture. Therefore, there is an urgent need to reduce GHG emissions and other negative environmental impacts in this sector of key importance through introduction of new technologies, practices, and business models in agriculture and food sector. The main EU goal in agriculture sector is to minimize the carbon and environmental footprint of the EU agriculture and food system, to increase resilience and agility of this sector, and to secure sustainable food supply at competitive prices by tapping into new opportunities provided by low-carbon transition pathways.

All main energy and land-use systems which are highly interlinked, such as power, industry, mobility, buildings, agriculture, forestry, and waste, have to undertake huge transformations to achieve carbon neutral society. All climate change mitigation actions in these systems must be harmonized across all sectors and layouts. The most important climate change mitigation actions include replacing fossil fuels by zero-carbon energy sources such as renewables or hydrogen; increasing energy production and use efficiency, employing circular economy principles; reducing production and consumption of carbon-intensive products, utilization, boosting carbon sinks, and utilizing carbon capture and storage. Agriculture and food sector is a main contributor to methane (CH4) coming from livestock digestion and manure management and nitrous oxide (N_2O) originating mainly from agricultural soils due to nitrogen fertilization processes also energy consumption in agriculture is an important sources of GHG emissions. Therefore, these climate change mitigation actions are very relevant agriculture sector though the decision makers were placing priorities on other sectors. In addition, the specifics of agriculture and food sector due to key importance of renewable energy production and use in agriculture by ensuring circular economy principles have been ignored for some time.

The potential of renewable energy production and use in agriculture and circular economy principles based on bioenergy potential of this sector are basically unexploited by farmers. As advanced bio-refineries producing bio-fertilisers, bio-chemicals and bioenergy can provide many opportunities for the low-carbon transition in EU including creation of new jobs and generating additional income and welfare for farmers and other employees in agriculture and food sectors. Farms have many options to achieve GHG emission reductions from livestock by initiating the renewable energy production from agricultural waste and residues in anaerobic digesters. Farmers also have the potential to collect and generate biogas from other agricultural waste and residues including food waste, wastewater, and sewage. Carbon sequestration and new farming practices linked to carbon removal provide new business opportunities for farmers and foresters by delivering also climate neutrality objectives. Farms have good conditions for investing in solar and wind parks as they require land and also provide a good source of income for farmers. Therefore, agriculture sector even can help other sectors to ensure decarbonization and to decarbonize the entire food chain.

Therefore, there is a huge need for new studies exploring the decarbonization options for agriculture and food sectors and assessing the socioeconomic and environmental effects of these options for agriculture sector. In addition, it is important to define how large might be the differences in these impacts of low-carbon transition of agriculture in different countries. The socioeconomic implications of low-carbon transition can vary across countries and regions therefore it is important to assess and compare low-carbon transition impacts on sustainability of agriculture in different countries.

The aim of this book is to analyze the main implications of low-carbon transition to sustainable agriculture development in European Union under current global crisis conditions by providing detailed analysis of policies and strategies aiming at low-carbon transition in agriculture, GHG mitigation and adaptations trends in EU agriculture sector and the impacts of low-carbon transition policies to sustainable agriculture and food systems development. For this purpose, the sustainable agriculture concepts, targets, measures, and indicators of sustainable agriculture development are analyzed and systematized, the importance of renewable energy and agricultural business in low-carbon transition and sustainable agriculture development is examined and the case study on sustainability assessment of low-carbon transition policies in agriculture is developed for Lithuania.

The book is constituted of six interconnected parts that sequentially reveal the main theoretical principles and policies of low-carbon transition and sustainable agriculture development, including concepts, targets, frameworks, indicators, and practical case study on assessment of low-carbon transition policies impacts on sustainable development of agriculture.

Chapter 1 provides the main theoretical issues of low-carbon transition including analysis of policies to ensure low-carbon transition in European Union agriculture and food sector.

In Chapter 2, the trends, sources, and structure of GHG emissions from agriculture in EU were analyzed and discussed during 1990–2022 year period.

In Chapter 3, sustainable agriculture concepts, goals, assessment frameworks, and indicator are scrutinized and systematized.

Chapter 4 examines the central role of renewable energy in low-carbon transition and delivering to sustainable agriculture development goals.

Chapter 5 analyzes the role of agriculture business and green agriculture initiatives in providing for low-carbon transition in agriculture sector.

Chapter 6 provides case study on assessment of low-carbon transition policies impacts on sustainable agriculture development in selected EU Member States.

Introduction

The aim of this book is to analyze the main implications of low-carbon transition to sustainable agriculture development in European Union by providing detailed analysis of policies and strategies aiming at low-carbon transition in agriculture, greenhouse gas (GHG) mitigation and adaptation trends in European Union (EU) agriculture sector, and the impacts of low-carbon transition policies to sustainable agriculture and food systems development. For this purpose, the sustainable agriculture concepts, targets, measures, and indicators of sustainable agriculture development are analyzed and systematized, the importance of renewable energy and agricultural business in low-carbon transition and sustainable agriculture development is examined, and the case study on sustainability assessment of low-carbon transition policies in agriculture is developed for selected EU member states.

Chapter 1 of the monograph introduces low-carbon transition concept by providing a theory for transition to carbon-neutral society. This chapter analyzes the international political support to climate change mitigation and low-carbon transition including major conventions and international agreements and their targets. The barriers of low-carbon transition and policies to promote low-carbon transition with emphasis on agriculture were analyzed and systematized. EU Green Deal policy areas are analyzed by emphasizing the targets set for agriculture. Cooperation for the transition to a carbon-neutral society among various actors and sectors and other initiatives relevant to agriculture are scrutinized by highlighting their importance for low-carbon transformations.

Chapter 2 surveys the dynamics and patterns in the GHG emission in the EU agriculture. The data from the United Nations Framework Convention on Climate Change (UNFCCC), Eurostat, and Farm Accountancy Data Network (FADN) are considered for the analysis. The trends, sources, and structure of GHG emissions from agriculture in EU is analyzed and discussed during 1990–2022 year period. The main drivers of GHG emissions from energy consumption in agriculture of EU member states are discussed.

Chapter 3 of the monograph analyzes sustainable agriculture concept based on sustainable development paradigm and discusses the main issue of sustainability of agriculture. Climate-smart agriculture concept is introduced and linkages with sustainable agriculture are investigated based on literature review. The barriers and enables of sustainable agriculture development are examined. The measurements of sustainable agriculture are analyzed and systematized. The single indicators and integrated frameworks of sustainable agriculture assessment are analyzed by revealing their strengths and shortages.

Chapter 4 of the monograph analyzes the role of renewables in ensuring sustainable agriculture development. The renewable energy production and its importance for achieving sustainable agriculture development are discussed. Renewable energy production and usage in agriculture can ensure implementation of cyclic economy principles and also allow to achieve low-carbon transition in agriculture as use of biomass linked to waste in agriculture can provide for sustainable energy usage in agriculture, reduction of waste, and GHG emissions in this sector. Agriculture sector is an important source of GHG emissions, and use of various agriculture waste for energy generation provides double benefit in terms of GHG emission reduction from agriculture and energy-related GHG emissions in this sector.

Chapter 5 of the book analyzes the role of green business and green agriculture initiatives in providing for low-carbon transition in European Union agriculture and food sector. The role of corporate social responsibility in agricultural business for supporting transition to low-carbon society is analyzed by providing good practices.

Chapter 6 presents case study on assessment of low-carbon transition impact on sustainable agriculture development in selected EU member states by applying multi-criteria decision aiding (MCDA) tools. Low-carbon energy transition challenges to sustainable agriculture are highlighted. The main policies to promote low-carbon energy transition in agriculture were selected and grouped for selected EU member state. The use of MCDM tools for assessment of sustainability and low-carbon transition success in agriculture is analyzed and the most suitable MCDA tools were selected for case study.

1 Low-Carbon Transition

Dalia Streimikiene

1.1 INTERNATIONAL POLITICAL SUPPORT TO CLIMATE CHANGE MITIGATION AND LOW-CARBON TRANSITION

Climate change is one of the most important challenges facing humanity. The effects of climate change have a significant negative impact on human health, the natural environment around us, and the economies of countries. Climate change is currently one of the most serious challenges facing planet Earth, as scientists unanimously agree that the planet is warming much faster than ever. Skeptics of the causes of climate change and climate change itself do not provide any evidence or even deny the fact that the temperature of the atmosphere is rising.

Climate change skeptics attempt to cast doubt on climate change itself as a phenomenon and only make factual claims that at some point in time, temperatures in one area or another are below or above normal (Škiudas et al., 2020). They also try to explain the phenomenon of climate change with the natural cyclical development of the natural environment. In this case, an attempt is made to prove that, for example, an unusually hot summer repeats itself every certain period of time and there is no need to worry, because the next period of the natural cycle will already be a cooler summer. It should be noted that undoubtedly the climate has changed and the temperature of the atmosphere has changed due to natural processes not related to human activity, such as volcanic eruptions and meteorite dust. At the same time, the concentration of carbon dioxide in the atmosphere has also changed over the millennia due to natural causes. However, there is no evidence that global warming can only be caused by natural changes in the natural environment (Trenberth & Fasullo, 2013; Marotzke & Forster, 2015). Scientists recommend to look at the analysis of the causes of climate change in a complex way, because the process of climate change itself is complex and includes different phenomena, areas of activity, and participants of this process.

After a detailed examination of the conducted studies, which are aimed at determining the causes of climate change, it can be recognized that, to date, most likely, not all the causes of climate change have been determined. The different effects of various external and internal factors on the climate are difficult to model, sometimes even impossible, because they are heterogeneous. Therefore, the global causes of climate change are often strengthened by regional ones and vice versa. The causes (factors) of regional and global climate change can either strengthen or weaken or neutralize each other.

It is also worth noting that climate fluctuations are long-term and slow enough, but many random climate change determinants can act in their background, which influence short-term climate changes. The latter are easy to spot but easily confused with global climate change phenomena.

It is necessary to mitigate the effects of climate change for several reasons:

1. **Environmental Reasons**: Climate change has many negative effects on the natural environment around us. Increased levels of atmospheric greenhouse gases (GHGs) such as carbon dioxide lead to global warming and climate change, which can have serious consequences for ecosystems, biodiversity, marine and aquatic resources, and air quality. Reducing the climate change process can reduce the negative impact on the environment.
2. **Social Causes**: Climate change can have significant impacts on human health and well-being. Increased heat, extreme weather, rising sea levels, and food supply problems can lead to food shortages, water shortages, economic problems, and increased social instability. Mitigating climate change can reduce these risks and ensure a safer, healthier, and more just society.
3. **Economic Causes**: Climate change can have significant economic impacts. Extreme weather conditions can cause production disruptions, infrastructure damage, and significant economic losses. However, mitigating climate change and transitioning to a sustainable, low-carbon economy can create new business opportunities, innovation, and jobs based on energy conservation, renewable energy sources, and green and sustainable development.

Factors promoting climate change mitigation can be singled out (Figure 1.1):

1. **Environmental Protection**: As mentioned, climate change is causing negative effects on our planet's environment. Increased GHG emissions contribute to rising temperatures, rising sea levels, extreme weather events, and loss of biodiversity. Acting on GHG emissions can reduce these negative impacts and contribute to environmental protection.
2. **Health and Well-being**: Climate change has a direct impact on human health and well-being. Rising levels of air pollution, extreme weather events, and food supply instability are just a few examples. Reducing GHG

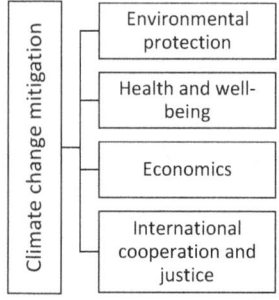

FIGURE 1.1 Drivers of climate change mitigation. (Created by authors.)

emissions can reduce these negative health factors and create a healthier and safer living environment.
3. **Economic Benefits**: Climate change mitigation can create economic benefits. Investments in renewable energy, energy efficiency, and other climate change mitigation measures drive job creation, technology development, and business innovation. In addition, reducing energy costs and dependence on carbon-emitting sources can achieve long-term economic sustainability.

Climate mitigation measures are actions taken to reduce GHG emissions and reduce the effects of climate change. These measures can cover various areas such as energy, transport, agriculture, forestry management, and construction. Climate mitigation measures play a crucial role in addressing the pressing challenge of climate change by curbing GHG emissions. They encompass a wide range of sectors, including energy, transport, agriculture, forestry management, and construction. In the energy sector, measures may involve transitioning to renewable energy sources, improving energy efficiency, and adopting cleaner technologies. In transportation, promoting public transport, electric vehicles (EVs), and sustainable urban planning are vital components of mitigation efforts. Additionally, sustainable agricultural practices, reforestation, and forest conservation are essential in the pursuit of climate mitigation goals. Furthermore, constructing energy-efficient buildings and employing low-carbon construction materials are also significant aspects of comprehensive mitigation strategies.

Climate change mitigation policy is associated with the Paris Agreement, which took place in 2015. In December, The Paris Agreement breaks new ground in international climate policy (Falkner, 2016). The Paris Climate Agreement aims to prevent the worst effects of climate change, with signatories agreeing to limit the rise in average global surface temperature to 2°C above pre-industrial levels. To achieve this goal, global GHG emissions would need to peak as soon as possible and then decline sharply.

The key basics of the Paris Agreement are as follows:

- **Long-Term Temperature Goal**: The primary aim of the agreement is to restrict the rise in global temperatures to significantly below 2°C above pre-industrial levels and to strive for limiting the temperature increase to 1.5°C. Achieving this goal is of utmost importance in preventing the most severe consequences of climate change.
- **Nationally Determined Contributions (NDCs)**: Each participating country is required to submit its own national climate action plan known as "Nationally Determined Contributions." These plans outline the country's efforts to reduce GHG emissions and adapt to climate change. The NDCs are intended to be ambitious and are subject to regular updates to increase their level of ambition over time.
- **Transparency and Accountability**: The Paris Agreement emphasizes the importance of transparency and accountability. Countries are expected to regularly report on their GHG emissions and progress toward meeting their

NDCs. This transparency is essential to track global progress and ensure that countries are fulfilling their commitments.
- **Climate Finance**: Developed countries have committed to providing financial support to developing nations to assist them in both mitigating their GHG emissions and adapting to the impacts of climate change. The agreement aims to mobilize significant funds from various sources to support climate action in developing countries.
- **Loss and Damage**: In the Paris Agreement, there is acknowledgment of the significance of tackling loss and damage resulting from the adverse impacts of climate change, with particular attention given to the most vulnerable countries and communities.
- **Global Stocktake**: The agreement mandates a global stocktake every five years to assess the collective progress toward achieving the agreement's goals. This process allows for course correction and increasing ambition based on the latest science and developments.
- **Cooperation and Support**: The Paris Agreement emphasizes the importance of international cooperation and support among countries to effectively tackle climate change. It encourages collaboration on technology development, capacity-building, and sharing best practices.

Rogel et al. (2016) conducted an analysis of the Paris climate change mitigation policy and presented the policy findings and its prospects and impacts.

The Intergovernmental Panel on Climate Change (IPCC) Sixth Assessment Report, published in 2021, states that changes in the earth's climate are now being observed in all regions of the world. Climate hazards are becoming increasingly complex and can affect crop yields and disrupt distribution networks and community livelihoods (IPCC, 2021).

Different academic papers present different perspectives on political support for climate change mitigation. Drews and van den Bergh (2016) provide empirical and experimental research on public attitudes toward climate policy, identifying socio-psychological factors, policy development, and contextual factors as key influences on policy support. Interactions among policies at diverse levels of governance have been explored and policies in the agriculture sector, nature conservation, and water sectors can provision or weaken adaptation to climate change (Urwin & Jordan, 2008). Norwegian researchers have analyzed a broad alliance of Norwegian civil society groups that have come together to make climate change mitigation a top priority in 2013 election campaign issue and found that although various groups joined the movement, their commitment was feeble (Nilsen et al., 2008). A large-scale UK study identifies barriers to UK public engagement on climate change, identifies individual and social barriers to engagement, and advocates for targeted information delivery and broader structural modification to enable peoples and communities to reduce carbon dependence (Lorenzoni et al., 2007). Many research papers indicate that political support for climate change mitigation is influenced by a variety of factors, including social and psychological factors, policy development, contextual factors, and barriers to engagement.

Different types of political support can have varying effectiveness for climate change mitigation. Bechtel and Scheve (2012) found that public support for global climate agreements depends on institutional design, with lower costs, fair distribution of costs, more countries involved, and small sanctions for non-compliance increasing support. Linde (2018) found that political communication can significantly influence public policy attitudes toward climate mitigation, but the effect varies across political contexts. Partisanship and party cues also affect public policy attitudes, even in a consensus context, according to Linde (2018). According to Nilsen's research in 2018, an extensive coalition of civil society groups focused on climate change mitigation might not be fully devoted to all political demands, suggesting that climate change mitigation could be considered a valence issue.

The shift toward a low-carbon economy plays a crucial role in mitigating climate change. This involves promoting the adoption of renewable energy sources, enhancing energy efficiency, and implementing innovative approaches in the transportation sector. These measures lead to a reduction in carbon dioxide emissions, actively contributing to climate change mitigation, and fostering the establishment of a sustainable economy in the long run. This allows us to reduce our dependence on fossil fuels and improve the state of the environment now and in the future.

The shift toward a low-carbon economy will result in significant structural transformations. Certain industries will experience a boost in their economic significance, whereas others, particularly those closely tied to the production and use of fossil fuels, will need to decrease their production levels (Semieniuk et al., 2020). Such a systemic shift could have major implications for the stability of financial systems through sudden asset revaluations, debt defaults, and bubbles.

The low-carbon transition refers to the global shift from an economy heavily reliant on carbon-intensive energy sources, such as fossil fuels, to one that primarily uses low-carbon or carbon-neutral alternatives. It is a response to the challenges posed by climate change and aims to mitigate GHG emissions to reduce their impact on the environment.

The low-carbon transition encompasses various sectors, including energy production, transportation, industry, agriculture, and buildings. Some scientists believe that heavy industry is the main source of CO_2 emissions and identify economic growth, urbanization, and energy efficiency as important factors in reducing emissions (Xu & Chen, 2021). Rissman et al. (2020) identified key supply side tools, such as energy efficiency, carbon capture, electrification, and zero-carbon hydrogen, as well as demand-side approaches, such as material-efficient design and circular economy interventions, that can achieve net zero industrial emissions. Sharmina et al. (2020) suggested that reducing demand for critical sectors, such as aviation, shipping, road freight, and industry, is necessary to meet emission reduction goals. In Gerres et al. (2019) research, essential avenues for reducing emissions were highlighted in every subsector of the energy-intensive industry. These included achieving decarbonization of low-temperature heat, adopting membrane technology in the (petro)-chemical industry, implementing carbon-neutral steelmaking processes, exploring alternative feedstock for cement production, and utilizing carbon capture and storage (CCS) methods. Overall, the papers suggest that a low-carbon transition requires a blend of supply side technologies and demand-side approaches across multiple sectors. It involves adopting cleaner and more sustainable energy sources, such as renewable energy (e.g., solar, wind, hydro, geothermal) and nuclear power, while reducing the reliance on fossil fuels. Additionally, it entails implementing energy efficiency measures and adopting technologies that minimize carbon emissions.

There are cutting-edge technologies accessible, providing opportunities to attain carbon neutrality and promote sustainable development. These technologies encompass renewable energy generation, transformation of food systems, valorization of waste, conservation of carbon sinks, and even carbon-negative production methods (Wang et al., 2021). Various technological solutions related to CCS can be used, which can significantly reduce CO_2 emissions (Olejnik & Sobiecka, 2017). However, the use of new technologies is associated with obstacles due to problematic to decarbonize services and procedures, counting likely technological keys and research and development (R&D) significances (Davis et al., 2018). There are currently several technologies that can reduce carbon emissions, including carbon sequestration, renewable energy generation, waste assessment, and CCS.

The transition to a low-carbon economy involves both policy changes and technological advancements. Governments around the world are implementing regulations, incentives, and carbon pricing mechanisms to promote the development and deployment of low-carbon technologies. These policies aim to drive funds in renewable energy infrastructure, promote energy efficiency in industries and buildings, and encourage sustainable practices in various sectors.

Technological advancements play a crucial role in facilitating the low-carbon transition. Researchers and innovators are developing new and improved technologies for energy generation, energy storage, and energy efficiency. For example, advancements in solar and wind technologies have made them more cost-effective and accessible, contributing to their rapid growth in recent years (Wilberforce et al., 2019; Wang, 2023). Similarly, developments in battery storage and smart grid systems are helping integrate intermittent renewable energy sources into the power grid more effectively (Kumar et al., 2023).

The low-carbon transition is a multi-faceted and complex process that requires collaboration among governments, businesses, communities, and individuals. It involves a combination of policy support, investment, technological innovation, and changes in behavior and consumption patterns. By transitioning to a low-carbon economy, societies aim to reduce GHG emissions, mitigate climate change, improve air quality, enhance energy security, and foster sustainable economic growth.

Political support for climate change mitigation and the low-carbon transition varies across countries and governments. Public opinion on climate change varies from country to country and is influenced by various factors:

- Views on climate change differ from views on other environmental issues, and political views play a significant role in determining views on complex questions such as climate change (Smith et al., 2017).
- In developing countries, public concern about climate change is higher, and vulnerability to climate risks has a strong influence on personal commitment and support for climate policy (Kim & Wolinsky-Nahmias, 2014).

- Richer and more educated countries are more aware of climate change, and perceived risk and human causes are also influenced by political orientation and vulnerability (Knight, 2016).
- Public preference for CCS technologies varies from country to country and is influenced by context (Ashworth et al., 2013).

Overall, public opinion about climate change is complex and influenced by a variety of factors, including political views, education, wealth, and vulnerability to climate change (Arıkan & Günay, 2021). Research on public attitudes toward climate change also distinguishes between individual and global threats. The influence of personal and biospheric threats on adapting to climate change can be identified. Personal threats are perceived as risks to oneself or one's family, while biospheric threats are seen as risks to other species and the overall health of ecosystems and the planet, resulting from environmental degradation (Arıkan & Günay, 2021; Helm et al., 2018). However, in recent years, there has been a growing recognition of the urgent need to address climate change, and many governments around the world have taken steps to support these efforts.

Here are some examples of political support for climate change mitigation and low-carbon transition (Figure 1.2):

1. **International Agreements**: The Paris Agreement, adopted in 2015, is a landmark global agreement that sets the framework for countries to combat climate change. It has been ratified by nearly all countries, demonstrating a high level of political support for climate action at the international level.

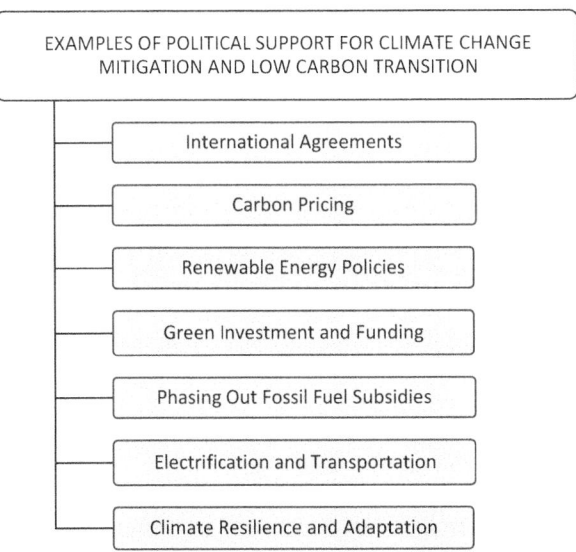

FIGURE 1.2 Examples of political support for climate change mitigation and low-carbon transition. (Created by authors.)

The Paris Agreement represents a novel approach to international climate policy, acknowledging the importance of domestic politics in addressing climate change. It grants countries the autonomy to determine their own level of ambition for climate change mitigation (Falkner, 2016). The agreement establishes a framework for voluntary pledges, encouraging international comparison and review, with the hope of enhancing global ambition through a mechanism of scrutiny and accountability (Falkner, 2016). Nevertheless, there are doubts about whether the treaty can effectively achieve the urgent goal of decarbonizing the global economy (Falkner, 2016). While the Paris Agreement marks a new beginning for the climate regime, its future direction and outcomes remain uncertain (Savaresi, 2016).

One of the noteworthy aspects of the Paris Agreement is its recognition of the intrinsic link between climate change actions, responses, and impacts with equitable access to sustainable development and the eradication of poverty (Seo, 2017).

Besides the Paris Agreement, several other international agreements and treaties are crucial in supporting climate change mitigation and the transition to a low-carbon economy. Some of the important ones include:

- *Kyoto Protocol.* The Kyoto Protocol was adopted in 1997 as an extension of the United Nations Framework Convention on Climate Change (UNFCCC). It established legally binding emission reduction targets for developed countries for the commitment period 2008–2012. Though it has been largely replaced by the Paris Agreement, it played a significant role in advancing international cooperation on climate change.
- *Montreal Protocol.* While primarily focused on protecting the ozone layer, the Montreal Protocol has also contributed to climate change mitigation. It phased out the production and consumption of substances responsible for ozone depletion, many of which were potent GHGs.
- *Kigali Amendment.* The Kigali Amendment is an extension of the Montreal Protocol, adopted in 2016. It aims to phase down the production and consumption of hydrofluorocarbons (HFCs), which are potent GHGs commonly used in refrigeration and air conditioning.
- *Sustainable Development Goals (SDGs).* While not exclusively focused on climate change, the SDGs, particularly Goal 13, emphasize the need for urgent action to combat climate change and its impacts. The SDGs provide a broader framework for global sustainability efforts, including climate mitigation.
- *The United Nations Convention on Biological Diversity (CBD).* Biodiversity conservation and climate change mitigation are interconnected. The CBD aims to conserve biodiversity, which also contributes to climate resilience and carbon sequestration.
- *The United Nations Convention to Combat Desertification (UNCCD).* Desertification and land degradation contribute to climate change. The UNCCD focuses on sustainable land management and plays a role in climate mitigation.

- *The International Solar Alliance (ISA).* The ISA is a treaty-based alliance of countries focused on promoting solar energy deployment globally. It aims to increase the use of solar energy and reduce reliance on fossil fuels.

    These agreements, along with the Paris Agreement, demonstrate the international community's efforts to address climate change through collective action and cooperation.

2. **Carbon Pricing**: Numerous nations have adopted carbon pricing mechanisms, including carbon taxes or cap-and-trade systems, to encourage the reduction of GHG emissions. These policies create economic incentives for industries and individuals to adopt low-carbon practices and technologies. It is important to evaluate the findings about the effectiveness of carbon pricing in reducing emissions.

    Carbon pricing has been found to have a limited effect on emissions, with most studies reporting total reductions of between 0% and 2% per year and limited average annual reductions under EU-emissions trading schemes (ETSs) (Green, 2021). Carbon taxes and emissions trading systems differ in many ways. Carbon taxes provide cost certainty, while emissions trading systems, in contrast, provide quantity certainty. In practice, the distinction between the two policies is sometimes blurred (Hepburn, 2006). Also, no empirical evidence is found that carbon pricing has led to the technological changes necessary for full decarbonization (Lilliestam et al., 2020). However, carbon pricing is important in the fight against climate change and can level the playing field between high- and low-carbon economic activities (Zechter et al., 2017) At the same time, carbon pricing can stimulate cost-effective climate change mitigation and innovation to address environmental challenges and can ensure that emissions are kept to target levels (Boyce, 2018).

    Successful implementation of carbon pricing measures depends on several key factors. These factors can vary depending on the specific context and political landscape of a country or region. Here are some common factors that contribute to the successful implementation of carbon pricing:

    - *Clear Policy Objectives.* Having well-defined and transparent policy objectives for carbon pricing is essential. These objectives should include specific targets for emissions reductions and a clear vision of how carbon pricing fits into the overall climate change mitigation strategy.
    - *Broad Political Support.* Carbon pricing is more likely to be successful when there is broad political support across different parties and stakeholders. Building consensus among policymakers, businesses, and civil society can help overcome resistance and ensure a stable policy framework.
    - *Credible and Predictable Carbon Price.* A credible and predictable carbon price provides certainty to businesses and investors, encouraging them to make long-term decisions that align with low-carbon strategies. A stable carbon price signal helps drive innovation and investments in cleaner technologies.

- *Gradual Phasing-In.* Gradually phasing in carbon pricing measures allows industries and consumers time to adjust to the new economic incentives. It helps avoid abrupt shocks to the economy and allows for a smoother transition.
- *Revenues and Use of Funds.* Designing a clear plan for the revenues generated from carbon pricing is crucial. Reinvesting the funds into renewable energy projects, sustainable infrastructure, or supporting vulnerable communities can enhance public support and demonstrate the positive impact of the policy.
- *International Cooperation.* Carbon pricing measures can be more effective when implemented in coordination with other countries or regions. International cooperation can help prevent carbon leakage (where emissions-intensive industries relocate to regions without carbon pricing) and promote a level playing field.
- *Monitoring, Reporting, and Verification.* Implementing a robust system to monitor emissions, report progress, and verify compliance with carbon pricing regulations is crucial. This transparency enhances the credibility and effectiveness of the policy.
- *Addressing Social Equity.* Carbon pricing can lead to higher costs for certain industries and consumers. Addressing social equity concerns and designing policies that protect vulnerable groups can garner broader public support.
- *Flexibility and Adaptability.* Carbon pricing policies should be designed to be flexible and adaptable to changing circumstances and evolving climate goals. Regular reviews and adjustments based on real-world outcomes and new data are essential for continuous improvement.
- *Public Awareness and Engagement.* Building public awareness and engaging citizens in discussions about carbon pricing can create a more informed and supportive public. Education campaigns can help dispel myths and misinformation and promote understanding of the benefits of carbon pricing.

Carbon pricing has distributional effects on different groups of people. The impact of carbon pricing on households depends on their income and spending patterns, and policymaking and the treatment of other government policies also influence the distributional impact (Rausch et al., 2011). A meta-analysis of empirical studies also shows that the distributional outcomes of carbon pricing vary depending on the country's income level and the sector targeted by the policy (Ohlendorf et al., 2020). It can be seen that carbon pricing has incremental distributional effects in low- and middle-income countries with per capita income below $15,000 per year (Dorband et al., 2019) and that this policy can be effective in reducing emissions, but distributional effects emerge equity issues that can be addressed by global dividends financed by carbon revenues (Boyce 2018).

3. **Renewable Energy Policies**: There are different types of renewable energy policies that aim to promote the development and deployment of renewable energy sources.

According to Beck and Martinot's (2004) study, there are six categories of policies that impact the advancement of renewable energy. These encompass policies for promoting renewable energy, encouraging transportation biofuels, reducing emissions, restructuring electricity systems, supporting distributed generation, and facilitating rural electrification. However, renewable energy policies are currently very diverse and mostly aim to remove various barriers to renewable energy development. These can include cost barriers, regulatory and institutional barriers, and barriers related to the perceived risks of renewable energy. Governments worldwide have introduced various policies to promote the deployment of renewable energy sources. This includes setting renewable energy targets, providing subsidies and incentives for renewable energy projects (Tryndina et al., 2022), and implementing feed-in tariffs to encourage the development of clean energy infrastructure. Renewable energy targets, which must be set based on energy system and resource costs (Keay, 2013), have both benefits and challenges.

Renewable energy sources provide various benefits, such as ensuring secure energy supply, granting access to energy, promoting social and economic progress, aiding in climate change mitigation, and lessening environmental and health repercussions (Owusu & Asumadu-Sarkodie, 2016). However, the sustainability of renewable energy sources faces hindrances, such as market failures, limited information availability, difficulties in accessing raw materials, and managing daily carbon footprints (Owusu & Asumadu-Sarkodie, 2016; Spillias et al., 2020). Renewable energy targets affect different EU member states and households in different income quintiles, and it has been found that renewable energy targets for electricity generation can make net exporters and net importers worse off (Landis & Heindl, 2019).
4. **Green Investment and Funding**: Governments are increasingly channeling financial resources toward climate change mitigation and low-carbon transition. This includes establishing green investment funds, providing grants and loans for clean energy projects, and supporting R&D of innovative technologies to reduce emissions.

The global green investment market has been growing rapidly in recent years (Naqvi et al., 2021). Pham (2016) found that since 2007 the green bond market grew by 50% annually, and in 2014 the issue of green bonds amounted to 36.6 billion and other macroeconomic and institutional factors are driving the growing issuance of green bonds that will finance climate and sustainability investments in the future. Between 2002 and 2017, green investment in China witnessed remarkable growth, soaring from 118.56 billion Chinese yuan to 950.86 billion Chinese yuan, representing an increase of over seven times (Sheng et al., 2021). This surge in green investment has emerged as the primary impetus behind the energy sector's development in

China. The surge is attributed to several factors, including robust economic growth, a stable financial system that supports low-interest rates, and soaring fuel prices, all of which are fueling the momentum toward ecological investment.

Implementing clean energy projects presents several challenges. In particular, there is the problem of scale and storage of clean energy, which can lead to economic failures and dependence on traditional energy sources (Bonnici et al., 2021). Another challenge is the availability, affordability, and efficiency of clean energy, including energy poverty, demand-supply mismatch, and great dependence on non-renewable energy sources (Singh & Ru, 2022).

In order to reduce the ever-increasing levels of pollution, promising innovative emission-reduction technologies are being developed. This could include the use of flue gas scrubbers and district cooling systems in combined heat and power generation to reduce emissions and save energy (Auvinen et al., 2020). The study identifies essential supply technologies, encompassing energy efficiency, carbon capture, electrification, carbon-free hydrogen, and other promising technologies tailored to the most environmentally detrimental industries (Rissman et al., 2020). Demand-side approaches include material efficiency, circular economy interventions, and low-carbon labeling.

5. **Phasing Out Fossil Fuel Subsidies**: Some countries have initiated efforts to phase out subsidies for fossil fuels, recognizing that such subsidies encourage the consumption of carbon-intensive energy sources and undermine climate change mitigation efforts. Redirecting these subsidies toward clean energy can accelerate the low-carbon transition.

   Phasing out fossil fuel subsidies can have both positive and negative effects on the economy, as phasing out subsidies could reduce GHG emissions and have short-term economic benefits, but it could also slow down the transition to renewable energy if other policies are not implemented. In addition, phasing out subsidies could improve macroeconomic performance, reduce inequality, and provide fiscal space for renewable energy policies (Monasterolo & Raberto, 2019). However, the economic impact of removing fossil fuel subsidies depends on the specific policies and context in which they are implemented.

   Phasing out fossil fuel subsidies is likely to have a limited impact on reducing emissions worldwide, except in high-income oil and gas exporting regions (Jewell et al., 2018). However, phasing out subsidies in high-income countries could help improve macroeconomic performance, reduce inequality, and support sustainable renewable energy policies (Monasterolo & Raberto, 2019). Li and Sun (2018) argue that removing fossil fuel subsidies alone will not achieve CO_2 reduction, as low-emission fuels would be replaced by carbon-intensive ones, and capital and labor would be replaced by energy.

6. **Electrification and Transportation**: Governments are promoting the electrification of transportation to reduce reliance on fossil fuels. This includes supporting the development of EV infrastructure, providing incentives for EV adoption, and investing in public transportation systems.

Electrification of transport has several advantages. Zou (2023) found that EVs have lower emissions, higher energy system efficiency, and lower costs compared to internal combustion and hydrogen fuel cell vehicles. However, lithium-ion batteries in EVs still have some limitations, such as limited driving distance and inability to meet the needs of use in extreme environments and found that the use of electric buses instead of diesel or compressed natural gas (CNG) in urban transit has positive environmental benefits, and the environmental benefits of operating an electric bus fleet are about $65 million per year in Los Angeles and over $10 million per year in six other MSAs (Holland et al., 2020). The advent of EVs has brought about substantial effects in various domains, particularly concerning the power grid. However, by implementing suitable charge management strategies, this challenge can be effectively addressed. Furthermore, integrating EVs into the smart grid presents numerous opportunities, especially when considering vehicle-to-grid technology as a means to tackle the issue of renewable energy intermittency.

7. **Climate Resilience and Adaptation**: Governments are also recognizing the importance of building resilience to climate change impacts. They are implementing policies and funding initiatives to support climate adaptation measures, such as improving infrastructure, enhancing disaster preparedness, and protecting vulnerable communities.

Climate adaptation measures can have multiple benefits. Investments in agricultural adaptation measures can have mitigation benefits that are inexpensive compared to many activities with the primary goal of mitigating climate change (Lobell et al., 2013). Owen (2020) classified 110 implemented adaptation initiatives and demonstrated their actual effectiveness levels. The study also identified shared characteristics among effective adaptation initiatives, which include the reduction of risk and vulnerability, the enhancement of resilient social systems, environmental improvements, increased economic resources, and advancements in governance and institutions. Successful adaptation to climate change requires effectiveness, efficiency, equity, and legitimacy, and each of these decision-making elements is implicit in currently formulated socio-economic future scenarios, both in terms of emission trajectories and adaptation. Adaptation can be effective in removing much of the potential damage from moderate climate change, but not short-term climate change (Adger et al., 2005; Massetti & Mendelsohn, 2018). Climate adaptation measures can have a range of benefits, including mitigation benefits, risk reduction, resilience building, environmental improvement, economic resource enhancement and management, and institution strengthening.

It's important to note that political support for climate change mitigation and low-carbon transition can vary, and some governments may be more proactive than others. However, overall, there is a global trend toward greater recognition and support for addressing climate change and transitioning to a low-carbon economy

1.2 BARRIERS OF LOW-CARBON TRANSITION AND POLICIES TO PROMOTE LOW-CARBON TRANSITION WITH EMPHASIS ON AGRICULTURE

The transition to a low-carbon economy has become a pressing global priority as the world grapples with the challenges of climate change and the need to mitigate its impact. The energy sector is the most successful in reducing GHG emissions, particularly through fuel switching and increasing penetration of renewable energy sources (Lamb et al., 2021). Lamb et al. (2021) found that energy systems in Europe and North America have decarbonized on average due to fuel substitution and increasing penetration of renewables. Gerber (2013) also found that the livestock sector is a significant emitter of GHGs but has the potential to significantly reduce emissions. However, Lamb et al. (2021) noted that emissions growth in the industrial, building, and transport sectors, particularly in East Asia, South Asia, and Southeast Asia, was driven by high demand for materials, floor space, energy services, and travel.

Meanwhile, emissions from the transport sector are stable or increasing (Lamb et al., 2021; Greene & Schäfer, 2003). Some countries have made sustained emissions reductions, particularly in the energy systems sector, suggesting that climate change mitigation is possible even with very different mitigation measures (Lamb et al., 2021). While several sectors contribute to GHG emissions, agriculture, in particular, holds significant potential for reducing its carbon footprint and promoting sustainable practices. Insights into sustainable practices in the agricultural sector can be found in research papers. According to Dicks' et al. (2018) research, there are 18 farm management practices in the UK that hold significant promise for sustainable intensification. These practices encompass precision farming and animal health diagnostics, among others. Syahruddin and Kalchschmidt (2012), Kalchschmidt and Syahruddin (2011) both delve into the topic of sustainable supply chain management (SSCM) within the agricultural sector, underscoring the significance of tackling environmental degradation, food safety considerations, and social and sustainability matters. Middelberg (2013) focuses on the challenges facing South Africa's agricultural sector, including the need to provide food for a growing population while addressing climate change. In most cases, it is recommended to develop an integrated strategy and take into account sustainable agricultural practices such as organic and sustainable agriculture.

Sustainable agriculture can have social benefits. Social benefits such as human health, employment, democratic participation, resilience, biological and cultural diversity, equity, and ethics can be provided by diversified farming systems (DFS) (Bacon et al., 2012). However, DFS practices are unevenly adopted and depend on the institutional environment that specifically promotes diversified farming practices. Decision making about sustainable agricultural development is also influenced by the general level of knowledge about different practices, and the increased area of transformed land and the number of social groups for sustainable agriculture and land management, mainly in less developed countries, improve environmental performance and agricultural productivity.

Sustainable agriculture offers not only social benefits but also economic advantages. Different agricultural sectors can experience diverse economic gains. For instance, integrated pest management (IPM) in cabbage production has demonstrated several economic benefits, including enhanced water quality, improved food safety, safer pesticide usage, and the long-term sustainability of pest management systems (Shamsudin et al., 2010). Sustainable agriculture and rice intensification systems can reduce farmers' cultivation costs, increase crop yields, and reduce chemical pollution of rivers and groundwater. The perceived costs and benefits of sustainable viticulture practices determine whether or not certain practices involve decisions about innovation or collaboration (Kassam & Brammer, 2013). In addition to all this, sustainable agriculture promotes new scientific achievements and the necessary economic incentives for the development and application of precision technologies. Overall, sustainable agriculture can provide economic benefits by reducing costs, increasing yields, and improving environmental quality.

It is worth noting that consumers are ready to pay a higher price for sustainable agricultural products. According to Melović et al. (2020), their study revealed that 30.8% of the respondents were willing to pay prices up to 20% higher for organic food in comparison to conventional food. Additionally, 39.4% of the respondents indicated a willingness to pay prices up to 40% higher for organic food. Cecchini et al. (2018) reviewed 41 studies and found that a large proportion of consumers are willing to pay a higher price for green and organic products.

The most important drivers of sustainable agricultural product price variation are quality assurance (QA) and social impact (SI), which are the most important drivers of online purchase intentions for sustainable agricultural products (Zia et al., 2022).

However, numerous barriers hinder the low-carbon transition in agriculture, requiring effective policies and measures to overcome these obstacles. This theoretical material aims to explore the barriers of low-carbon transition in agriculture and propose policies that can facilitate and promote the adoption of low-carbon practices in the sector.

The agricultural sector is a significant contributor to global GHG emissions, responsible for approximately 10%–12% of total emissions. The primary sources of these emissions are diverse and include enteric fermentation (digestive processes in livestock leading to methane release), manure management practices, methane emissions from rice cultivation in flooded paddies, emissions from the use of synthetic fertilizers, and energy consumption in various agricultural operations. Addressing emissions from the agricultural sector is vital for effective climate change mitigation efforts.

Indeed, the conversion of forest to agricultural land can lead to the loss of soil carbon, which can have adverse effects on both the environment and agricultural productivity. Additionally, studies evaluating the economic impact of transitioning to a low-carbon economy have shown that ambitious mitigation targets may have negative implications for global agriculture, potentially affecting agricultural production. Balancing climate change mitigation goals with sustainable agricultural practices is crucial to ensure both environmental protection and food security in the face of climate challenges (Jensen et al., 2019; Murty et al., 2002).

It is important to mention that rural communities have an important role to play in this process, and they have an important role to play in the transition to a low-carbon society (Markantoni & Woolvin, 2015). A low-carbon society refers to a socio-economic system that significantly reduces its GHG emissions, particularly carbon dioxide (CO_2), with the goal of mitigating climate change and minimizing its environmental impact. In a low-carbon society, energy generation, transportation, industry, and other sectors adopt sustainable practices and technologies that minimize the use of fossil fuels and prioritize renewable energy sources. The objective is to achieve a substantial reduction in carbon emissions while promoting economic growth, social well-being, and environmental sustainability. This transition often involves a combination of energy efficiency measures, renewable energy adoption, sustainable transportation systems, waste management strategies, and changes in consumption patterns. A low-carbon society recognizes the importance of sustainable development, balance between economic growth and environmental protection, and the urgent need to combat climate change.

Transitioning to a low-carbon future in agriculture, especially moving away from livestock production, requires the right transition (Blattner, 2020).

The figure shows the reasons why low-carbon technologies in agriculture will be very important in the future (Figure 1.3).

A low-carbon future in agriculture is crucial for several reasons (Liu et al., 2020; Zhao & Zhou, 2021; Stevanović et al., 2016) (Figure 1.3):

- **Climate Change Mitigation**. Agriculture is a significant contributor to GHG emissions, primarily through the production and use of synthetic fertilizers, livestock emissions, and deforestation for agricultural expansion. By adopting low-carbon practices, such as precision farming, organic farming, and agroforestry, emissions can be reduced, helping to mitigate climate change.

    Precision farming can reduce the environmental costs associated with agriculture. Finger (2019) argues that precision farming can reduce waste

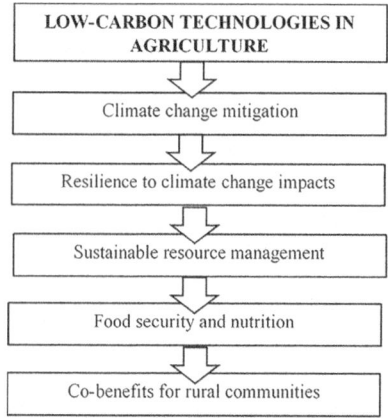

FIGURE 1.3 The benefits of low-carbon technologies in agriculture. (Created by authors.)

and agrochemical residues, thereby reducing private variable costs and environmental costs. Precision agriculture can use remote sensing technologies to optimize agricultural inputs, including nutrient use and disease and pest management, thereby reducing environmental impact (Sishodia et al., 2020). However, factors such as farm size, yield, profitability, and income influence farmers' perceptions of environmental quality improvement through precision farming technologies, suggesting that precision farming may not always be environmentally beneficial. This goes against the main goal of reducing environmental impact, but precision farming can meet both ecological and social requirements, preserving environmental biodiversity and improving the economy (Larkin et al., 2005; Abobatta, 2022; Finco et al., 2023).

Global trends in organic farming indicate that the global market for organic food is growing. Willer and Lernoud (2018) in Global Trends in Organic Agriculture report that approximately 2.7 million farmers organically manage 57.8 million hectares of agricultural land, and global organic food and beverage sales reached $90 billion in 2016. In the active arena of organic agriculture, access to good-quality data on organic farming is crucial. Reliable data helps measure success or failure in achieving Sustainable Development Goals and is a source of sound analysis and informed decision making by researchers, policymakers, industry, and other stakeholders across the value chain (Willer et al., 2023).

- **Resilience to Climate Change Impacts**: The agricultural sector is exceptionally susceptible to the effects of climate change, including occurrences like droughts, floods, and extreme weather events. Transitioning to low-carbon practices can enhance the resilience of agricultural systems, making them better able to adapt to changing climatic conditions and reducing the risk of crop failures and food shortages.
- **Sustainable Resource Management**: Agriculture relies on finite resources such as water, soil, and biodiversity. Unsustainable farming practices, such as intensive monocultures and excessive use of chemical inputs, degrade soil quality, deplete water resources, and harm biodiversity. Low-carbon approaches, such as agroecology and regenerative agriculture, prioritize sustainable resource management, promoting soil health, efficient water use, and biodiversity conservation.

Agroecology is a multidimensional concept that encompasses scientific, social, and cultural aspects (Wezel et al., 2023). Wezel et al. (2009) explain that agroecology has evolved over time, with different meanings and interpretations in different countries. Altieri (1995) emphasizes that agroecology is a science that provides ecological principles for designing and managing agroecosystems that are sustainable, socially just, and economically viable. Gliessman (2007) provides an introduction to agroecology, covering topics such as the ecology of sustainable food systems, plant and abiotic factors of the environment, biotic factors, and the energetics of agroecosystems. Altieri and Toledo (2011) argues that an agroecological revolution is taking place in Latin America, where peasants, NGOs, and some government and academic institutions are applying blended agroecological science and indigenous knowledge systems to enhance food security, conserve natural resources, and empower local communities. Agroecology is a holistic approach to agriculture that considers ecological, social, and cultural factors, and it has the potential to promote sustainable and equitable food systems. Agroecology is a holistic concept that encompasses various interpretations, intentions, and realities that depend on the country's issues, context, history, stakeholders, and socio-political environment (Wezel et al., 2023).

Regenerative agriculture has gained popularity in recent years, but a widely accepted definition of the term is yet to be established. Newton et al. (2020) discovered multiple definitions and descriptions of regenerative agriculture, based on various processes or outcomes, or a combination of both. According to Giller et al. (2021), the renewed interest in regenerative agriculture signifies a re-framing of two distinct approaches to agricultural futures, agroecology, and sustainable intensification, unified under the same concept. LaCanne and Lundgren (2018) observed that regenerative farming systems offer greater ecosystem services and profitability for farmers compared to an input-intensive model of corn production. However, Ikerd (2021) cautions that regenerative agriculture faces similar risks as sustainable agriculture if it lacks a clear definition that ensures agricultural sustainability.

- **Food Security and Nutrition**: With a growing global population, ensuring food security and nutrition is a pressing challenge. Low-carbon agriculture practices can improve productivity and efficiency, making food production more sustainable and resilient. By reducing waste and promoting diverse and nutritious crops, a low-carbon future in agriculture can contribute to meeting global food demand while minimizing environmental impacts.
- **Co-benefits for Rural Communities**: Transitioning to low-carbon agriculture can bring multiple co-benefits to rural communities. It can create new job opportunities, improve farmers' livelihoods, enhance rural development, and promote social equity. By supporting sustainable farming practices, low-carbon agriculture fosters a more inclusive and resilient agricultural sector.

Creating an all-encompassing value chain is vital for promoting inclusivity in the agricultural sector. This involves ensuring the active involvement of smallholder farmers in burgeoning markets for agricultural products, necessitating a comprehensive evaluation of opportunities and obstacles. Integrating value chain strategies with other innovative and rural livelihood systems approaches becomes essential to address the complexities of the situation effectively (Devaux et al., 2018). It can be seen and does provide a critical assessment of inclusive business in agriculture, as it is important to assess who, where, and under what conditions are operating (Chamberlain & Anseeuw 2017; German et al., 2018).

Overall, a low-carbon future in agriculture is essential to address climate change, build resilience, promote sustainable resource management, ensure food security, and create a more equitable and sustainable agricultural sector. It is a crucial component of achieving global sustainability goals and creating a resilient and prosperous future for both people and the planet.

By transitioning to low-carbon practices, the agricultural sector can:

- substantially reduce its environmental impact,
- mitigate climate change,
- enhance sustainability, and
- improve resilience in the face of climate-related challenges.

Transitioning to a low-carbon economy in the agricultural sector presents unique challenges, but it also offers significant opportunities for sustainable development. This section provides both basic and advanced understanding, encompassing key definitions, barriers, and policy measures.

There are many barriers to the adoption of low-carbon practices in agriculture, and addressing these barriers will require a multi-pronged approach that includes policy interventions, education and training, and the use of social media to promote adoption (Kushnarenko et al., 2021; Van Hoof, 2023). Wreford et al. (2017) identify a number of potential barriers, including farm-level constraints, sector-level constraints, and policy-related barriers. Derpsch et al. (2010) argue that adoption of no-till farming, a low-carbon practice, has been slow due to barriers such as lack of knowledge, lack of access to equipment, and resistance to change. Yang et al. (2021) find that participation in social media can positively influence low-carbon farming practices and that adoption of these practices can increase household income of farmers. Zheng et al. (2022) identify factors such as government support, farmers' familiarity, social capital, and personal and family characteristics that may influence farmers' behavior in adopting low-carbon agricultural technologies.

Kushnarenko et al. (2021) identified market failures, policy failures, and climate uncertainties as the primary hindrances to adopting the low-carbon adaptation model. According to Gomes and Reidsma (2021), a lack of awareness among farmers regarding their soil GHG production, economic challenges, personal mindset, on-farm complications, and the need to reconcile different stakeholders' rates of adoption were critical barriers. Burbi et al. (2016) found that limited financial capital, distrust in government action, and uncertainty about the effectiveness of farm advice on mitigation were the main obstacles to innovation. The papers suggest that overcoming these barriers may involve policy reforms, government intervention, the implementation of various adaptation approaches, such as early warning systems, disaster risk management, climate-smart agriculture, and insurance systems, and engaging a broader range of farmers and farming systems.

Figure 1.4 presents the most common barriers to a smooth transition to low-carbon agriculture.

Overcoming the barriers to low-carbon transition in agriculture requires a multi-faceted approach that combines technological innovation, supportive policies, capacity-building, and stakeholder engagement. Governments, international organizations, research institutions, and agricultural communities play a crucial role in creating an enabling environment for sustainable agricultural practices.

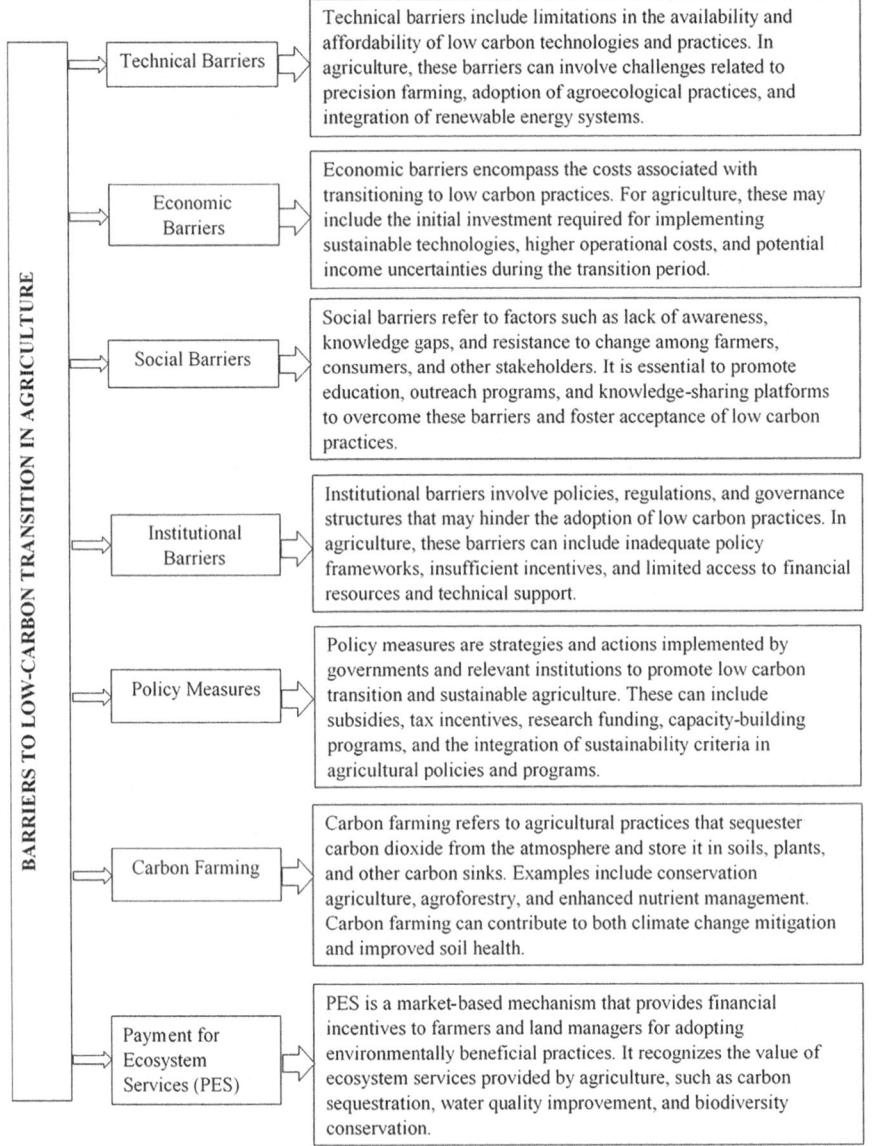

FIGURE 1.4 Barriers to low-carbon transition in agriculture. (Created by authors.)

By promoting policies that incentivize low-carbon practices, fostering innovation, and addressing the challenges faced by farmers and stakeholders, we can accelerate the transition to a sustainable and climate-resilient agricultural system. This will not only mitigate climate change but also contribute to food security, environmental conservation, and the well-being of rural communities.

Transitioning to a low-carbon economy in the agricultural sector presents unique challenges, but it also offers significant opportunities for sustainable development.

A closer look reveals the following barriers to the transition to low-carbon technologies in agriculture:

1. **Financial Constraints**: Many low-carbon technologies and practices in agriculture require significant investments. Farmers often face financial constraints and limited access to capital, making it challenging for them to adopt these technologies and practices. The high upfront costs of renewable energy systems, precision agriculture equipment, or organic farming methods can be prohibitive for many farmers, especially small-scale farmers.

    The papers suggest that farmers perceive financial risks and opportunities when making decisions about agricultural practices. Duarte et al. (2023) found that family farming managers perceive risk situations and have a low willingness to take risks. Lane et al. (2018, 2019) found that farmers and advisors in New York and Wisconsin are concerned about climate change impacts on agriculture, but other business pressures such as profitability, market conditions, government regulations, and labor availability often affect their decision making more. Käyhkö (2019) found that Nordic farmers and agricultural extension officers have different risk responses when making adaptation decisions, including risk aversive, opportunity-seeking, and experimental. Overall, farmers consider financial risks and opportunities when making decisions about agricultural practices, but other factors such as climate change impacts and business pressures also play a significant role in their decision making.

2. **Lack of Awareness and Information**: Farmers may lack awareness of low-carbon practices and technologies or have limited access to information about their benefits. It is crucial to provide farmers with the knowledge and resources necessary to understand and adopt low-carbon practices effectively. Educational programs, training workshops, and extension services can play a vital role in disseminating information and raising awareness among farmers.

    Awareness of climate change among farmers can affect their decisions. Priyanto and Mulyo (2021) found that farmers from a Proklim location had a higher awareness of climate change compared to non-Proklim farmers. Raghuvanshi and Ansari (2017) found that all farmers in their study were aware of climate change and perceived its impact on agriculture productivity. Farmers in Sub-Saharan Africa are aware of climate change and have adopted various adaptation strategies to cope with its adverse effects (Sani, 2016). Access to agricultural extension programs, radio, mobile phones, newspapers, education, and participation in developmental and farmer organizations positively impact climate change awareness among communal farmers in Zimbabwe (Mudombi et al., 2014; Ricart et al., 2022).

    Thus, farmers' understanding of climate change influences their decisions on how to manage their land. Some examples:
    - Prolonged wet, hot, and dry weather conditions affect the resource efficiency and investment decisions of smallholder farmers in Zimbabwe (Mutekwa, 2009).

- In Northwest Ethiopia, various factors play a significant role in farmers' decisions to adopt sustainable land management as a climate change adaptation strategy. These factors include their perception of climate change, the effectiveness of adaptation methods, education level, awareness of land degradation, slope of the land, vulnerability to degradation, number of plots owned, income from crop enterprise, farm size, distance to the farm, size of the economically active family, and the specific agro-ecological context (Asrat & Simane, 2017).
- Farmers in New York and Pennsylvania are concerned about climate impacts on their farms, but other business pressures such as profitability, market conditions, labor availability, or government regulations often influence their decision making (Lane et al., 2018).
- A positive temperature deviation from the medium-term average affects farmland values in Pakistan, and farmers' perceptions of climate change influence their active and prospective farm adaptation strategies (Arshad et al., 2017).

3. **Technical Knowledge and Skills**: Implementing low-carbon practices often requires specific technical knowledge and skills. Farmers may lack the necessary training or expertise to adopt new practices or operate advanced technologies. Training programs and capacity-building initiatives can help farmers acquire the skills needed for successful implementation of low-carbon practices.

    Farmers have several avenues to acquire the technical knowledge and skills required for implementing low-carbon practices. Enhancing the efficiency of extension services and broadening farmers' Internet access, along with improving their ability to utilize online resources for technology application, can significantly contribute to developing farmers' capabilities (Iskandar et al., 2021). Paustian (2013) proposes that incentivizing farmers to provide management information and ensuring timely and accurate reporting can effectively engage farmers from developing countries in GHG accounting. Parker (2022) proposes targeted nitrogen and irrigation education in California's nitrate-impacted regions to address nitrogen loss from agricultural soils via leaching and volatilization. Providing farmers with access to information and resources can help them acquire the technical knowledge and skills needed to implement low-carbon practices.

4. **Infrastructure Limitations**: Inadequate infrastructure, such as storage facilities, irrigation systems, or transportation networks, can pose challenges to the adoption of low-carbon practices. For example, the lack of proper storage facilities may result in post-harvest losses, discouraging farmers from adopting sustainable agricultural practices. Improving infrastructure is essential to support the low-carbon transition in agriculture.

    Infrastructure, government support, social capital, and social media participation are important factors influencing the adoption of low-carbon technologies in agriculture. Government support and farmer familiarity had a positive effect on adoption behavior, and social capital played an important role in promoting adoption (Zheng et al., 2022). Chen and Chen (2020) found that government extension service is an important factor influencing adoption, while education level is the main determinant of the probability of adoption. Latynskiy et al. (2014) discussed the use of multi-agent system models for policy evaluation, which can inform policy evaluators about the modeled results and social media participation has a positive effect on the intensity of adoption of low-carbon agricultural practices and the adoption of such practices can increase the income of farm households (Yang et al., 2021).

5. **Policy and Regulatory Barriers**: The absence of supportive policies or the presence of conflicting regulations can impede the adoption of low-carbon practices. Governments need to establish clear policies that promote and incentivize sustainable agriculture. This can include financial incentives, subsidies, or tax breaks for farmers who adopt low-carbon practices. Removing existing policy barriers and creating a favorable policy environment is crucial for encouraging widespread adoption.

Policy and regulatory barriers refer to obstacles and limitations within the legal and regulatory frameworks that hinder the adoption and implementation of low-carbon practices in agriculture. These barriers can vary across different countries and regions. Here are some examples of policy and regulatory barriers in the context of low-carbon transition in agriculture:

- *Lack of Supportive Policies.* The absence of clear policies and regulations that prioritize and promote low-carbon agriculture can be a significant barrier. If there are no specific targets, incentives, or support mechanisms in place, farmers may be less motivated to adopt sustainable practices. Governments need to establish policies that provide a supportive framework for low-carbon agriculture, such as setting emission reduction targets, promoting renewable energy use, or allocating R&D funds for sustainable farming practices.
- *Conflicting Regulations.* Sometimes existing regulations can impede the adoption of low-carbon practices. For example, there might be conflicting regulations or standards that make it difficult for farmers to transition to sustainable farming methods. Harmonizing regulations across different sectors and ensuring consistency in environmental, agricultural, and energy policies can help overcome these barriers.
- *Inflexible Land-Use Regulations.* Land-use regulations that restrict the use of agricultural land for low-carbon activities can hinder the transition. For instance, zoning laws that prioritize conventional agriculture over sustainable practices like agroforestry or organic farming can limit farmers' choices. Flexible land-use policies that allow for diversification and integration of low-carbon practices can be instrumental in promoting sustainable agriculture.
- *Lack of Financial Incentives.* The absence of financial incentives can deter farmers from adopting low-carbon practices. Governments can play a role in providing subsidies, grants, or tax breaks to farmers who adopt sustainable farming methods. Financial support can help offset the initial costs and make low-carbon practices more economically viable.

- *Insufficient Research and Development Support.* Inadequate funding and support for R&D in low-carbon agriculture can hinder innovation and the development of new technologies and practices. Governments and funding agencies should invest in R&D programs focused on sustainable agriculture to encourage the development of innovative solutions and provide farmers with access to cutting-edge technologies and knowledge.
- *Limited Access to Credit and Financing.* Farmers often face challenges in accessing credit and financing for adopting low-carbon practices. Financial institutions may perceive sustainable agriculture as riskier and less profitable, making it difficult for farmers to secure loans or investment capital. Establishing specialized financing programs, guarantee mechanisms, or partnerships between financial institutions and agricultural stakeholders can help address this barrier.
- *Inadequate Monitoring and Reporting Mechanisms.* Effective monitoring and reporting of GHG emissions and the impact of low-carbon practices are essential for evaluating progress and ensuring compliance. However, the lack of standardized monitoring methodologies, data collection systems, and reporting frameworks can hinder the implementation of low-carbon practices. Governments can establish monitoring and reporting requirements and provide support to farmers in complying with these requirements.

Addressing policy and regulatory barriers requires a comprehensive and coordinated effort involving policymakers, government agencies, agricultural organizations, and stakeholders. It involves the development of supportive policies, regulatory frameworks, and financial mechanisms that incentivize and facilitate the adoption of low-carbon practices in agriculture.

6. **Market Challenges**: The agricultural market dynamics can present challenges to the low-carbon transition. For instance, low-carbon products may face market competition from conventionally produced goods, which may be cheaper due to economies of scale or subsidies. There is a need to create market mechanisms and incentives that recognize and reward low-carbon agriculture, such as eco-labeling or carbon pricing initiatives.

Market challenges refer to obstacles and complexities within the agricultural market that can impede the transition to a low-carbon agriculture system. These challenges can affect the adoption, scalability, and economic viability of low-carbon practices and technologies. Some key market challenges in the context of low-carbon transition in agriculture are presented in Figure 1.5.

Addressing market challenges requires a multi-stakeholder approach involving farmers, industry players, consumers, policymakers, and agricultural organizations. Governments can play a crucial role in creating supportive market conditions through targeted policies, awareness campaigns, market incentives, and public–private collaborations. Strengthening consumer demand for sustainable products, improving market access, and

Low-Carbon Transition

MARKET CHALLENGES IN THE CONTEXT OF LOW-CARBON TRANSITION IN AGRICULTURE

- **Market Competition**: Low-carbon agricultural products may face competition from conventionally produced goods that are often cheaper due to economies of scale or subsidies. This can create a barrier for farmers who adopt low-carbon practices, as they may struggle to compete on price. Market mechanisms that recognize and reward the environmental benefits of low-carbon agriculture, such as eco-labeling or certification programs, can help create a competitive advantage for sustainable products.

- **Limited Consumer Demand**: The level of consumer demand for low-carbon agricultural products can impact the market viability of sustainable farming practices. If consumers are not aware of the environmental benefits or are not willing to pay a premium for sustainable products, farmers may be reluctant to invest in low-carbon practices. Educating consumers about the environmental impact of their food choices and promoting the value of sustainable agriculture can help drive demand for low-carbon products.

- **Supply Chain Constraints**: Complex supply chains in agriculture can pose challenges to the adoption of low-carbon practices. Farmers need to ensure that their low-carbon products can be efficiently transported, stored, and distributed without compromising their environmental credentials. Strengthening supply chain infrastructure, including storage facilities, cold chain systems, and transportation networks, can help facilitate the transition to low-carbon agriculture.

- **Market Access and Distribution Channels**: Accessing appropriate distribution channels and markets can be challenging for farmers adopting low-carbon practices. They may face barriers in terms of market entry, certification requirements, or limited access to distribution networks. Supportive policies and initiatives that facilitate market access for sustainable products, such as preferential procurement programs, direct marketing opportunities, or partnerships with retailers, can help overcome these challenges.

- **Market Volatility and Uncertainty**: Agricultural markets are prone to price volatility and uncertainty, which can make it difficult for farmers to plan and invest in low-carbon practices. Fluctuating commodity prices, weather-related risks, and changing consumer preferences can create uncertainties and perceived risks for farmers. Risk-sharing mechanisms, such as insurance programs or contract farming arrangements, can help mitigate the financial risks associated with low-carbon agriculture and provide stability to farmers.

- **Knowledge and Information Asymmetry**: Limited access to accurate information and knowledge about low-carbon practices, technologies, and market opportunities can be a barrier for farmers. Lack of awareness about the benefits, costs, and potential market advantages of sustainable farming methods may deter farmers from adopting low-carbon practices. Knowledge dissemination programs, extension services, and farmer-to-farmer networks can help bridge this information gap and empower farmers with the necessary knowledge and resources.

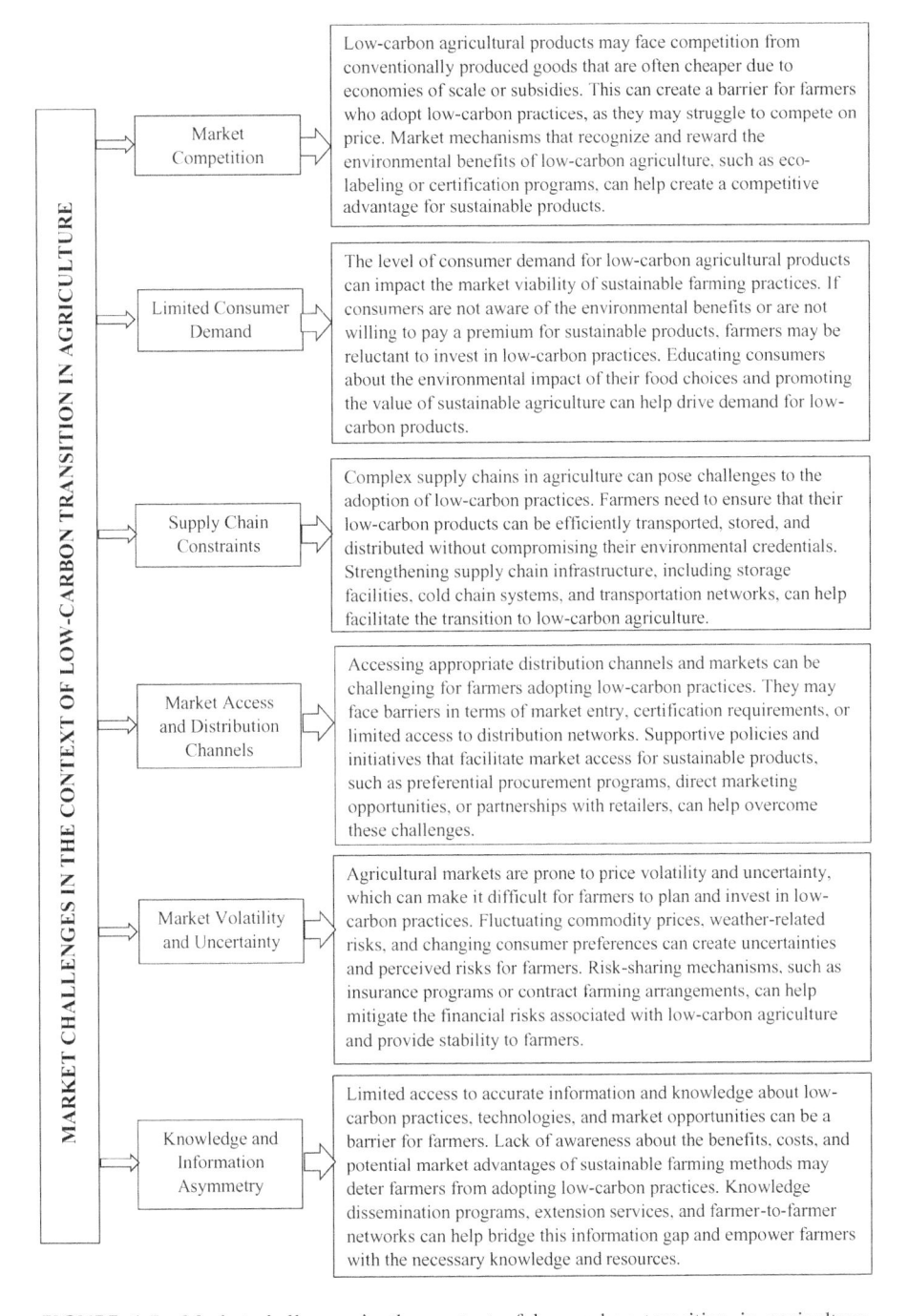

FIGURE 1.5 Market challenges in the context of low-carbon transition in agriculture. (Created by authors.)

ensuring fair pricing mechanisms can foster the growth of low-carbon agriculture in the market.
7. **Risk and Uncertainty**: Transitioning to low-carbon agriculture often involves a degree of uncertainty and perceived risk. Farmers may be concerned about the potential financial risks, productivity losses, or market acceptance of new practices. Providing support mechanisms, such as insurance programs, risk-sharing arrangements, or pilot projects, can help mitigate these concerns and encourage farmers to adopt low-carbon practices.

Risk and uncertainty are important factors that can hinder the adoption of low-carbon practices in agriculture. They refer to the potential for negative outcomes or unknowns associated with transitioning to sustainable farming methods (Morton et al., 2017; Findlater et al., 2019). Here's a further explanation of risk and uncertainty in the context of the low-carbon transition in agriculture (Iqbal et al., 2020; Waldman et al., 2020; Hanger-Kopp et al., 2019; Nozharov & Nikolova, 2020; Carbone et al., 2021):

- *Financial Risk.* Transitioning to low-carbon practices often requires upfront investments in technologies, equipment, or infrastructure. There is a financial risk involved if farmers are uncertain about the returns on their investment or face uncertainty regarding the market demand for sustainable products. The potential for lower yields or increased production costs during the transition phase can also contribute to financial risks.
- *Productivity and Performance Risk.* Adopting new farming practices or technologies may involve a learning curve and uncertainties about their impact on productivity and performance. Farmers may be concerned that low-carbon practices could lead to lower yields or quality compared to conventional methods. The uncertainty surrounding the productivity and performance of sustainable practices can deter farmers from making the transition.
- *Market Acceptance and Demand Uncertainty.* Farmers may be unsure about the market acceptance and demand for low-carbon agricultural products. Shifting consumer preferences and market dynamics can create uncertainty about the willingness of consumers to pay a premium for sustainable products. Farmers may hesitate to invest in low-carbon practices if they are uncertain about the profitability and marketability of their products.
- *Policy and Regulatory Uncertainty.* Uncertainty about future policy directions and regulatory changes can create challenges for farmers considering a low-carbon transition. Changes in government policies, subsidies, or carbon pricing mechanisms can impact the economic viability of sustainable farming practices. Farmers may be reluctant to invest in low-carbon practices if they are unsure about the long-term support and stability of the policy environment.
- *Climate and Environmental Risks.* Climate change-related risks, such as extreme weather events, water scarcity, or pest and disease outbreaks,

can affect the success of low-carbon agriculture. Farmers may face uncertainties about how their low-carbon practices will perform under changing climatic conditions or how resilient their farming systems will be to climate-related challenges. Climate risks can contribute to hesitation and concerns about the effectiveness and stability of low-carbon practices.

Addressing these barriers requires a multi-faceted approach involving stakeholders from the agricultural sector, policymakers, researchers, and financial institutions. By implementing supportive policies, providing financial incentives, improving access to information and resources, and fostering collaboration, it is possible to overcome these barriers and facilitate a successful low-carbon transition in agriculture. Achieving a low-carbon transition in agriculture is crucial for addressing climate change and promoting sustainable development. Despite the barriers, such as knowledge gaps, financial constraints, and policy challenges, there are several policy measures that can facilitate the adoption of low-carbon practices. By enhancing knowledge dissemination, providing financial support, and developing coherent policy frameworks, governments and stakeholders can promote sustainable agriculture, reduce GHG emissions, and contribute to a more resilient and environmentally conscious future. The low-carbon transition in agriculture requires collaborative efforts, innovative solutions, and a shared commitment to building a sustainable future.

1.3 GREEN DEAL POLICY AREAS WITH EMPHASIS ON AGRICULTURE

The third subtopic of this teaching material focuses on the Green Deal policy areas, with an emphasis on agriculture. The Green Deal is an ambitious policy framework established by the European Union (EU) to combat climate change, achieve sustainability, and foster economic growth. This section provides both basic and advanced understanding, encompassing key definitions, policy areas, and their implications for the agricultural sector.

The Green Deal is a comprehensive policy framework launched by the EU to transform Europe into the world's first climate-neutral continent by 2050. It aims to reconcile environmental sustainability with economic growth, fostering a just and inclusive transition.

The European Green Deal (EGD), introduced by the European Commission in 2019, is a comprehensive policy package designed to tackle climate change and facilitate a fair and inclusive transition. Its objective is to attain carbon neutrality in Europe's production systems and address environmental sustainability, structural transformation, and economic fairness in the region (Pianta & Lucchese, 2020). The Green Deal encompasses various policy domains, including climate and energy, agriculture and fisheries, products and services, and trade and foreign policy. It also incorporates the EU climate law to ensure climate neutrality by 2050 (Krämer, 2020).

The Green Deal presents a roadmap for a transformative shift toward a resource-efficient economy decoupled from resource consumption, with the ultimate goal of achieving net-zero emissions by 2050 (Kazak, 2022). However, criticisms have been raised against the Green Deal, including its lack of a clear vision for a just, post-carbon economy in Europe, inadequate resources allocated to achieve the stated objectives, and limited implementation tools (Pianta & Lucchese, 2020). Furthermore, it has been scrutinized for its dependence on global finance and for not providing enough policy space for systemic changes or ambitious green macroeconomics and green industrial policies (Storm, 2020).

The EGD outlines a path to ensure the sustainability of the EU economy by transforming climate and environmental challenges into opportunities across all policy domains, while ensuring a just and inclusive transition for all (Kazak, 2022). It seeks to achieve climate neutrality for the European Union by 2050, aiming to reduce GHG emissions by 65% compared to 1990 levels (Hainsch et al., 2020). To establish a strong foundation for current national and EU-level economic stimulus packages, the Green Deal demands substantial investments, with around €3,000 billion needed for renewable energy initiatives (Hainsch et al., 2020). Additionally, a successful transition requires annual investment requirements of up to €855 billion in the EU27 (Wildauer et al., 2020). Furthermore, the Green Deal aims to tackle chemical pollution comprehensively, involving improved substance design and remediation options to eliminate chemicals from the environment (Van Dijk et al., 2021).

The EGD is an ambitious initiative launched by the European Commission in 2019. It aims to make Europe the world's first climate-neutral continent by 2050, while also promoting sustainable economic growth.

The implementation plan of the EGD encompasses a wide range of policies and measures across various sectors (Figure 1.6):

1. **Climate Law and Emissions Reduction Targets**: The EGD is underpinned by the European Climate Law, which sets the objective of reaching net-zero GHG emissions by 2050. The plan includes a significant increase in the 2030 emissions reduction target, aiming for a minimum reduction of 55% compared to 1990 levels.
2. **Clean Energy Transition**: The EU aims to accelerate the transition to clean energy sources and increase energy efficiency. Measures include supporting renewable energy generation, improving energy storage capabilities, promoting energy efficiency in buildings, and developing smart grids.
3. **Sustainable Mobility**: The plan focuses on promoting sustainable and clean transportation. Initiatives include expanding EV infrastructure, promoting alternative fuels, encouraging public transport, and developing sustainable mobility strategies in urban areas.
4. **Circular Economy**: The EGD aims to transition to a circular economy, where resources are used efficiently and waste is minimized. Measures include promoting recycling and waste reduction, encouraging sustainable product design, and establishing a framework for a more sustainable use of resources.

Low-Carbon Transition   31

- Climate Law and Emissions Reduction Targets
- Clean Energy Transition
- Sustainable Mobility
- Circular Economy
- Biodiversity and Sustainable Agriculture
- Just Transition and Social Dimension
- Research, Innovation, and Investment
- International Cooperation

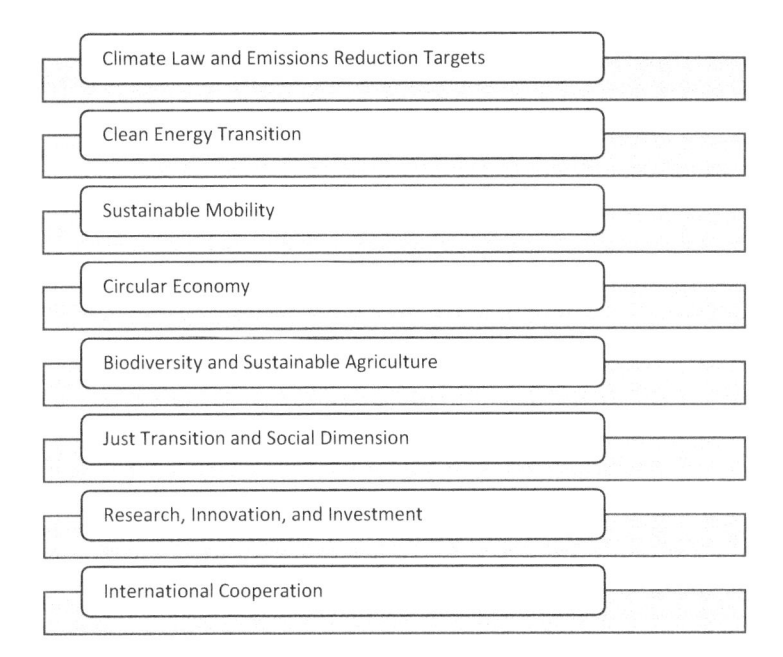

FIGURE 1.6 A simplified implementation plan for the European Green Deal. (Created by authors.)

5. **Biodiversity and Sustainable Agriculture**: The EU seeks to protect and restore biodiversity, promote sustainable agriculture practices, and reduce the use of chemical pesticides. Initiatives include the Farm to Fork Strategy, which aims to ensure a fair, healthy, and environmentally friendly food system.
6. **Just Transition and Social Dimension**: The plan emphasizes the need for a just transition, ensuring that no one is left behind during the transformation to a sustainable economy. Measures include supporting affected regions and sectors, providing retraining and reskilling opportunities, and investing in green jobs.
7. **Research, Innovation, and Investment**: The EGD emphasizes the importance of research, innovation, and investment to drive the transition to a sustainable future. Funding programs, such as Horizon Europe, aim to support research and innovation in areas such as clean energy, sustainable mobility, and circular economy solutions.
8. **International Cooperation**: The EU aims to lead global efforts in addressing climate change and promoting sustainability. It seeks to work closely with international partners to achieve global climate goals and integrate sustainability into trade and foreign policy.

The EU Green Deal is a strategic program initiated by the European Commission with the objective of promoting environmentally friendly practices and reorienting policies and regulations in various areas, including climate and energy, agriculture and fisheries, products and services, and trade and foreign policy. A significant milestone is the adoption of an EU climate law, which ensures the achievement of EU climate neutrality by 2050 (Krämer, 2020). The Green Deal's impact extends globally and significantly influences the EU's external actions, encompassing diplomatic efforts, cooperation, and trade instruments (Ciot, 2022). It not only seeks to promote EU priorities and values but also aims to advance the global public good. By incorporating climate goals into the global agenda, the EU demonstrates its commitment to inclusive governance (Larionova, 2021). Moreover, the Green Deal is envisioned as a mechanism for reallocating resources, encouraging investment shifts and labor substitution in critical economic sectors, while providing support to the most vulnerable segments of society throughout the decarbonization process.

However, despite its initiatives, the EGD has faced criticism on various fronts. Pianta and Lucchese (2020) argue that the EGD lacks a clear vision of a just, post-carbon economy for Europe, and the available resources are insufficient to achieve the stated objectives. Wildauer et al. (2020) suggest that the EGD's current assumptions regarding the necessary investments to reduce GHG emissions by 40% by 2030 may be underestimated, and more ambitious targets would require even greater investments. Ossewaarde and Ossewaarde-Lowtoo (2020) question whether the EGD represents a distinct third alternative to the growth-based green economy and the degrowth model, suggesting that specific interpretations and implementations of the EGD could potentially make it an alternative to both. Fleming and Mauger (2021) discuss recent developments related to the "green" and "just" aspects of the EGD, concluding with critical remarks on how these aspects are being integrated into energy and climate law. Overall, the EGD exhibits some shortcomings and limitations, indicating that further action is necessary to ensure a just and effective transition to a post-carbon economy (Domorenok & Graziano, 2023; Eckert & Kovalevska, 2021).

The EGD would have significant impacts on different sectors. Lamenta and Grzybowska (2023) found that businesses are taking actions to implement the guidelines of the EGD but face both hindering and motivating factors. Leonard et al. (2021) discuss the geopolitical aspects of the EGD, including relationships with important neighboring countries and global players, and propose seven actions to promote the deal's success. Bezdek (2020) focuses on the jobs impact of the Green New Deal in the USA, estimating that it would generate more than 18.3 million jobs throughout the economy, with a concentration in sectors such as manufacturing and professional, scientific, and technical services. The EGD and Green New Deal would have significant impacts on different sectors, including job creation and geopolitical relationships.

The implementation of the Green Deal is expected to bring about substantial economic and environmental consequences for the agriculture sector. According to Beckman et al. (2020), the proposed input reductions under the Green Deal could result in decreased agricultural production and competitiveness, affecting both domestic and export markets. This could lead to higher global food prices, which may have adverse effects on consumer budgets. Leonard et al. (2021) delve into the geopolitical implications of the Green Deal, particularly concerning relationships with neighboring countries and global players, and provide seven suggested actions to ensure its successful implementation. Given these factors, the Green Deal's impact on the agriculture sector requires thorough consideration of economic, environmental, and geopolitical factors.

Within the EGD, there are several policy areas that specifically address agriculture and aim to promote sustainability in the sector. Here are some key policy areas with an emphasis on agriculture:

1. **Farm to Fork Strategy**: The Farm to Fork Strategy is a central component of the EGD that focuses on making the food system more sustainable, from production to consumption. It aims to promote healthy and sustainable food, reduce the use of chemical pesticides and fertilizers, and improve animal welfare.

   Tengilimoğlu (2021) explores the strategic focal points of key European documents and initiatives related to future sustainable agri-food systems, considering human health and well-being, and encompassing ecological, economic, and social aspects of sustainability. Abdalla et al. (2022) investigate the utilization of nanomaterials in various agricultural practices for vegetables to align with the Farm to Fork strategy. Abbasi et al. (2023) analyze the value chain of indigenous chicken micro-farming in Malaysia and devise an initial integrated model to align with the Farm to Fork strategy. Giannou (2022) examines the interconnectedness of the EU's approach to sustainable food systems and the broader context of a sustainable agricultural and economic future.

   Together, these studies suggest that the Farm to Fork Strategy's primary objectives involve establishing an equitable, healthy, and eco-friendly food system that is sustainable and supports local farmers, while also reducing food waste and promoting the production of nutritious food.

   Dobrin et al. (2022) discovered that the implementation of the strategy aimed at reducing pesticides and fertilizers would not have a detrimental impact on Romania's agricultural productivity. Heyl et al. (2023) examined how the Farm to Fork Strategy's proposed measures, such as digital precision fertilization and sustainable agricultural practices, are being put into action through the Common Agricultural Policy (CAP) in Germany and Saxony. The findings indicate that while Germany does offer some support for sustainable agricultural practices and digital precision fertilization, the overall level of support is relatively limited. On the other hand, Moschitz et al. (2021) criticize the Farm to Fork Strategy for its heavy emphasis on technical innovations while neglecting the social and structural aspects of transforming food systems. The paper proposes that Member States need to adequately equip their agricultural knowledge and innovation systems (AKIS) and provide education to advisors, researchers, knowledge brokers, and others to bring about the necessary change in attitudes and practices.

   At the heart of the EGD lies the Farm to Fork Strategy, which seeks to revolutionize the food system to become more sustainable, nutritious, and adaptable. This strategy encompasses the entire food supply chain, from production to consumption, with the goal of tackling the environmental,

Sustainable Food Production
- The Farm to Fork Strategy promotes sustainable farming practices that reduce the environmental impact of agriculture. It encourages the adoption of agroecology, organic farming, and precision agriculture techniques to minimize the use of chemical inputs, enhance soil health, and protect biodiversity.

Reduction of Chemical Pesticides and Fertilizers
- The strategy sets targets to reduce the use and risk of chemical pesticides and fertilizers in agriculture. It aims to promote integrated pest management, natural pest control methods, and the development of safe and sustainable alternatives to chemical inputs.

Enhancing Biodiversity and Ecosystem Services
- The Farm to Fork Strategy recognizes the importance of biodiversity in agriculture and aims to preserve and restore ecosystems. It promotes measures such as creating ecological focus areas, preserving habitats, and promoting pollinator-friendly practices to support biodiversity and enhance ecosystem services in agricultural landscapes.

Sustainable Aquaculture
- The strategy addresses the sustainability challenges of aquaculture, aiming to ensure that fish and seafood production is environmentally responsible and socially inclusive. It encourages the adoption of sustainable aquaculture practices, reduces the use of antibiotics and chemicals, and promotes responsible fish feed sourcing.

Healthy and Sustainable Diets
- The Farm to Fork Strategy aims to promote healthy and sustainable diets for European citizens. It supports education, awareness, and information campaigns to promote balanced and nutritious food choices, reduce food waste, and encourage plant-based diets with lower environmental footprints.

Food Loss and Waste Reduction
- The strategy sets targets to reduce food loss and waste along the entire food supply chain. It promotes measures to improve food storage, processing, and distribution, and encourages collaboration among stakeholders to address the causes of food loss and waste.

Transparent and Traceable Food Systems
- The Farm to Fork Strategy aims to enhance transparency and traceability in the food system. It promotes digital technologies and data-sharing platforms to improve food traceability, provide consumers with reliable information, and ensure the safety and authenticity of food products.

Sustainable Food Procurement
- The strategy encourages public institutions, such as schools and hospitals, to adopt sustainable food procurement practices. It promotes the inclusion of sustainability criteria in public food procurement contracts to support sustainable farming practices and contribute to local and regional food systems.

FIGURE 1.7 Farm to Fork strategy within the context of the European Green Deal. (Created by authors.)

social, and economic issues connected to the food system. A summary of the Farm to Fork Strategy within the context of the EGD is presented in Figure 1.7.

In essence, the Farm to Fork Strategy, as part of the EGD, strives to advance sustainable food production, safeguard the environment, boost biodiversity, and enhance the health and well-being of European citizens. It acknowledges the necessity for a holistic and cohesive approach to revolutionize the food system, making it more sustainable, robust, and socially accountable.

2. **Sustainable Agriculture Practices**: The Green Deal encourages the adoption of sustainable farming practices that reduce environmental impact. Sustainable agriculture practices refer to a range of farming methods and techniques that aim to minimize negative environmental impacts, preserve natural resources, promote biodiversity, and ensure long-term food production. These practices prioritize the sustainability of agricultural systems by considering ecological, economic, and social factors. Figure 1.8 shows some key components of sustainable agriculture practices.

There are several types of sustainable agriculture. Szekely and Jijakli (2022) describe "bioponics" as an emerging form of hydroponics that recycles organic waste into a nutrient-rich solution for plant growth. Khorramnia et al. (2014) discuss precision agriculture as a tool for achieving sustainable agriculture, using sensors, positioning and navigation systems, GIS, software, and variable rate technology. Bon et al. (2011) highlight the importance of urban agriculture in developing countries, which can provide fresh and nutritious food while also cleansing and opening up urban

COMPONENTS OF SUSTAINABLE AGRICULTURE PRACTICES

- Conservation Tillage
- Crop Rotation and Diversification
- Agroforestry
- Integrated Pest Management (IPM)
- Organic Farming
- Water Conservation and Irrigation Management
- Livestock Management
- Nutrient Management
- Climate-Smart Agriculture

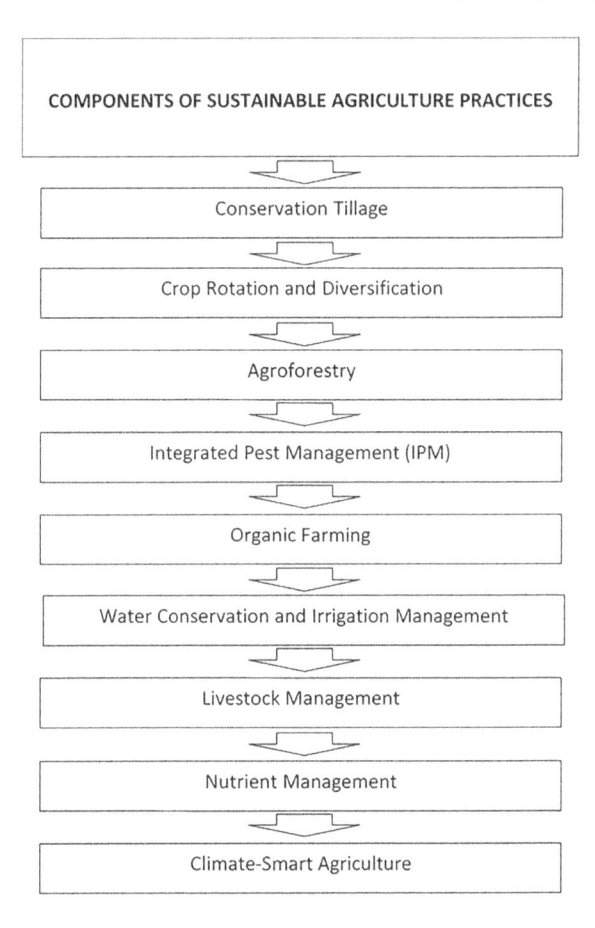

FIGURE 1.8 Key components of sustainable agriculture practices. (Created by authors.)

spaces. AlShrouf (2017) discusses hydroponics, aeroponics, and aquaponics as innovative methods that conserve water and reduce the use of agrichemicals. Sustainable agriculture can take many forms, including those that recycle waste, use precision technology, and integrate with urban environments (Komatsuzaki & Ohta, 2013).

As we can see from Figure 1.8, some key components of sustainable agriculture practices are (Velten et al., 2015; Dubey et al., 2015; Sollen-Norrlin et al., 2020; Jamnadass et al., 2013; Beillouin et al., 2020, 2021; Goodell et al., 2014; Niggli, 2014; Fonteh et al., 2013):

- *Conservation Tillage.* Conservation tillage practices minimize soil disturbance by reducing or eliminating plowing. This helps prevent soil erosion, improves soil structure, retains moisture, and enhances organic matter content, promoting overall soil health and fertility.
- *Crop Rotation and Diversification.* Crop rotation involves the systematic sequencing of different crops in the same field over time. It helps break disease cycles, manage pests and weeds, and improve soil nutrient balance. Diversifying crops can enhance ecosystem resilience, support pollinators, and reduce the reliance on specific crops, contributing to more sustainable agriculture.
- *Agroforestry.* Agroforestry integrates trees and shrubs with crops or livestock in agricultural systems. It provides multiple benefits, such as improved soil fertility, increased biodiversity, carbon sequestration, and reduced water runoff. Agroforestry systems can enhance ecological resilience while supporting sustainable agricultural production.
- *Integrated Pest Management (IPM).* IPM focuses on managing pests using an ecosystem-based approach that minimizes the use of chemical pesticides. It combines various pest control methods, including biological control, crop rotation, and cultural practices, to reduce pest populations while minimizing environmental impacts and preserving natural enemies of pests.
- *Organic Farming.* Organic farming avoids the use of synthetic fertilizers, pesticides, and genetically modified organisms (GMOs). It promotes soil health through the use of organic matter, crop rotation, and biological pest control. Organic farming also emphasizes animal welfare, biodiversity conservation, and the exclusion of artificial additives in food production.
- *Water Conservation and Irrigation Management.* Sustainable agriculture practices emphasize efficient water use and conservation. This includes implementing technologies like drip irrigation and precision irrigation systems, managing irrigation schedules based on crop needs, and adopting water-saving techniques such as mulching and soil moisture monitoring.
- *Livestock Management.* Sustainable livestock management practices prioritize animal welfare, minimize environmental impacts, and optimize resource use. This may involve rotational grazing systems, proper waste

management, responsible use of antibiotics, and the integration of livestock with crop systems to utilize by-products and enhance nutrient cycling.
- *Nutrient Management.* Sustainable agriculture practices aim to optimize nutrient use efficiency and reduce nutrient runoff. This includes proper soil testing, balanced fertilizer application, use of organic fertilizers, and practices such as cover cropping and conservation tillage that enhance nutrient cycling and reduce the risk of water pollution.
- *Climate-Smart Agriculture.* Climate-smart agriculture focuses on reducing GHG emissions, enhancing climate resilience, and adapting to climate change impacts. It includes practices such as carbon sequestration through agroforestry or cover cropping, improving water management in response to changing rainfall patterns, and optimizing crop varieties and planting dates to match changing climatic conditions.

Sustainable agriculture practices prioritize the overall ecological and socio-economic sustainability of farming systems, including factors such as farmer livelihoods, social equity, and community resilience, whereas traditional agriculture practices may have a narrower focus on maximizing yields and short-term economic gains without considering broader sustainability dimensions.

3. **Biodiversity Protection**: The Green Deal recognizes the importance of protecting and restoring biodiversity in agricultural landscapes. It aims to integrate biodiversity considerations into farming practices, promote the preservation of pollinators, and protect habitats and ecosystems that provide valuable services for agriculture.

Biodiversity protection within the EGD recognizes the critical role of biodiversity in maintaining healthy ecosystems and promoting sustainable agriculture. It encompasses a range of measures aimed at preserving and restoring biodiversity in agricultural landscapes. Some key aspects of biodiversity protection within the EGD are presented in Figure 1.9.

Discussing the aspects in more detail (Figure 1.9):
- *Habitat Preservation.* Biodiversity protection involves preserving and restoring habitats within agricultural landscapes. This includes maintaining natural areas such as hedgerows, woodlands, wetlands, and meadows, which provide crucial habitats for a wide range of plant and animal species. Preserving these habitats helps support biodiversity and ecosystem functioning.
- *Agroecological Practices.* The EGD promotes the adoption of agroecological practices that prioritize biodiversity conservation. These practices include diversifying crops, implementing crop rotations, intercropping, and integrating trees and shrubs within agricultural fields. By creating a more diverse and complex agricultural landscape, agroecological practices support a variety of species and promote ecological resilience.
- *Pollinator Protection.* Pollinators, such as bees, butterflies, and other insects, play a vital role in agricultural ecosystems by facilitating plant reproduction and promoting crop yields. Biodiversity protection

```
┌─────────────────────────────────────────┐
│ KEY ASPECTS OF BIODIVERSITY PROTECTION WITHIN THE │
│         EUROPEAN GREEN DEAL             │
└─────────────────────────────────────────┘
         ├──── Habitat Preservation
         ├──── Agroecological Practices
         ├──── Pollinator Protection
         ├──── Ecosystem Services
         ├──── Genetic Diversity
         ├──── Cross-Compliance and Greening Measures
         └──── Ecological Focus Areas
```

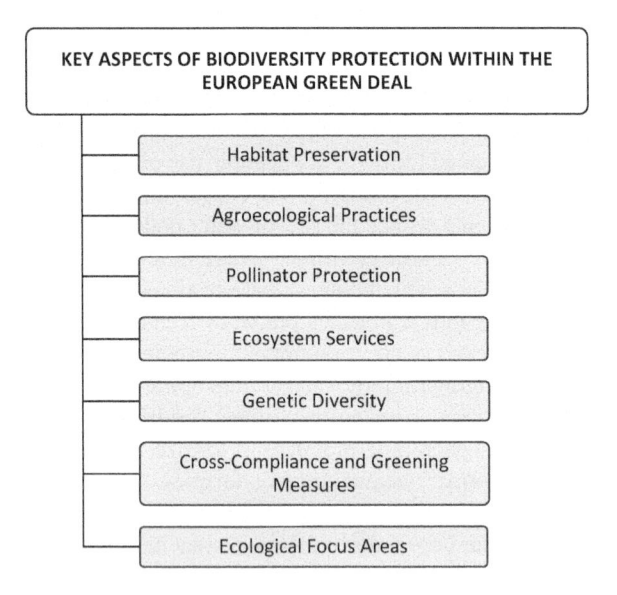

FIGURE 1.9 Key aspects of biodiversity protection within the European Green Deal. (Created by authors.)

measures within the EGD aim to safeguard pollinators by providing suitable habitats, reducing the use of harmful pesticides, and promoting flowering plant diversity.

- *Genetic Diversity.* Preserving genetic diversity is crucial for agricultural resilience and sustainability. The EGD emphasizes the conservation of crop genetic resources and the promotion of diverse crop varieties. Protecting and using a wide range of genetic resources helps maintain the adaptability of crops to changing environmental conditions and supports the development of more resilient and sustainable agricultural systems.
- *Ecosystem Services.* Biodiversity provides various ecosystem services that are essential for agricultural productivity and sustainability. These services include pest control, nutrient cycling, soil fertility, and water regulation. By protecting and promoting biodiversity, the EGD aims to harness these ecosystem services to support sustainable agriculture and reduce the reliance on synthetic inputs.
- *Ecological Focus Areas.* The EGD encourages the establishment of ecological focus areas (EFAs) within agricultural landscapes. EFAs are areas set aside for biodiversity enhancement and ecological benefits. They can include features such as buffer strips, field margins, and flower-rich habitats. These areas contribute to biodiversity protection by providing refuge for wildlife, supporting pollinators, and improving overall landscape connectivity.
- *Cross-Compliance and Greening Measures.* The CAP of the European Union integrates biodiversity protection through cross-compliance and

greening measures. These measures link the provision of direct payments to farmers with the adoption of environmentally friendly practices, including maintaining permanent grasslands, protecting habitats, and implementing crop diversification.

By integrating biodiversity protection into agricultural practices, the EGD aims to conserve and enhance the natural capital that underpins sustainable food production, resilient ecosystems, and the well-being of both wildlife and people. It recognizes that agriculture has a crucial role to play in biodiversity conservation and seeks to promote a harmonious coexistence between agriculture and nature.

4. **Reduction of Chemical Pesticides and Fertilizers**: The EGD seeks to reduce the use of chemical pesticides and fertilizers in agriculture. It encourages the development and adoption of alternative pest and disease management strategies, as well as the promotion of organic and IPM practices.

   Exposure to chemical pesticides and fertilizers can have negative health effects on humans. Exposure to agrichemicals can result in acute and chronic health effects, including neurotoxicity, lung damage, chemical burns, infant methemoglobinemia, and various cancers (Weisenburger, 1993), exposure to pesticides can lead to short-term and chronic impacts, such as skin and eye irritation, headaches, dizziness, nausea, cancer, asthma, and diabetes (Kim et al., 2017). Toxicity of pesticides in humans is influenced by various factors, such as age, gender, and health status, children are at greater risk than adults and farmers who are routinely exposed to pesticides are at risk of developing a wide variety of diseases, ranging from respiratory effects to various types of cancer (Dhananjayan & Ravichandran, 2018). Eco-friendly pesticide alternatives and organic farming may be viable solutions to reduce the toxic effects of chemicals.

   By reducing the reliance on chemical pesticides and fertilizers, agricultural systems can minimize environmental pollution, protect beneficial organisms, safeguard human health, and promote long-term sustainability. The goal is to adopt more holistic and integrated approaches that balance productivity, profitability, and environmental responsibility in agriculture.

5. **Soil Health and Carbon Sequestration**: The Green Deal promotes soil health and the sequestration of carbon in agricultural soils. It supports measures such as conservation agriculture, cover cropping, and agroforestry, which enhance soil fertility, reduce erosion, and contribute to carbon storage.

   Healthy soils have numerous benefits for sustainable agriculture, environmental quality, and human health. Handayani and Hale (2022) emphasize that soil health is the foundation of crop health and productivity, and that soil health assessments should consider physical, chemical, and biological properties. Tahat et al. (2020) note that soil microorganism diversity and activity are key components of soil health, and that sustainable agriculture depends on healthy soils. Wall et al. (2015) highlight the benefits of soil biodiversity for human health, including suppressing disease-causing soil organisms and providing clean air, water, and food. Liang et al. (2020)

present a framework for future soil management that combines research, industry, and policymakers to ensure high-quality outputs that are transferable to industry and end-users. Healthy soils are essential for sustainable food production, environmental quality, and human health.

6. **Sustainable Water Management**: The Green Deal emphasizes the importance of sustainable water management in agriculture. It promotes efficient irrigation techniques, water-saving practices, and the protection of water resources to ensure their availability for agricultural activities while minimizing water pollution.

Several factors (Figure 1.10) play a crucial role in determining sustainable water management practices. These factors vary depending on the specific context, but here are some key considerations (Velasco-Muñoz et al, 2022; Chaminé & Gómez-Gesteira, 2019; Richter et al., 2003; Ding & Ghosh, 2017; Alexandratos et al., 2019; Mastrorilli & Zucaro, 2019; Russo et al., 2014; Setegn, 2015; Boelee et al., 2019; Cai et al., 2015; Bonetti et al., 2022):

- *Water Availability and Accessibility*. Assessing the quantity and quality of water resources, understanding seasonal variations, and considering the needs of different sectors are essential for effective water management planning.

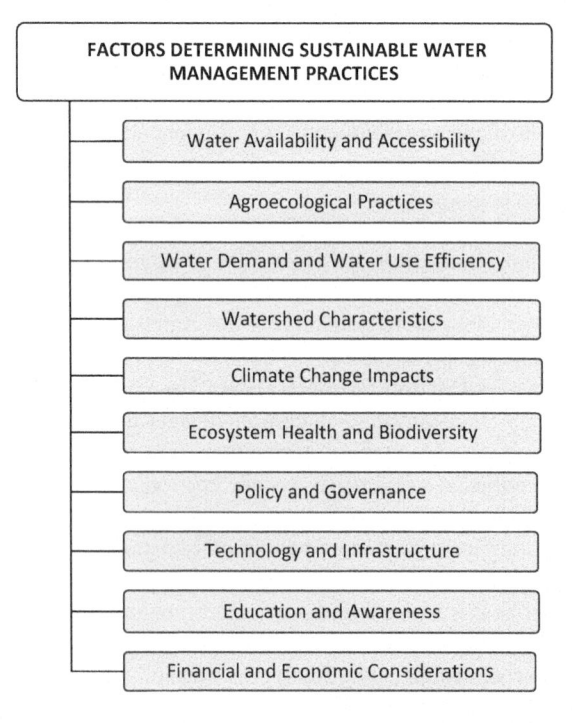

FIGURE 1.10 Factors determining sustainable water management practices. (Created by authors.)

- *Water Demand and Water Use Efficiency.* Understanding water demands from various sectors, including agriculture, industry, and domestic use, is essential for sustainable water management. Promoting water use efficiency through the adoption of efficient technologies, practices, and behavior change helps optimize water allocation and reduce wastage.
- *Watershed Characteristics.* The characteristics of a watershed, such as topography, soil type, and vegetation cover, influence water availability and quality. Sustainable water management considers these factors to assess the impacts of land use practices, erosion potential, and the need for conservation measures like reforestation or wetland protection.
- *Climate Change Impacts.* Climate change affects water availability, precipitation patterns, and water demand. Sustainable water management takes into account climate change projections to anticipate and plan for potential shifts in water resources, such as changes in rainfall patterns, increased evapotranspiration, or more frequent droughts or floods.
- *Ecosystem Health and Biodiversity.* The health and functioning of ecosystems are closely linked to water resources. Sustainable water management considers the impacts on aquatic habitats, wetlands, and riparian zones to ensure the preservation of biodiversity, maintain ecosystem services, and support the resilience of natural water systems.
- *Policy and Governance.* Effective policies, regulations, and governance frameworks are critical for sustainable water management. Clear water allocation mechanisms, equitable access, stakeholder participation, and coordinated decision making among different levels of government and stakeholders support sustainable water management practices.
- *Technology and Infrastructure.* Appropriate technology and infrastructure play a significant role in sustainable water management. Efficient irrigation systems, water treatment facilities, water storage and distribution networks, and monitoring systems enable better water management, conservation, and effective utilization of water resources.
- *Education and Awareness.* Education and awareness among water users, communities, and stakeholders are essential for sustainable water management. Promoting water literacy, encouraging responsible water use, and raising awareness about the importance of water conservation foster a culture of sustainable water practices.
- *Financial and Economic Considerations.* Sustainable water management requires adequate financial resources and economic incentives. Investing in water infrastructure, incentivizing water-efficient technologies and practices, and integrating the cost of water use and pollution into economic systems can encourage sustainable water management practices.

By considering these factors and adopting integrated approaches, societies can enhance their capacity to manage water sustainably, ensure water security, and minimize the impacts of water scarcity and pollution on both human and ecological systems.

7. **Support for Agroecological Transition**: The EGD provides support and incentives for farmers to transition to more sustainable and agroecological practices. This includes financial support, advisory services, and knowledge-sharing platforms to facilitate the adoption of sustainable farming methods.
8. **Digitalization and Precision Agriculture**: The Green Deal acknowledges the potential of digital technologies and precision agriculture in promoting sustainable agriculture. It encourages the use of digital tools, remote sensing, and data analytics to optimize resource use, improve farm management practices, and reduce environmental impact.

Digitalization and precision agriculture have several benefits. Bolfe et al. (2020) found that Brazilian farmers perceive increased productivity as the main benefit of using digital technologies in their production systems. Molin et al. (2020) discussed how digital technologies, such as the Internet of Things and Artificial Intelligence, can enhance agricultural management by predicting attributes of the crop and soil using advanced data-driven tools. Buick (1997) discussed how precision agriculture can assist with environmental compliance and management record keeping through automated record keeping. Digitalization and precision agriculture can lead to increased productivity, improved environmental compliance, and better management of spatial variability in fields.

These policy areas demonstrate the commitment of the EGD to transform the agricultural sector into a more sustainable and environmentally friendly industry. By promoting sustainable farming practices, biodiversity conservation, and resource efficiency, the aim is to ensure a resilient and sustainable food system for the future.

1.4 COOPERATION FOR THE TRANSITION TO A CARBON-NEUTRAL SOCIETY AMONG VARIOUS ACTORS AND SECTORS WITH EMPHASIS FOR AGRICULTURE

The fourth subtopic of this teaching material focuses on the cooperation required for the transition to a carbon-neutral society, emphasizing the agricultural sector. Achieving a sustainable and low-carbon future necessitates collaboration among multiple actors, including governments, businesses, farmers, and civil society. This section provides both basic and advanced understanding, encompassing key definitions, the importance of cooperation, and examples of collaboration in the agricultural sector.

A carbon-neutral society is a concept where the net carbon emissions produced by human activities are balanced or offset by removing an equivalent amount of carbon dioxide from the atmosphere. The goal is to achieve a state where the amount of carbon emissions released into the atmosphere is equal to the amount of carbon removed or absorbed.

Creating a carbon-neutral society requires a combination of strategies, including carbon reduction technologies, negative emission technologies, carbon trading, and carbon taxes. Chen et al. (2022) suggest plans such as change from fossil fuels to renewable energy, development of low-carbon technologies, low-carbon agriculture, changing dietary habits, and increasing the value of food and agricultural waste. Wheeler (2017) argues that developers and experts need to consider planned activities to change social ecology and climate policy, such as putting a price on carbon and developing education and social marketing programs for behavior change. Cheng (2021) identifies global trends in decarbonization actions embedded in countries' sustainable development agendas for the foreseeable future, including accelerating mature carbon solutions and promoting carbon-neutral technologies. Williams et al. (2021) present several ways by 2050 to achieve net-zero and net-negative CO_2 emissions, which meet all projected US energy needs and 2050 accounted for 0.2% to 1.2% of GDP using only commercial or near-commercial technologies and on demand commercial technologies, and requiring no early retirement of existing infrastructure.

The papers suggest that achieving carbon neutrality is an urgent and challenging task that requires significant societal transformation. Huang and Zhai (2021) note that to achieve the Paris Agreement temperature goals, carbon neutrality must be achieved by the middle of the century, with far-reaching transitions in all sectors of society. Chen et al. (2022) propose various strategies to achieve carbon neutrality, including shifting away from fossil fuels, developing low-carbon technologies, and increasing the value of food and agricultural waste. Zhao (2022) discusses China's goal of achieving carbon neutrality before 2060 and the challenges and potential solutions to achieving this goal. Wang et al. (2021) review innovative technologies that offer solutions to achieving carbon neutrality and sustainable development, including renewable energy production, waste valorization, and C-negative manufacturing. Achieving carbon neutrality is a complex and multifaceted task that requires significant societal transformation and the adoption of innovative technologies.

To create a carbon-neutral society, several key strategies and actions can be implemented (Chen et al., 2022; Cheon, 2022; Bookhart, 2008; Zhao, 2021; Gan & Zhang, 2022):

1. **Renewable Energy**: Transitioning to renewable energy sources such as solar, wind, hydro, and geothermal power can significantly reduce carbon emissions from the energy sector. By phasing out fossil fuel-based power generation, we can greatly reduce our carbon footprint.
2. **Energy Efficiency**: Improving energy efficiency in industries, buildings, transportation, and appliances can help reduce energy consumption and lower carbon emissions. This includes adopting energy-efficient technologies, better insulation, efficient transportation systems, and smart grid solutions.
3. **Electrification**: Shifting from fossil fuel-powered vehicles to EVs can have a significant impact on reducing carbon emissions in the transportation sector. Promoting EV adoption, expanding charging infrastructure, and incentivizing clean modes of transportation are essential steps.
4. **Sustainable Agriculture and Land Use**: Implementing sustainable agricultural practices, such as precision farming and organic methods, can help

reduce emissions from livestock, synthetic fertilizers, and deforestation. Conserving forests and promoting reforestation efforts also aid in carbon sequestration.
5. **Carbon Capture and Storage (CCS)**: Developing and deploying technologies for capturing and storing carbon dioxide from power plants and industrial facilities can help offset emissions that are challenging to eliminate completely. CCS can involve capturing CO_2 and either storing it underground or using it for other purposes like enhanced oil recovery or the production of building materials.
6. **Transitioning to a Circular Economy**: Promoting a circular economy that focuses on reducing waste, reusing materials, and recycling can reduce the carbon emissions associated with resource extraction, manufacturing, and waste management. It involves designing products for longevity and recyclability.
7. **Education and Awareness**: Raising awareness about the importance of a carbon-neutral society and providing education about sustainable practices can foster behavioral change and encourage individuals, communities, and businesses to make environmentally conscious decisions.
8. **Policy and Regulation**: Governments can play a crucial role by implementing policies and regulations that incentivize and support the transition to a carbon-neutral society. This includes setting emissions reduction targets, implementing carbon pricing mechanisms, providing subsidies for renewable energy, and promoting sustainable practices.

Achieving a carbon-neutral society requires a collective effort involving governments, businesses, communities, and individuals. Collaboration, technological advancements, and a commitment to sustainability are key to successfully mitigating climate change and creating a more environmentally friendly future.

A carbon-neutral society is a transformative and necessary response to the challenges posed by climate change. It offers environmental, economic, social, and health benefits while paving the way for a sustainable and resilient future (Figure 1.11).

As we can see from Figure 1.11, a carbon-neutral society is significant for several reasons (Chen et al., 2022; Zhang et al., 2021; Mi et al., 2023; Qin et al., 2021; Wang et al., 2021; Carrasco, 2014; Millot et al., 2018):

- **Climate Change Mitigation**: Carbon neutrality is a crucial step in mitigating climate change. By balancing or offsetting carbon emissions, a carbon-neutral society aims to reduce the concentration of GHGs in the atmosphere, which is a major driver of global warming and climate-related impacts. Agriculture is a significant contributor to GHG emissions, primarily through methane from livestock, nitrous oxide from synthetic fertilizers, and carbon dioxide from deforestation and land-use change. Achieving carbon neutrality in agriculture is crucial for mitigating climate change by reducing these emissions and preventing further environmental degradation.
- **Environmental Stewardship**: Achieving carbon neutrality demonstrates a commitment to environmental stewardship. It signifies a responsible

Low-Carbon Transition 45

```
THE IMPORTANCE OF A CARBON-NEUTRAL SOCIETY
    ├── Climate Change Mitigation
    ├── Environmental Stewardship
    ├── Sustainable Development
    ├── Energy Independence
    ├── Innovation and Technological Advancements
    ├── Health and Well-being
    ├── International Cooperation
    └── Positive Reputation and Leadership
```

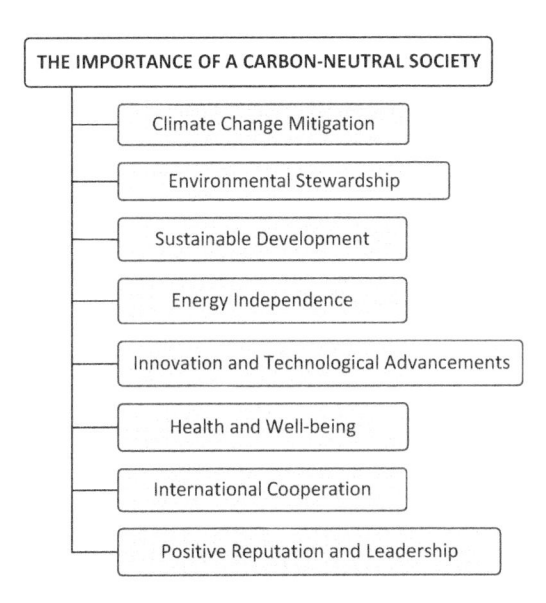

FIGURE 1.11 The importance of a carbon-neutral society. (Created by authors.)

approach to managing resources and minimizing the ecological footprint associated with human activities. This includes preserving ecosystems, protecting biodiversity, and ensuring a sustainable future for generations to come.

- **Sustainable Development**: Transitioning to a carbon-neutral society encourages sustainable development practices. It promotes the integration of economic growth, social well-being, and environmental protection, ensuring that progress is achieved without compromising the ability of future generations to meet their needs.
- **Energy Independence**: Shifting toward renewable energy sources as part of carbon neutrality reduces reliance on fossil fuels. This enhances energy independence, reducing vulnerability to geopolitical tensions and price fluctuations associated with fossil fuel extraction and consumption.
- **Innovation and Technological Advancements**: Pursuing carbon neutrality stimulates innovation and the development of new technologies. It drives investment and research in renewable energy, energy efficiency, CCS, and other clean technologies. This fosters economic growth, job creation, and the transition to a more sustainable and low-carbon economy.
- **Health and Well-being**: A carbon-neutral society can have significant benefits for human health. By reducing air and water pollution associated with carbon-intensive activities, it improves air quality, reduces respiratory illnesses, and enhances overall well-being. Additionally, promoting sustainable agriculture and reducing deforestation can contribute to food security and healthier ecosystems.
- **International Cooperation**: Achieving carbon neutrality requires global collaboration and cooperation. It encourages nations to work together to

address climate change collectively. International agreements like the Paris Agreement provide a framework for countries to set emissions reduction targets and support each other in transitioning to a low-carbon future.
- **Positive Reputation and Leadership**: A carbon-neutral society demonstrates leadership and responsible citizenship on the global stage. Countries, cities, businesses, and individuals that prioritize carbon neutrality gain a positive reputation and inspire others to take similar actions. It can attract investment, foster partnerships, and position entities as leaders in sustainability and environmental responsibility.

In order to achieve the intended goals, cooperation is necessary, which refers to working together toward a common goal or objective. The transition to a carbon-neutral society involves collaboration between different actors and sectors to share knowledge, resources, and efforts to implement sustainable practices and reduce GHG emissions.

Collaboration is a key concept that involves the collective efforts and cooperation of many actors, organizations, and stakeholders working together toward a common goal. It involves pooling knowledge, resources, expertise, and effort to achieve results that would be difficult or impossible to achieve individually.

Figure 1.12 presents the key aspects of collaboration according to Lawson (2004), Camarihna-Matos and Afsarmanesh (2008), Baker (2015), Graham and Barter (1999), Littler et al. (1998), Inshakov (2013), Thompson (2001), Walker and Elberson (2005), Byrne and Hansberry (2007), Riaz and Din (2023), Chia (2006), Camarinha-Matos et al. (2017).

KEY ASPECTS OF COLLABORATION
- Shared Vision and Goals
- Trust and Relationship Building
- Inclusive and Participatory Approach
- Clear Roles and Responsibilities
- Effective Communication
- Resource Sharing and Mobilization
- Conflict Resolution and Consensus Building
- Continuous Learning and Adaptation
- Monitoring, Evaluation, and Accountability
- Long-Term Sustainability

FIGURE 1.12 The key aspects of collaboration. (Created by authors.)

One of the key aspects is *shared vision and goals*, as collaboration begins with establishing a shared vision and goals that align with the desired outcome. Collaborators identify a common goal, whether it's achieving carbon neutrality, implementing sustainable practices, or solving a specific challenge. This shared vision is the unifying basis for cooperation.

Another aspect is *trust and building relationships*. Trust is essential in collaboration, as it is essential to build and nurture relationships based on mutual respect, transparency, and open communication. Collaborators must trust each other's intentions, capabilities, and commitments in order to work effectively together. Trust fosters an environment of collaboration, creativity, and problem-solving. Equally important is an *inclusive and participatory approach*, as collaboration is inclusive and includes diverse perspectives, knowledge, and experiences. This includes involving stakeholders from different sectors, fields, and disciplines, ensuring their meaningful participation in decision-making processes. This involvement improves the quality of outcomes and promotes ownership and buy-in by all involved.

In collaboration, *clear roles and responsibilities* must be established. Collaboration requires clearly defining the roles, responsibilities, and expectations of participants. Each collaborator should understand their specific contribution and how it fits into the overall collaborative effort. Clear roles and responsibilities reduce confusion, improve coordination, and enable effective results. The aspect of *effective communication* refers to clear, timely, and open channels of communication that facilitate information sharing, understanding, and alignment among co-workers. It encourages active listening, constructive dialogue, and exchange of ideas, enabling effective decision making and problem-solving. *Resource sharing and mobilization* are also important in cooperation. Resources in this case may include financial support, technical expertise, research results, infrastructure, and human capital. Collaborators pool their resources, leverage complementary strengths, and bridge gaps to maximize their collective impact. In the modern world, crisis management in various activities is becoming relevant, so another aspect is distinguished – *conflict resolution and consensus-building*, because cooperation often involves navigating different perspectives, priorities, and possible conflicts. Colleagues must be skilled in resolving conflicts, negotiating, and finding a common language. Consensus-building techniques such as brainstorming, mediation, and compromise are used to reconcile differences and reach mutually acceptable solutions.

Without *continuous learning and adaptation*, a dynamic collaborative process will not be effective, as collaborators adhere to a culture of continuous learning, evaluating results, and adjusting strategies based on feedback and new information. Learning from successes and failures helps improve methods, improve efficiency, and drive innovation.

Monitoring, evaluation and accountability is another aspect that includes mechanisms for monitoring progress, evaluating results, and ensuring accountability. Collaborators establish metrics, indicators, and feedback loops to measure the effectiveness of their collaboration. Regular monitoring and evaluation allow for course correction and continuous improvement toward common goals.

Cooperation can also ensure *long-term sustainability*, when it is not limited to short-term initiatives and aims for systemic changes and long-term solutions.

Collaborators consider the long-term implications of their actions, identify opportunities for scaling, and encourage the creation of networks and structures that can support collaborative efforts beyond the initial stages.

Collaboration leverages the collective power and capabilities of various actors, fostering synergy and collective intelligence. It empowers stakeholders to work together toward a shared vision, fostering innovation, fostering trust, and delivering results that address complex challenges more effectively than individual efforts could ever do.

Cooperation types among various actors and sectors are essential for the successful transition to a carbon-neutral society, with specific emphasis on agriculture:

1. **Multi-Stakeholder Engagement**: Cooperation involves engaging a wide range of stakeholders in the agricultural sector, including farmers, agricultural associations, research institutions, policymakers, consumers, and environmental organizations. Each stakeholder brings unique perspectives, expertise, and resources to the table. Cooperation allows for effective collaboration, knowledge sharing, and coordinated action to drive sustainable practices and reduce emissions in agriculture.

    In agriculture, multi-stakeholders refer to the diverse range of actors and organizations that have an interest or are affected by agricultural activities. Figures 1.13 and 1.14 present stakeholders, who vary depending on the specific context and location (Häring et al., 2009; Dumont et al., 2017; Neef & Neubert, 2011; Babu et al., 2021; Konefal et al., 2017; Prager & Freese, 2009; Ebrahimi et al., 2021; MacLeod et al., 2021).

    These stakeholders interact and collaborate to drive sustainable agriculture practices, mitigate climate change impacts, promote food security, and achieve a carbon-neutral agriculture sector. Multi-stakeholder engagement ensures that their diverse perspectives and interests are considered in decision-making processes, creating a more inclusive and comprehensive approach to sustainable agriculture.

2. **Public–Private Partnerships**: Cooperation between the public and private sectors plays a crucial role in advancing sustainability in agriculture. Public–private partnerships (PPPs) foster collaboration and resource-sharing between government entities, businesses, and other stakeholders. These partnerships can facilitate the development of policies, funding mechanisms, and initiatives that incentivize sustainable farming practices and support the adoption of low-carbon technologies in agriculture.

    PPPs are a composite and multi-layered concept that has increased popularity globally. Hodge and Greve (2007) argue that assessments of PPPs have produced inconsistent results concerning their effectiveness, and greater care is needed to strengthen upcoming assessments. A methodical and combined method to PPP financing in the framework of public policy can clarify the project financing approaches used for this resolution. Linder (1999) disapprovingly observes the multiple connotations that the term "public-private partnership" accepts in modern debates, exposing their

Low-Carbon Transition

	Stakeholder	Description
THE MULTI-STAKEHOLDERS IN AGRICULTURE (1)	Farmers and Farmer Organizations	Farmers are primary stakeholders in agriculture. They include small-scale farmers, large-scale farmers, family farmers, and agricultural cooperatives. Farmer organizations, such as farmer associations, cooperatives, and unions, represent the collective interests of farmers and advocate for their needs.
	Agricultural Producers and Agribusinesses	Agricultural producers, including crop growers, livestock farmers, and fishers, are important stakeholders. Agribusinesses, such as food processors, distributors, and retailers, are also key actors in the agricultural value chain. These stakeholders play a crucial role in implementing sustainable practices and driving market-driven solutions.
	Government Agencies and Policymakers	Government agencies at various levels, such as ministries of agriculture, environment, and rural development, are significant stakeholders. Policymakers formulate and implement agricultural policies, regulations, and programs that influence farming practices, land use, and sustainability initiatives.
	Research Institutions and Academia	Research institutions, universities, and agricultural colleges contribute to knowledge generation, innovation, and capacity building in the agricultural sector. They play a vital role in conducting research, developing technologies, and providing scientific expertise to inform sustainable agricultural practices.
	Non-Governmental Organizations (NGOs)	NGOs, including environmental organizations, development organizations, and farmer support groups, are important stakeholders in agriculture. They work towards promoting sustainable farming practices, advocating for social and environmental issues, providing technical assistance, and supporting farmers' welfare.
	Consumers and Consumer Organizations	Consumers have a stake in agriculture as they influence market demand and purchasing decisions. Their preferences for sustainably produced, locally sourced, and environmentally friendly food products drive changes in the agricultural sector. Consumer organizations advocate for fair trade, organic farming, and sustainable food systems.
	Financial Institutions	Financial institutions, such as banks, impact investors, and microfinance institutions, are stakeholders in agriculture. They provide financing and investment opportunities to farmers and agribusinesses engaged in sustainable practices, innovative technologies, and climate-smart agriculture

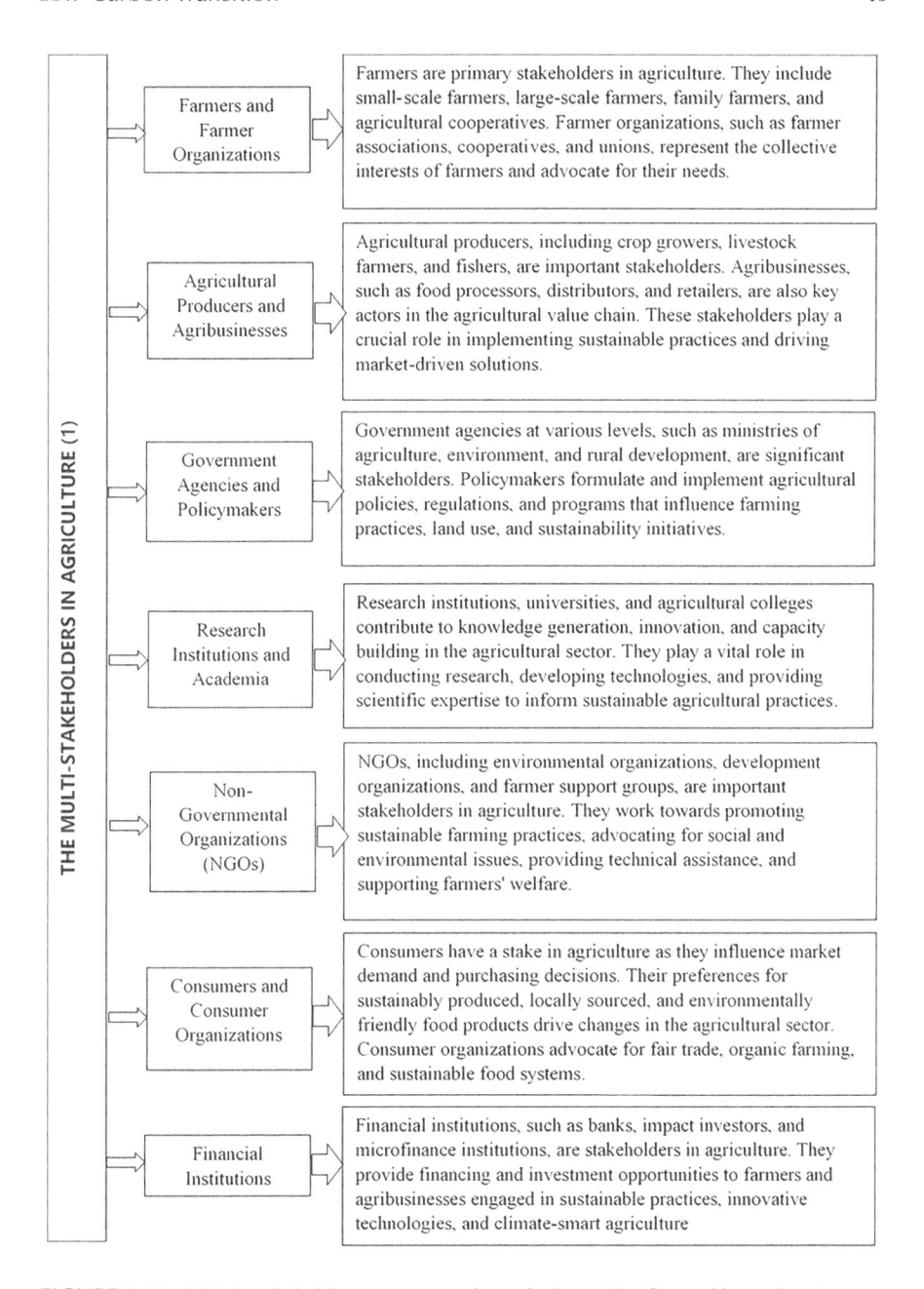

FIGURE 1.13 Multi-stakeholder engagement in agriculture (1). (Created by authors.)

fundamental premises and ideological promises. Broadbent and Laughlin (2003) provide a suggestion of PPPs as an existing allowance of the "new public management" schedule for variations in the way public services are provided and highlight the need for additional investigation on the wildlife, rule, pre-decision analysis, and post-project assessment of PPPs. PPPs are

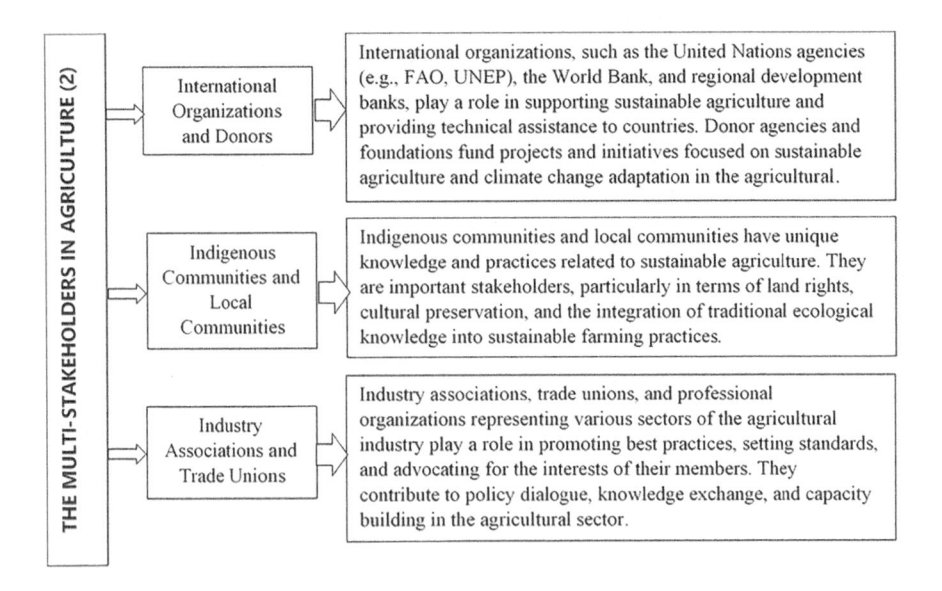

THE MULTI-STAKEHOLDERS IN AGRICULTURE (2)	International Organizations and Donors	International organizations, such as the United Nations agencies (e.g., FAO, UNEP), the World Bank, and regional development banks, play a role in supporting sustainable agriculture and providing technical assistance to countries. Donor agencies and foundations fund projects and initiatives focused on sustainable agriculture and climate change adaptation in the agricultural.
	Indigenous Communities and Local Communities	Indigenous communities and local communities have unique knowledge and practices related to sustainable agriculture. They are important stakeholders, particularly in terms of land rights, cultural preservation, and the integration of traditional ecological knowledge into sustainable farming practices.
	Industry Associations and Trade Unions	Industry associations, trade unions, and professional organizations representing various sectors of the agricultural industry play a role in promoting best practices, setting standards, and advocating for the interests of their members. They contribute to policy dialogue, knowledge exchange, and capacity building in the agricultural sector.

FIGURE 1.14 Multi-stakeholder engagement in agriculture (2). (Created by authors.)

a multifaceted and evolving idea that needs careful assessment and reflection of their fundamental premises and conceptual commitments (Torchia et al., 2015).

PPPs are important for the agriculture sector in relations of distribution incomes, technology, and benefits. Hall (2006) found that PPP can empower farmers, especially women, through knowledge management, capacity building, and market promotion. Shukla et al. (2016) discussed how PPP can accelerate development in various areas of agribusiness and infrastructure, such as crop diversification, marketing infrastructure, and contract farming. Dustmurodov et al. (2020) emphasized the importance of improving the mechanism of PPP in agriculture in Uzbekistan. However, Spielman and Grebmer (2006) highlighted that PPP between public research agencies and private firms are constrained by different incentive structures, competition, and risk. Overall, PPP can play a significant role in the agriculture sector, but there are challenges that need to be addressed to ensure successful collaboration between public and private sectors (Kolomoiets et al., 2021).

PPPs in agriculture refer to collaborations between public sector entities, such as governments or government agencies, and private sector actors, including businesses, corporations, and non-profit organizations, to address agricultural challenges and achieve common goals. These partnerships leverage the strengths and resources of both sectors to promote sustainable agriculture, enhance food security, and drive innovation. Here are some key aspects and benefits of PPPs in agriculture (Figure 1.15).

- *Shared Resources and Expertise.* PPPs contribute to food security and improved nutrition by enhancing agricultural productivity and supply chain efficiency. These collaborations can focus on increasing crop

Low-Carbon Transition 51

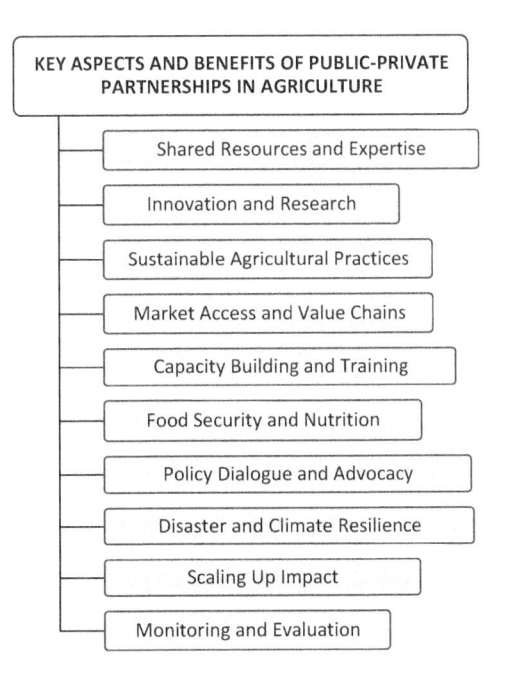

KEY ASPECTS AND BENEFITS OF PUBLIC-PRIVATE PARTNERSHIPS IN AGRICULTURE
- Shared Resources and Expertise
- Innovation and Research
- Sustainable Agricultural Practices
- Market Access and Value Chains
- Capacity Building and Training
- Food Security and Nutrition
- Policy Dialogue and Advocacy
- Disaster and Climate Resilience
- Scaling Up Impact
- Monitoring and Evaluation

FIGURE 1.15 Key aspects and benefits of public-private partnerships in agriculture. (Created by authors.)

yields, reducing post-harvest losses, and fortifying food products to address malnutrition. By combining efforts, PPPs can address food insecurity at local and global levels.

- *Innovation and Research.* PPPs promote innovation in agriculture by combining R&D efforts from both sectors. Private companies often invest in research and technological advancements, while public research institutions contribute scientific knowledge and expertise. Together, they can develop and deploy cutting-edge technologies and practices that improve agricultural productivity and sustainability.
- *Sustainable Agricultural Practices.* PPPs play a crucial role in promoting and implementing sustainable agricultural practices. By working together, stakeholders can support the adoption of climate-smart farming techniques, resource-efficient methods, and environmentally friendly practices. These partnerships contribute to reducing the environmental footprint of agriculture and enhancing its resilience to climate change.
- *Market Access and Value Chains.* PPPs facilitate the integration of smallholder farmers and rural communities into formal markets and value chains. Private sector partners often provide market linkages, technical training, and access to improved inputs and technologies. This empowers farmers to produce higher-quality products, access fair prices, and improve their livelihoods.
- *Capacity Building and Training.* PPPs support capacity building and training programs for farmers and other stakeholders in the agricultural

sector. Public sector agencies and non-profit organizations can deliver extension services, agricultural education, and knowledge transfer. Private partners can offer training on sustainable farming practices, technology use, and business development, fostering the development of a skilled workforce.
- *Food Security and Nutrition.* PPPs contribute to food security and improved nutrition by enhancing agricultural productivity and supply chain efficiency. These collaborations can focus on increasing crop yields, reducing post-harvest losses, and fortifying food products to address malnutrition. By combining efforts, PPPs can address food insecurity at local and global levels.
- *Policy Dialogue and Advocacy.* PPPs facilitate policy dialogue and advocacy efforts for sustainable agriculture. Collaboration between government and private sector stakeholders allows for the development of policies and regulations that support innovation, investment, and the adoption of sustainable practices. PPPs can also advocate for the inclusion of sustainable agriculture in national and international agendas.
- *Disaster and Climate Resilience.* PPPs contribute to building resilience in agriculture by jointly addressing disaster risk reduction and climate change adaptation. These partnerships can develop early warning systems, promote resilient crop varieties, and implement climate-smart practices to help farmers cope with the impacts of extreme weather events.
- *Scaling Up Impact.* PPPs have the potential to scale up successful pilot projects and initiatives. Once proven effective, these collaborations can attract additional funding and investment, leading to the wider adoption of sustainable practices and technologies across different regions and farming systems.
- *Monitoring and Evaluation.* PPPs emphasize monitoring and evaluation to assess the effectiveness of joint initiatives and track progress toward shared goals. Regular assessment helps identify areas for improvement, measure the impact of interventions, and inform evidence-based decision making for future projects.

Overall, PPPs in agriculture offer a powerful mechanism for addressing complex challenges, driving innovation, and achieving sustainable agricultural development. By leveraging the complementary strengths of both sectors, these collaborations contribute to a more inclusive, resilient, and environmentally conscious agricultural sector.

3. **Research and Innovation Collaboration**: Cooperation in research and innovation is key to developing sustainable solutions for agriculture. Collaborative efforts between research institutions, universities, and agricultural organizations can accelerate the development and deployment of climate-smart technologies, precision agriculture techniques, and improved farming practices. Sharing research findings, data, and resources enables stakeholders to make informed decisions and drive innovation in the sector.

Successful research and innovation collaboration in agriculture depends on several critical factors that enable effective partnerships and the achievement of desired outcomes. According to Velten et al. (2021), Lambrecht et al. (2015), Hoffmann et al. (2007), Begum & Sami (1988), Moser (2019), Hall (2012), Hermans et al. (2015), Ingram et al. (2020), Williams et al. (2020), Manjula and Rengalakshmi (2021), there are the key factors determining successful research and innovation collaboration in agriculture:

- *Shared Vision and Goals.* A clear and shared vision for the collaboration is essential. All partners must align their goals and objectives, ensuring that their efforts are directed toward common targets. Establishing a collective understanding of the desired outcomes fosters commitment and unity among collaborators.
- *Complementary Expertise.* Successful collaboration in agriculture requires partners with diverse and complementary expertise. Research institutions bring scientific knowledge and technical expertise, while agricultural organizations and businesses contribute practical insights and on-the-ground experience. This combination of expertise enhances problem-solving and innovation.
- *Trust and Communication.* Building trust among collaborators is fundamental. Open and transparent communication is vital for effective collaboration. Trust fosters honest exchange of ideas, allows partners to share challenges and risks openly, and facilitates constructive feedback.
- *Defined Roles and Responsibilities.* Clearly defined roles and responsibilities for each partner are critical for a smooth collaboration. This includes delineating tasks, contributions, and expectations. Clarity in roles minimizes confusion, reduces redundancy, and ensures efficient progress toward shared goals.
- *Resource Commitment.* Successful collaboration requires a commitment of resources from all partners. This includes financial investment, access to facilities, research materials, and human resources. Adequate resource allocation ensures that the collaboration can pursue its objectives effectively.
- *Effective Leadership and Coordination.* Strong leadership and effective coordination are crucial for guiding the collaboration. Designating a project leader or coordinator who can facilitate communication, manage timelines, and address conflicts helps keep the collaboration on track.
- *Flexibility and Adaptability.* Collaboration in agriculture may face unforeseen challenges and changing circumstances. Being flexible and adaptable to new information, emerging opportunities, and evolving priorities is essential for the success of the partnership.
- *Data Sharing and Intellectual Property Rights.* Collaborators must agree on data-sharing protocols and address intellectual property rights. Ensuring that data is shared in a timely and transparent manner promotes collaboration and enables all partners to benefit from the research.

- *Evaluation and Feedback Mechanisms.* Establishing mechanisms for regular evaluation and feedback enhances collaboration. Periodic assessments help track progress, identify strengths and weaknesses, and make necessary adjustments to improve outcomes.
- *Engagement of Stakeholders.* Engaging relevant stakeholders throughout the collaboration process is vital for ensuring that research and innovation address real-world challenges and meet the needs of end-users. Involving farmers, policymakers, and other stakeholders provides valuable insights and increases the relevance and impact of the collaboration's outputs.
- *Sustainability and Long-Term Commitment.* Long-term commitment to the collaboration is crucial for achieving lasting impact. Sustainable collaborations foster ongoing learning, knowledge exchange, and joint problem-solving beyond individual projects.

By paying attention to these factors, research and innovation collaboration in agriculture can be effectively harnessed to develop and implement solutions that address pressing agricultural challenges, drive sustainable practices, and promote food security and environmental conservation.

4. **Knowledge Sharing and Capacity Building**: Cooperation facilitates knowledge-sharing and capacity-building initiatives. Stakeholders can collaborate on educational programs, training workshops, and extension services to disseminate knowledge about sustainable agricultural practices, carbon sequestration methods, and climate change mitigation strategies (Srivastava, 2020; Coulibaly et al., 2021; Singh et al., 2021). This cooperation strengthens the capacity of farmers and other actors in the agricultural sector to adopt and implement sustainable practices effectively.
5. **Supply Chain Cooperation**: Cooperation along the agricultural supply chain is crucial for achieving carbon neutrality. Collaboration between farmers, processors, distributors, retailers, and consumers allows for the development of sustainable supply chain practices. Stakeholders can work together to reduce waste, improve energy efficiency, promote sustainable sourcing, and implement carbon accounting measures. Cooperation enables the traceability and transparency necessary to achieve a carbon-neutral supply chain.

Cooperation in the agricultural supply chain is essential for optimizing efficiency, reducing waste, and promoting sustainability. Successful cooperation depends on several key factors that enable effective collaboration and value creation for all stakeholders involved. Figure 1.16 shows success factors for cooperation in the agricultural supply chain.

According to Figure 1.16, we can see these success factors for cooperation in the agricultural supply chain (Yu, 2011; Zhu et al., 2016, 2021; Usuga et al., 2012; Huo et al., 2022; Fang & Meng, 2009; Ouhna, 2019; Settle et al., 2017; Zaraté et al., 2020; Linsner et al., 2022; Anantraj, 2021; Rumble, 2015; Brezuleanu et al., 2015; Chaponnière et al., 2012):

```
┌─────────────────────────────────────┐
│ THE SUCCESS FACTORS FOR COOPERATION IN │
│    THE AGRICULTURAL SUPPLY CHAIN    │
└─────────────────────────────────────┘
   ├── Clear Communication
   ├── Shared Goals and Objectives
   ├── Trust and Reliability
   ├── Collaborative Planning
   ├── Transparent Information Sharing
   ├── Win-Win Relationships
   ├── Flexibility and Adaptability
   ├── Performance Evaluation and Continuous Improvementesilience
   ├── Responsible and Sustainable Practices
   └── Conflict Resolution Mechanisms
```

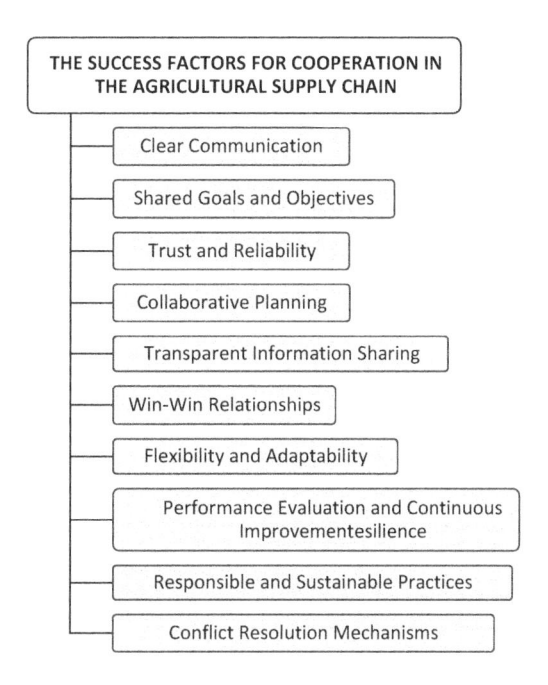

FIGURE 1.16 Success factors for cooperation in the agricultural supply chain. (Created by authors.)

- *Clear Communication.* Effective communication is crucial for successful cooperation in the agricultural supply chain. All stakeholders, including farmers, suppliers, processors, distributors, retailers, and consumers, must communicate openly and transparently. Clear communication ensures that information, orders, and expectations are conveyed accurately, reducing misunderstandings and delays.
- *Shared Goals and Objectives.* Cooperation thrives when all participants share common goals and objectives. Stakeholders in the agricultural supply chain should have a shared commitment to sustainable practices, quality standards, and customer satisfaction. Aligning goals helps create a sense of purpose and fosters a cooperative spirit among all involved.
- *Trust and Reliability.* Building and maintaining trust is fundamental for effective cooperation. Each stakeholder in the supply chain should trust others to fulfill their responsibilities and obligations. Reliability in delivering products on time, adhering to agreed-upon quality standards, and meeting contractual commitments is essential for fostering trust.
- *Collaborative Planning.* Collaboration in the agricultural supply chain requires collaborative planning and coordination. Stakeholders should work together to develop efficient supply chain processes, including forecasting, inventory management, and distribution strategies. Collaborative planning reduces inefficiencies, stockouts, and surplus, optimizing the overall supply chain performance.

- *Transparent Information Sharing.* Transparency in sharing information is critical for ensuring the smooth flow of products and data across the supply chain. Timely sharing of information on inventory levels, demand forecasts, market trends, and pricing helps all stakeholders make informed decisions and respond proactively to changes.
- *Win-Win Relationships.* Cooperation in the agricultural supply chain should aim for win–win relationships. All stakeholders should benefit from the collaboration, whether through improved profitability, reduced costs, enhanced market access, or increased customer satisfaction. A win–win approach encourages long-term commitment to the partnership.
- *Flexibility and Adaptability.* The agricultural supply chain can be influenced by various external factors, such as weather conditions, market fluctuations, and policy changes. Successful cooperation requires flexibility and adaptability to respond to unexpected events. Being able to adjust supply chain processes and strategies helps mitigate risks and capitalize on emerging opportunities.
- *Performance Evaluation and Continuous Improvement.* Regularly evaluating the performance of the cooperative efforts is essential. Monitoring key performance indicators (KPIs) and seeking feedback from stakeholders enable the identification of areas for improvement. Continuous improvement initiatives help enhance the efficiency and effectiveness of the supply chain collaboration over time.
- *Responsible and Sustainable Practices.* Cooperation in the agricultural supply chain should promote responsible and sustainable practices. This includes environmentally friendly farming methods, ethical sourcing, resource conservation, and social responsibility. Emphasizing sustainability enhances the reputation of the supply chain and attracts environmentally conscious consumers and partners.
- *Conflict Resolution Mechanisms.* Despite the best efforts, conflicts may arise in the agricultural supply chain. Establishing conflict resolution mechanisms allows stakeholders to address disputes promptly and constructively. Effective resolution ensures that cooperation continues smoothly and that relationships are not negatively affected.

By embracing these success factors, stakeholders in the agricultural supply chain can foster strong and effective collaboration. Cooperation in the supply chain leads to enhanced efficiency, sustainability, and overall performance, benefiting all stakeholders involved, from producers to consumers.

6. **Policy Alignment and Harmonization**: Cooperation among policymakers and regulatory bodies is necessary to develop coherent policies and regulations that support the transition to a carbon-neutral agricultural sector. Collaboration ensures that policies are aligned and consistent across different levels of governance. This includes setting targets for emissions reduction, implementing carbon pricing mechanisms, providing financial incentives for sustainable farming practices, and creating a supportive

policy environment that encourages innovation and adoption of low-carbon technologies (Fahamsyah & Chansrakaeo, 2022).

Successful policy alignment and harmonization in the agricultural sector depend on several key factors. Some of the crucial factors that contribute to effective policy coordination and integration include (Lu et al., 2020; Wirén-Lehr, 2001; Jordan et al., 2020; Raišienė et al., 2019; Pe'er et al., 2020; Neef & Neubert, 2011; Konefal et al., 2017; Larsen & Powell, 2013; Muscat et al., 2021; Brooks, 2014; Darnhofer et al., 2010; Adukova et al., 2022):

- *Clear Objectives and Shared Vision.* Having a common understanding of the goals and objectives among all stakeholders involved in the agricultural sector is essential. A shared vision helps align policies toward a unified direction and facilitates cooperation.
- *Political Will and Commitment.* Strong political will and commitment from governments and policymakers are vital for successful policy alignment. It requires leadership and dedication to overcome challenges and implement coordinated policies.
- *Multi-Stakeholder Engagement.* Involving various stakeholders, such as farmers, industry representatives, researchers, and civil society, in the policy-making process fosters inclusivity and ensures that policies address diverse needs and concerns.
- *Cross-Sectoral Collaboration.* Agriculture is interconnected with various sectors, such as environment, trade, health, and rural development. Effective policy alignment requires collaboration and coherence between different sectors to avoid conflicting policies.
- *Policy Coherence and Consistency.* Policymakers must ensure that agricultural policies do not contradict or undermine other related policies, such as environmental, climate, or trade policies. Coherence ensures a more integrated and effective approach.
- *Capacity Building and Knowledge Sharing.* Enhancing the capacity of relevant institutions and stakeholders in understanding and implementing policies is crucial. Knowledge sharing and capacity building promote better policy implementation.
- *Data and Evidence-based Decision Making.* Policies should be grounded in sound data and evidence to ensure that they are relevant, efficient, and effective in achieving their objectives.
- *Flexibility and Adaptability.* The agricultural sector is subject to various uncertainties, including climate change, market fluctuations, and technological advancements. Policies should be flexible and adaptable to respond to changing circumstances.
- *Policy Evaluation and Feedback Loops.* Regular evaluation of policy outcomes and impact is essential to assess effectiveness and identify areas for improvement. Establishing feedback loops allows policymakers to learn from experiences and refine policies accordingly.
- *International Cooperation and Harmonization.* In an increasingly interconnected world, international cooperation and harmonization

of policies can enhance agricultural trade, mitigate cross-border challenges, and foster sustainable development.

Overall, successful policy alignment and harmonization in the agricultural sector require a combination of effective governance, stakeholder engagement, data-driven decision making, and adaptability to create a conducive policy environment that supports sustainable agriculture and rural development.
7. **Financial and Technical Support**: Cooperation involves mobilizing financial resources and technical support from various sources to support sustainable agriculture. Collaboration between financial institutions, development agencies, and agricultural stakeholders can facilitate access to funding, loans, and grants for implementing sustainable practices. Technical support can include sharing expertise, providing training on sustainable techniques, and facilitating the adoption of climate-smart technologies in agriculture.
8. **Scaling Up Successful Models**: Cooperation enables the scaling up of successful sustainable agricultural models and practices. By sharing experiences, lessons learned, and best practices, stakeholders can replicate and adapt successful initiatives in different contexts. Cooperation allows for the dissemination of knowledge and experiences to reach a broader audience, accelerating the adoption of sustainable practices and achieving a wider impact on carbon neutrality in agriculture.
9. **Advocacy and Awareness Campaigns**: Cooperation is essential for collaborative advocacy and awareness campaigns. Stakeholders can join forces to raise awareness about the importance of a carbon-neutral agriculture sector, promote sustainable farming practices, and engage policymakers, consumers, and the public in the transition. Cooperation allows for the pooling of resources, expertise, and networks to amplify messaging and drive positive change at a larger scale.

By fostering collaboration and cooperation, we can accelerate the transition to a carbon-neutral agricultural sector, ensuring food security, promoting rural development, and mitigating climate change impacts. Together, we can create a more sustainable and resilient future for agriculture and our planet.

1.5 CONCLUSIONS AND RECOMMENDATIONS

The most important conclusions are provided below:

1. **Urgency of Low-Carbon Transition**: The urgency to address climate change and transition to a low-carbon economy is evident. The agricultural sector, as a significant contributor to GHG emissions, must actively participate in this transition to ensure a sustainable future.
2. **International Political Support**: International political support is critical for effective climate change mitigation and low-carbon transition. Agreements such as the Paris Agreement provide a framework for collaboration and

collective action, emphasizing the importance of reducing emissions and promoting sustainable practices in agriculture.
3. **Barriers and Policy Measures**: Transitioning to a low-carbon agricultural sector faces various barriers, including technical, economic, social, and institutional challenges. Overcoming these barriers requires the implementation of supportive policies, financial incentives, capacity-building programs, and knowledge sharing initiatives to promote sustainable agricultural practices.
4. **Green Deal Policies**: The Green Deal, as a comprehensive policy framework, offers significant opportunities for sustainable agriculture. Policy areas such as the Farm to Fork Strategy, Biodiversity Strategy, circular economy, and support for agroecology and digitalization in agriculture contribute to the overall goal of achieving a climate-neutral society.
5. **Cooperation and Collaboration**: Collaboration among various actors and sectors is essential for the low-carbon transition in agriculture. Multi-stakeholder cooperation, PPPs, knowledge sharing, and international collaborations enable the exchange of ideas, resources, and expertise to implement sustainable practices and drive innovation.

The main recommendations are summarized below:

1. **Strengthen Policy Support**: Governments should continue developing and implementing supportive policies, financial incentives, and regulatory frameworks that promote sustainable agricultural practices, renewable energy integration, and carbon sequestration.
2. **Enhance Collaboration and Cooperation**: Stakeholders, including governments, farmers, researchers, NGOs, and businesses, should foster collaboration and cooperation through PPPs, knowledge-sharing platforms, and research networks to accelerate the low-carbon transition in agriculture.
3. **Invest in Research and Innovation**: Increased investment in R&D is crucial to drive innovation and develop climate-smart solutions for agriculture. This includes advancing precision agriculture, developing climate-resilient crop varieties, and exploring sustainable energy solutions for farming operations.
4. **Promote Education and Awareness**: Education and awareness programs should be encouraged to disseminate knowledge about the importance of sustainable agriculture, climate change impacts, and the benefits of low-carbon practices. This can empower farmers and stakeholders to adopt and advocate for sustainable approaches.
5. **Support Rural Communities**: Transitioning to a low-carbon agricultural sector should include support for rural communities, ensuring their socio-economic well-being, and providing access to training, financial resources, and market opportunities for sustainable agricultural practices.

By implementing these recommendations and fostering a collaborative and supportive environment, we can accelerate the low-carbon transition in agriculture, mitigate climate change impacts, promote sustainable practices, and ensure food security for future generations. It is our collective responsibility to actively engage in the transition toward a sustainable and carbon-neutral agricultural sector, contributing to a healthier planet and a more resilient future.

REFERENCES

Abbasi, I. A., Ashari, H., Ariffin, A. S., & Yusuf, I. (2023). Farm to Fork: Indigenous chicken value chain modelling using system dynamics approach. *Sustainability*, *15*(2), 1402. https://doi.org/10.3390/su15021402

Abdalla, Z. F., El-Ramady, H., Omara, A. E., Elsakhawy, T., Bayoumi, Y. A., Shalaby, T. A., & Prokisch, J. (2022). From farm-to-fork: A pictorial mini review on nano-farming of vegetables. *Environment, Biodiversity and Soil Security*, *6*, 149–163. https://doi.org/10.21608/jenvbs.2022.145977.1180

Abobatta, W. F. (2022). Importance of smart agriculture. *Open Journal of Biotechnology and Bioengineering Research*, *24*(1), 25–28. https://doi.org/10.37871/ojbbr.id19

Adger, W. N., Arnell, N. W., & Tompkins, E. L. (2005). Successful adaptation to climate change across scales. *Global Environmental Change: Human and Policy Dimensions*, 15, 77–86. https://doi.org/10.1016/j.gloenvcha.2004.12.005

Adukova, A. N., Adukov, R. K., & Zakharov, R. V. (2022). *Harmonious development of integration and cooperation in agriculture as a factor in ensuring food security*. IOP Conference Series: Earth and Environmental Science, 949. https://doi.org/10.1088/1755-1315/949/1/012145

Alexandratos, S. D., Barak, N. A., Bauer, D., Davidson, F. T., Gibney, B. R., Hubbard, S. S., Taft, H. L., & Westerhof, P. V. (2019). Sustaining water resources: Environmental and economic impact. *ACS Sustainable Chemistry & Engineering*, *7*(3), 2879–2888. https://doi.org/10.1021/ACSSUSCHEMENG.8B05859

AlShrouf, A. (2017). Hydroponics, aeroponic and aquaponic as compared with conventional farming. *American Scientific Research Journal for Engineering, Technology, and Sciences*, *27*, 247–255. https://core.ac.uk/download/pdf/235050152.pdf

Altieri, M. A. (1995). *Agroecology: The science of sustainable agriculture*, Second Edition. Boulder, CO: Westview Press.

Altieri, M. A., Toledo, V. M. (2011). The agroecological revolution in Latin America: rescuing nature, ensuring food sovereignty and empowering peasants. *The Journal of Peasant Studies*, *38*(3), 587–612, https://doi.org/10.1080/03066150.2011.582947

Anantraj, I. (2021). Secure data sharing in agriculture using blockchain technology. *Turkish Journal of Computer and Mathematics Education*, *12*(11), 1925–1933. https://www.turcomat.org/index.php/turkbilmat/article/view/6147/5114

Arıkan, G., & Günay, D. (2021). Public attitudes towards climate change: A cross-country analysis. *The British Journal of Politics and International Relations*, *23*(1), 158–174. https://doi.org/10.1177/1369148120951013

Arshad, M., Kächele, H., Krupnik, T.J., Amjath-Babu, T. S., Aravindakshan, S., Abbas, A., Mehmood, Y., & Müller, K. (2017). Climate variability, farmland value, and farmers' perceptions of climate change: Implications for adaptation in rural Pakistan. *International Journal of Sustainable Development & World Ecology*, *24*, 532–544. https://doi.org/10.1080/13504509.2016.1254689

Ashworth, P., Einsiedel, E. F., Howell, R., Brunsting, S., Boughen, N., Boyd, A. D., Shackley, S., Bree, B. V., Jeanneret, T., Stenner, K., Medlock, J., Mabon, L., Feenstra, C., & Hekkenberg, M. (2013). Public preferences to CCS: How does it change across countries? *Energy Procedia*, *37*, 7410–7418. https://doi.org/10.1016/J.EGYPRO.2013.06.683

Asrat, P., & Simane, B. (2017). Adapting smallholder agriculture to climate change through sustainable land management practices: Empirical evidence from north-west Ethiopia. *Journal of Agricultural Science & Technology A*, 7, 289–301. https://doi.org/10.17265/2161-6256/2017.05.001

Auvinen, H., Santti, U., & Happonen, A. (2020). *Technologies for reducing emissions and costs in combined heat and power production*. E3S Web of Conferences. 2019 7th International Conference on Environment Pollution and Prevention (ICEPP 2019). https://doi.org/10.1051/e3sconf/202015803006

Babu, S. C., Franzel, S. C., Davis, K., & Srivastava, N. (2021). Drivers of youth engagement in agriculture: Insights from Guatemala, Niger, Nigeria, Rwanda, and Uganda. SRPN: Farming & Agriculture (Topic). https://doi.org/10.2499/P15738COLL2.134328

Bacon, C. M., Getz, C., Kraus, S., Montenegro, M., & Holland, K. (2012). The social dimensions of sustainability and change in diversified farming systems. *Ecology and Society*, 17, 41. https://doi.org/10.5751/ES-05226-170441

Baker, M. (2015). Collaboration in collaborative learning. *Interaction Studies*, 16, 451–473. https://doi.org/10.1075/IS.16.3.05BAK

Bechtel, M.M., & Scheve, K.F. (2012). Mass support for global climate agreements depends on institutional design. *Proceedings of the National Academy of Sciences*, 110, 13763–13768.

Beck, F., & Martinot, E. (2004). Renewable energy policies and barriers. *Encyclopedia of Energy*, 5, 365–383 https://doi.org/10.1016/B0-12-176480-X/00488-5

Begum, K. J., & Sami, L. K. (1988). Research collaboration in agricultural science. *International Library Review*, 20, 57–63. https://doi.org/10.1016/0020-7837(88)90044-1

Beillouin, D., Ben-Ari, T., Malézieux, É., Seufert, V., & Makowski, D. (2020). Benefits of crop diversification for biodiversity and ecosystem services. bioRxiv. https://doi.org/10.1101/2020.09.30.320309

Beillouin, D., Ben-Ari, T., Malézieux, É., Seufert, V., & Makowski, D. (2021). Positive but variable effects of crop diversification on biodiversity and ecosystem services. *Global Change Biology*, 27, 4697–4710. https://doi.org/10.1111/gcb.15747

Bezdek, R. H. (2020). The jobs impact of the USA New Green Deal. *American Journal of Industrial and Business Management*, 10(6), 1085–1106. https://doi.org/10.4236/ajibm.2020.106072

Blattner, C. E. (2020). Just transition for agriculture? A critical step in tackling climate change. *The Journal of Agriculture, Food Systems, and Community Development*, 9(3), 1–6. https://doi.org/10.5304/jafscd.2020.093.006

Boelee, E., Geerling, G. W., van der Zaan, B. M., Blauw, A. N., & Vethaak, A. D. (2019). Water and health: From environmental pressures to integrated responses. *Acta Tropica*, 193, 217–226 https://doi.org/10.1016/j.actatropica.2019.03.011

Bolfe, É. L., Jorge, L., Sanches, I. D., Luchiari Júnior, A., da Costa, C. C., Victoria, D. D., Inamasu, R. Y., Grego, C. R., Ferreira, V. R., & Ramirez, A. R. (2020). Precision and digital agriculture: Adoption of technologies and perception of Brazilian farmers. *Agriculture*, 10(12), 653. https://doi.org/10.3390/agriculture10120653

Bon, H. D., Parrot, L., & Moustier, P. (2011). Sustainable urban agriculture in developing countries. A review. *Agronomy for Sustainable Development*, 30, 21–32. https://doi.org/10.1051/agro:2008062

Bonetti, S., Sutanudjaja, E. H., Mabhaudhi, T., Slotow, R., & Dalin, C. (2022). Climate change impacts on water sustainability of South African crop production. *Environmental Research Letters*, 17, 084017. https://doi.org/10.1088/1748-9326/ac80cf

Bonnici, M., Greene, H., & Bonnici, I. (2021). Barriers for clean energy projects. *Journal of Clean Energy Technologies*, 9(2), 24–27. https://doi.org/10.18178/jocet.2021.9.2.256

Bookhart, D. (2008). Strategies for carbon neutrality. *Sustainability: The Journal of Record*, 1, 34–40. https://doi.org/10.1089/SUS.2008.9991

Boyce, J. K. (2018). Carbon pricing: Effectiveness and equity. *Ecological Economics, 150*, 52–61. https://doi.org/10.1016/J.ECOLECON.2018.03.030

Brezuleanu, S., Brezuleanu, C. O., Brad, I., Iancu, T., & Ciani, A. (2015). Performance Assessment in business of agricultural companies using balanced scorecard model. *Cercetari Agronomice in Moldova, 48*, 109–120. https://doi.org/10.1515/cerce-2015-0035

Broadbent, J. M., & Laughlin, R. C. (2003). Public private partnerships: An introduction. *Making Policy Happen, 16*(3), 332–341. https://doi.org/10.1108/09513570310482282

Brooks, J. (2014). Policy coherence and food security: The effects of OECD countries' agricultural policies. *Food Policy, 44*, 88–94. https://doi.org/10.1016/j.foodpol.2013.10.006

Buick, R. (1997). Precision agriculture: an integration of information technologies with farming. https://doi.org/10.30843/nzpp.1997.50.11310

Burbi, S., Baines, R. N., & Conway, J. S. (2016). Achieving successful farmer engagement on greenhouse gas emission mitigation. *International Journal of Agricultural Sustainability, 14*, 466–483. https://doi.org/10.1080/14735903.2016.1152062

Byrne, A., & Hansberry, J. (2007). Collaboration: Leveraging resources and expertise. *New Directions for Youth Development, 114*, 75–84. https://doi.org/10.1002/YD.214

Cai, X., Zhang, X., Noël, P. H., & Shafiee-Jood, M. (2015). Impacts of climate change on agricultural water management: A review. *Wiley Interdisciplinary Reviews: Water, 2*, 439–455. https://doi.org/10.1002/wat2.1089

Camirhna-Matos, L. M., & Afsarmanesh, H. (2008). Concept of collaboration. https://doi.org/10.4018/978-1-59904-885-7.CH041

Camarinha-Matos, L. M., Afsarmanesh, H., & Fornasiero, R. (2017). Collaboration in a data-rich world. IFIP Advances in Information and Communication Technology. https://doi.org/10.1007/978-3-319-65151-4

Campuzano, L. R., HincapiéLlanos, G. A., Zartha Sossa, J. W., Orozco Mendoza, G. L., Palacio, J. C., & Herrera, M. (2023). Barriers to the adoptionof innovations for sustainable development in the agricultural sector-systematic literature review (SLR). *Sustainability, 15*, 4374. https://doi.org/10.3390/su15054374

Carbone, S., Giuzio, M., Kapadia, S., Kapadia, S., Krämer, J. S., Nyholm, K., and Vozian, K. (2021). The low-carbon transition, climate commitments and firm credit risk (December, 2021). ECB Working Paper No. 2021/2631. https://doi.org/10.2139/ssrn.3991358

Carrasco, J. (2014). The challenge of changing to a low-carbon economy: A brief overview. *Low Carbon Economy, 5*, 1–5. https://doi.org/10.4236/LCE.2014.51001

Cecchini, L., Torquati, B., & Chiorri, M. (2018). Sustainable agri-food products: A review of consumer preference studies through experimental economics. *Agricultural Economics (Zemědělská ekonomika), 64*(12), 554–565. https://doi.org/10.17221/272/2017-AGRICECON

Chamberlain, W., & Anseeuw, W. (2017). *Inclusive businesses in agriculture: What, how and for whom? Critical insights based on South African cases.* Stellenbosch: Sun Press, 282 p.

Chaminé, H. I., & Gómez-Gesteira, M. (2019). Sustainable resource management: Water practice issues. *Sustainable Water Resources Management, 5*, 3–9. https://doi.org/10.1007/s40899-019-00304-7

Chaponnière, A., Marlet, S., Perret, S. R., Bouleau, G., & Zaïri, A. (2012). Methodological pathways to improvements of evaluation approaches: The case of irrigated agriculture performance evaluation. *Journal of MultiDisciplinary Evaluation, 8*(18), p-47. https://doi.org/10.56645/jmde.v8i18.339

Chen, L., Msigwa, G., Yang, M., Osman, A. I., Fawzy, S., Rooney, D., & Yap, P. (2022). Strategies to achieve a carbon neutral society: A review. *Environmental Chemistry Letters, 20*, 2277–2310. https://doi.org/10.1007/s10311-022-01435-8

Chen, Z., & Chen, F. (2020). Socio-economic factors influencing the adoption of low carbon technologies under rice 1 production systems in China 2. Corpus ID: 251306417

Cheng, M. H. (2021). Towards a carbon-neutral state: International progress, national risks, and coping strategies. *International Journal of Scientific Research and Management*, 9(10), 790–802. https://doi.org/10.18535/ijsrm/v9i10.sh03

Cheon, Y. (2022). Review of global carbon neutral strategies and technologies. *Journal of the Korean Society of Mineral and Energy Resources Engineers*, 59(1), 99–112. https://doi.org/10.32390/ksmer.2022.59.1.099

Chia, S. H. (2006). *Collaboration in the 21st century: Driving-innovation and differentiation*. 2006 IEEE International Symposium on Semiconductor Manufacturing, xxxiii–xlviii. https://doi.org/10.1109/ISSM.2006.4493001

Ciot, M. (2022). Implementation perspectives for the European Green Deal in central and eastern Europe. *Sustainability*, 14(7), 3947. https://doi.org/10.3390/su14073947

Coulibaly, T. P., Du, J., & Diakité, D. (2021). Sustainable agricultural practices adoption. *Agriculture (Pol'nohospodárstvo)*, 67, 166–176. https://doi.org/10.2478/agri-2021-0015

Darnhofer, I., Bellon, S., Dedieu, B., & Milestad, R. (2010). Adaptiveness to enhance the sustainability of farming systems. A review. *Agronomy for Sustainable Development*, 30, 545–555. https://doi.org/10.1051/agro/2009053

Davis, S. J., Lewis, N. S., Shaner, M., Aggarwal, S., Arent, D. J., Azevedo, I. M., Benson, S. M., Bradley, T. H., Brouwer, J., Chiang, Y., Clack, C. T., Cohen, A., Doig, S. J., Edmonds, J., Fennell, P. S., Field, C. B., Hannegan, B., Hodge, B. S., Hoffert, M. I., Ingersoll, E., Jaramillo, P., Lackner, K. S., Mach, K. J., Mastrandrea, M. D., Ogden, J., Peterson, P. F., Sanchez, D. L., Sperling, D., Stagner, J., Trancik, J. E., Yang, C., & Caldeira, K. (2018). Net-zero emissions energy systems. *Science*, 360, eaas9793. https://doi.org/10.1126/science.aas9793

Derpsch, R., Friedrich, T., Kassam, A. H., & Hongwen, L. (2010). Current status of adoption of no-till farming in the world and some of its main benefits. *International Journal of Agricultural and Biological Engineering*, 3, 1–25. https://doi.org/10.25165/IJABE.V3I1.223

Devaux, A., Torero, M., Donovan, J. and Horton, D. (2018). Agricultural innovation and inclusive value-chain development: A review. *Journal of Agribusiness in Developing and Emerging Economies*, 8(1), 99–123. https://doi.org/10.1108/JADEE-06-2017-0065

Dhananjayan, V., & Ravichandran, B. (2018). Occupational health risk of farmers exposed to pesticides in agricultural activities. *Current Opinion in Environmental Science & Health*, 4, 31–37. https://doi.org/10.1016/J.COESH.2018.07.005

Dicks, L. V., Rose, D. C., Ang, F., Aston, S., Birch, A. N., Boatman, N. D., Bowles, E. L., Chadwick, D. R., Dinsdale, A., Durham, S., Elliott, J., Firbank, L. G., Humphreys, S., Jarvis, P., Jones, D., Kindred, D., Knight, S. M., Lee, M. R., Leifert, C., Lobley, M., Matthews, K., Midmer, A., Moore, M. A., Morris, C., Mortimer, S. R., Murray, T. C., Norman, K., Ramsden, S. J., Roberts, D. A., Smith, L., Soffe, R. J., Stoate, C., Taylor, B., Tinker, D. B., Topliff, M., Wallace, J. R., Williams, P., Wilson, P., Winter, M., & Sutherland, W. J. (2018). What agricultural practices are most likely to deliver "sustainable intensification" in the UK? *Food and Energy Security*, 8, e00148. https://doi.org/10.1002/FES3.148

Ding, G. K., & Ghosh, S. (2017). Sustainable water management – A strategy for maintaining future water resources. *Encyclopedia of Sustainable Technologies*, 2017, 91–103. https://doi.org/10.1016/B978-0-12-409548-9.10171-X

Dobrin, C., Croitoru, E. O., Dinulescu, R., & Marin, I., (2022). The impact of pesticide and fertiliser use on agricultural productivity in the context of the "Farm to Fork" Strategy in Romania and the European Union. *Amfiteatru Economic*, 24(60), 346–360. https://doi.org/10.24818/ea/2022/60/346

Domorenok, E., & Graziano, P. (2023). Understanding the European Green Deal: A narrative policy framework approach. *European Policy Analysis*, 9, 9–29. https://doi.org/10.1002/epa2.1168

Dorband, I. I., Jakob, M., Kalkuhl, M., & Steckel, J.C. (2019). Poverty and distributional effects of carbon pricing in low- and middle-income countries – A global comparative analysis. *World Development*, *115*, 246–257. https://doi.org/10.1016/J.WORLDDEV.2018.11.015

Drews, S., & van den Bergh, J. C. J. M. (2016). What explains public support for climate policies? A review of empirical and experimental studies. *Climate Policy*, *16*, 855–876.

Duarte, H. J., Morais, E. D., Siqueira, E. S., Nobre, F. C., & Nobre, L. H. (2023). Perception and tolerance to the risk of family agriculture managers. Corpus ID: 259767268.

Dubey, R. K., Tripathi, V., & Abhilash, P. (2015). Book review: Principles of plant-microbe interactions: Microbes for sustainable agriculture. *Frontiers in Plant Science*, *6*. https://doi.org/10.3389/fpls.2015.00986

Dumont, E. S., Bonhomme, S., Pagella, T., & Sinclair, F. L. (2017). Structured stakeholder engagement leads to development of more diverse and inclusive agroforestry options. *Experimental Agriculture*, *55*, 252–274. https://doi.org/10.1017/S0014479716000788

Dustmurodov, G. G., Yunusov, I., Ahmedov, U. K., Murodov, S. M., & Iskandarov, S. T. (2020). The mechanism for the development of public-private partnerships in agriculture (on the example of the Republic of Uzbekistan). *E3S Web of Conferences*, *224*, 04042. https://doi.org/10.1051/e3sconf/202022404042

Ebrahimi, H., Schillo, R. S., & Bronson, K. (2021). Systematic stakeholder inclusion in digital agriculture: A framework and application to Canada. *Sustainability*, *13*(12), 6879. https://doi.org/10.3390/su13126879

Eckert, E., & Kovalevska, O. (2021). Sustainability in the European Union: Analyzing the discourse of the European Green Deal. *Journal of Risk and Financial Management*, *14*(2), 80. https://doi.org/10.3390/JRFM14020080

Fahamsyah, E., & Chansrakaeo, R.. (2022). The legal politics harmonization of sustainable agricultural policy. *Fiat Justisia: Jurnal Ilmu Hukum*, *16*(2), 171–192. https://doi.org/10.25041/fiatjustisia.v16no2.2635

Falkner, R. (2016). The Paris Agreement and the new logic of international climate politics. *International Affairs*, *92*(5), 1107–1125. https://doi.org/10.1111/1468-2346.12708

Fang, L., & Meng, X. (2009). Research on collaborative structures of agricultural supply chains. *2009 International Workshop on Intelligent Systems and Applications*, 1–4. https://doi.org/10.1109/IWISA.2009.5073062

Finco, A., Bentivoglio, D., Belletti, M., Chiaraluce, G., Fiorentini, M., Ledda, L., & Orsini, R. (2023). Does precision technologies adoption contribute to the economic and agri-environmental sustainability of Mediterranean wheat production? An Italian case study. *Agronomy*, *13*, 1818. https://doi.org/10.3390/agronomy13071818

Findlater, K. M., Satterfield, T., & Kandlikar, M. (2019). Farmers' risk-based decision making under pervasive uncertainty: Cognitive thresholds and hazy hedging. *Risk Analysis*, *39*, 1755–1770. https://doi.org/10.1111/risa.13290

Fleming, R., & Mauger, R. (2021). Green and Just? An update on the 'European Green Deal'. *Journal for European Environmental & Planning Law*, *18*, 164–180. https://doi.org/10.1163/18760104-18010010

Fonteh, M. F., Tabi, F. O., Wariba, A. M., & Zie, J. (2013). Effective water management practices in irrigated rice to ensure food security and mitigate climate change in a tropical climate. *Agriculture and Biology Journal of North America*, *4*, 284–290. https://doi.org/10.5251/ABJNA.2013.4.3.284.290

Gan, S. Y., & Zhang, W. (2022). Building carbon neutrality goals break down strategies for sustainable energy development. *International Journal of Emerging Electric Power Systems*, *23*, 899–911. https://doi.org/10.1515/ijeeps-2022-0152

German, L., Cotula, L., Gibson, K., Locke, A., Bonanno, A. M., & Quan, J. (2018). *Land governance and inclusive business in agriculture: advancing the debate*. London: Overseas Development Institute (ODI). Available: https://www.econstor.eu/bitstream/10419/193655/1/104559637X.pdf

Gerres, T., Ávila, J. P., Llamas, P. L., & Román, T. G. (2019). A review of cross-sector decarbonisation potentials in the European energy intensive industry. *Journal of Cleaner Production*, *210*, 585–601. https://doi.org/10.1016/J.JCLEPRO.2018.11.036

Giannou, C. (2022). Farm to Fork: EU's strategy for a sustainable food system. *HAPSc Policy Briefs Series*, *3*(1), 189–198. https://doi.org/10.12681/hapscpbs.31008

Giller, K. E., Hijbeek, R., Andersson, J. A., & Sumberg, J. (2021). Regenerative agriculture: An agronomic perspective. *Outlook on Agriculture*, *50*, 13–25. https://doi.org/10.1177/0030727021998063

Gliessman, S. (2007). *Agroecology: The ecology of sustainable food systems*. Boca Raton: CRC Press.

Gomes, A. P., & Reidsma, P. (2021). Time to transition: Barriers and opportunities to farmer adoption of soil GHG mitigation practices in Dutch agriculture. *Frontiers in Sustainable Food Systems*, *5*, 706113. https://doi.org/10.3389/fsufs.2021.706113

Goodell, P. B., Zalom, F. G., Strand, J. F., Wilen, C. A., & Windbiel-Rojas, K. (2014). Maintaining long-term management: Over 35 years, integrated pest management has reduced pest risks and pesticide use. *California Agriculture*, *68*, 153–157. https://doi.org/10.3733/CA.V068N04P153

Graham, J. R., & Barter, K. (1999). Collaboration: A social work practice method. *Families in Society: The Journal of Contemporary Social Services*, *80*, 13–16. https://doi.org/10.1606/1044-3894.634

Green, J. F. (2021). Does carbon pricing reduce emissions? A review of ex-post analyses. *Environmental Research Letters*, *16*, 043004. https://doi.org/10.1088/1748-9326/abdae9

Greene, D., & Schäfer, A.W. (2003). Reducing Greenhouse gas emissions from U.S. Transportation. Corpus ID: 128928115

Hainsch, K., Göke, L., Kemfert, C., oei, P., & Hirschhausen, C. V. (2020). European Green Deal: Using ambitious climate targets and renewable energy to climb out of the economic crisis. Corpus ID: 224815567.

Hall, A. (2006). Public private sector partnerships in an agricultural system of innovation: concepts and challenges. *Cognition & Emotion*, *5*(1), 3–20. https://doi.org/10.1386/IJTM.5.1.3/1

Hall, A. (2012). *Partnerships in agricultural innovation: Who puts them together and are they enough?* Improving Agricultural Knowledge and Innovation Systems: OECD Conference Proceedings, OECD Publishing, Paris. https://doi.org/10.1787/9789264167445-18-en.

Handayani, I. P., & Hale, C. (2022). *Healthy soils for productivity and sustainable development in agriculture*. IOP Conference Series: Earth and Environmental Science, 1018. https://doi.org/10.1088/1755-1315/1018/1/012038

Hanger-Kopp, S., Lieu, J., & Nikas, A. (2019). Narratives of low-carbon transitions. https://doi.org/10.4324/9780429458781

Häring, A. M., Vairo, D., Dabbert, S., & Zanoli, R. (2009). Organic farming policy development in the EU: What can multi-stakeholder processes contribute? *Food Policy*, *34*, 265–272. https://doi.org/10.1016/J.FOODPOL.2009.03.006

Helm, S. V., Pollitt, A., Barnett, M. A., Curran, M. A., & Craig, Z. R. (2018) Differentiating environmental concern in the context of psychological adaption to climate change. *Global Environmental Change*, *48*, 158–167. https://doi.org/10.1016/j.gloenvcha.2017.11.012

Hepburn, C. (2006). Regulation by prices, quantities, or both: A review of instrument choice. *Oxford Review of Economic Policy*, *22*(2), 226–247.

Hermans, F., Klerkx, L., & Roep, D. (2015). Structural conditions for collaboration and learning in innovation networks: Using an innovation system performance lens to analyse agricultural knowledge systems. *The Journal of Agricultural Education and Extension*, *21*, 35–54. https://doi.org/10.1080/1389224X.2014.991113

Heyl, K., Ekardt, F., Roos, P., & Garske, B. (2023). Achieving the nutrient reduction objective of the Farm to Fork strategy. An assessment of CAP subsidies for precision fertilization and sustainable agricultural practices in Germany. *Frontiers in Sustainable Food Systems*, 7, 1088640. https://doi.org/10.3389/fsufs.2023.1088640

Hodge, G., & Greve, C. (2007). Public-private partnerships: An international performance review. *Public Administration Review*, 67, 545–558. https://doi.org/10.1111/J.1540-6210.2007.00736.X

Hoffmann, V., Probst, K., & Christinck, A. (2007). Farmers and researchers: How can collaborative advantages be created in participatory research and technology development? *Agriculture and Human Values*, 24, 355–368. https://doi.org/10.1007/S10460-007-9072-2

Holland, S. P., Mansur, E. T., Muller, N. Z., & Yates, A. (2020). The environmental benefits from transportation electrification: Urban buses. NBER Working Paper Series. https://doi.org/10.3386/w27285

Huang, M., & Zhai, P. (2021). Achieving Paris Agreement temperature goals requires carbon neutrality by middle century with far-reaching transitions in the whole society. *Advances in Climate Change Research*, 12(2), 281–286. https://doi.org/10.1016/J.ACCRE.2021.03.004

Huo, Y., Wang, J., Guo, X., & Xu, Y. (2022). The collaboration mechanism of agricultural product supply chain dominated by farmer cooperatives. *Sustainability*, 14(10), 5824. https://doi.org/10.3390/su14105824

Ikerd, J. (2021). THE ECONOMIC PAMPHLETEER: Realities of regenerative agriculture. *Journal of Agriculture, Food Systems, and Community Development*, 10(2), 7–10. https://doi.org/10.5304/JAFSCD.2021.102.001

Ingram, J., Gaskell, P., Mills, J., & Dwyer, J. C. (2020). How do we enact co-innovation with stakeholders in agricultural research projects? Managing the complex interplay between contextual and facilitation processes. *Journal of Rural Studies*, 78, 65–77. https://doi.org/10.1016/j.jrurstud.2020.06.003

Inshakov, O. V. (2013). Collaboration as a form of knowledge-based economy organization. *Economy of Region*, 1, 45–52. https://doi.org/10.17059/2013-3-3

IPCC (2021). Sixth assessment report. https://www.ipcc.ch/report/ar6/wg1/

Iqbal, M. A., Abbas, A., Naqvi, S. A., Rizwan, M., Samie, A., & Ahmed, U. I. (2020). Drivers of farm households' perceived risk sources and factors affecting uptake of mitigation strategies in Punjab Pakistan: Implications for sustainable agriculture. *Sustainability*, 12(23), 9895. https://doi.org/10.3390/su12239895

Iskandar, E. A., Amanah, S., Hubeis, A. V., & Sadono, D. (2021). *Structural equation model of Farmer competence and cocoa sustainability in Aceh, Indonesia*. IOP Conference Series: Earth and Environmental Science, 644. https://doi.org/10.1088/1755-1315/644/1/012008

Jamnadass, R., Torquebiau, E. F., Iiyama, M., Kehlenbeck, K., Masters, E. T., & McMullin, S. (2013). Agroforestry for food and nutritional security. https://doi.org/10.5716/wp13054.pdf

Jayson Beckman, M. I., Baquedano, F. G., & Scott, S. G. (2020). Economic and food security impacts of agricultural input reduction under the European Union Green Deal's Farm to Fork and biodiversity strategies. https://doi.org/10.22004/AG.ECON.307277

Jensen, H. G., Pérez Domínguez, I., Fellmann, T., Lirette, P., Hristov, J., & Philippidis, G. P. (2019). Economic impacts of a low carbon economy on global agriculture: The Bumpy Road to Paris. *Sustainability*, 11(8), 2349. https://doi.org/10.3390/SU11082349

Jewell, J., Mccollum, D., Emmerling, J., Bertram, C., Gernaat, D.E., Krey, V., Paroussos, L., Berger, L., Fragkiadakis, K., Keppo, I., Saadi, N., Tavoni, M., Vuuren, D. P., Vinichenko, V., & Riahi, K. (2018). Limited emission reductions from fuel subsidy removal except in energy-exporting regions. *Nature*, 554, 229–233. https://doi.org/10.1038/nature25467

Jordan, N., Gutknecht, J. L., Bybee-Finley, K. A., Hunter, M., Krupnik, T. J., Pittelkow, C. M., Prasad, P. V., & Snapp, S. (2020). To meet grand challenges, agricultural scientists must engage in the politics of constructive collective action. *Crop Science, 61*, 24–31. https://doi.org/10.1002/csc2.20318

Kalchschmidt, M., & Syahruddin, N. (2011). Towards sustainable supply chain management in agricultural sector. Corpus ID: 55109824.

Kassam, A. H., & Brammer, H. (2013). Combining sustainable agricultural production with economic and environmental benefits. *The Geographical Journal, 179*, 11–18. https://doi.org/10.1111/J.1475-4959.2012.00465.X

Käyhkö, J. (2019). Climate risk perceptions and adaptation decision-making at Nordic farm scale – A typology of risk responses. *International Journal of Agricultural Sustainability, 17*, 431–444. https://doi.org/10.1080/14735903.2019.1689062

Kazak, T. (2022). European Green Deal. *Yearbook of the Law Department, 9*(10), 304–315. https://doi.org/10.33919/yldnbu.20.9.12

Keay, M. (2013). Oxford Energy Comment February 2013 renewable energy targets: The importance of system and resource costs. https://www.oxfordenergy.org/wpcms/wp-content/uploads/2013/02/Renewable-energy-targets-the-importance-of-system-and-resource-costs.pdf

Khorramnia, K., Shariff, A. B., Rahim, A. A., & Mansor, S. (2014). Toward Malaysian sustainable agriculture in 21st century. *IOP Conference Series: Earth and Environmental Science, 18*, 012142. https://doi.org/10.1088/1755-1315/18/1/012142

Kim, K., Kabir, E., & Jahan, S. A. (2017). Exposure to pesticides and the associated human health effects. *The Science of the Total Environment, 575*, 525–535. https://doi.org/10.1016/j.scitotenv.2016.09.009

Kim, S. Y., & Wolinsky-Nahmias, Y. (2014). Cross-National Public Opinion on climate change: The effects of affluence and vulnerability. *Global Environmental Politics, 14*, 79–106. https://doi.org/10.1162/GLEP_a_00215

Knight, K. W. (2016). Public awareness and perception of climate change: A quantitative cross-national study. *Environmental Sociology, 2*, 101–113. https://doi.org/10.1080/23251042.2015.1128055

Kolomoiets, T., Galitsina, N., Sharaia, A. A., Kachuriner, V., & Danylenko, O. (2021). International experience of public-private partnership in agriculture. *Amazonia Investiga, 10*(41), 160–168. https://doi.org/10.34069/ai/2021.41.05.16

Komatsuzaki, M., & Ohta, H. (2013). Sustainable agriculture practices. https://doi.org/10.18356/4F898E93-EN

Konefal, J. T., Hatanaka, M., & Constance, D.H. (2017). Multi-stakeholder initiatives and the divergent construction and implementation of sustainable agriculture in the USA. *Renewable Agriculture and Food Systems, 34*, 293–303. https://doi.org/10.1017/S1742170517000461

Krämer, L. (2020). Planning for climate and the environment: the EU Green Deal. *Journal for European Environmental & Planning Law, 17*, 267–306. https://doi.org/10.1163/18760104-01703003

Kumar, N., Singh, G., & Kebede, H. (2023). An optimized framework of the integrated renewable energy and power quality model for the smart grid. *International Transactions on Electrical Energy Systems, 2023*, 11. Article ID 6769690. https://doi.org/10.1155/2023/6769690

Kushnarenko, T., Makeev, V., & Debesai, M. G. (2021). *A review of low-carbon adaptation model to climate change in rural farming households.* IOP Conference Series: Earth and Environmental Science, 937. https://doi.org/10.1088/1755-1315/937/2/022026

LaCanne, C. E., & Lundgren, J. G. (2018). Regenerative agriculture: Merging farming and natural resource conservation profitably. *PeerJ, 6*, e4428. https://doi.org/10.7717/peerj.4428

Lamb, W. F., Wiedmann, T. O., Pongratz, J., Andrew, R. M., Crippa, M., Olivier, J. G., Wiedenhofer, D., Mattioli, G., Khourdajie, A. A., House, J., Pachauri, S., Figueroa, M. J., Saheb, Y. P., Slade, R., Hubacek, K., Sun, L., Ribeiro, S. K., Khennas, S., de la rue du Can, S., Chapungu, L., Davis, S. J., Bashmakov, I. A., Dai, H., Dhakal, S., Tan, X., Geng, Y., Gu, B., & Minx, J. C. (2021). A review of trends and drivers of greenhouse gas emissions by sector from 1990 to 2018. *Environmental Research Letters*, *16*(7), 073005. https://doi.org/10.1088/1748-9326/abee4e

Lambrecht, E., Kühne, B., & Gellynck, X. (2015). Success factors of innovation networks: Lessons from agriculture in Flanders. https://doi.org/10.18461/PFSD.2015.1532

Lamenta, Z. A., & Grzybowska, K. (2023). Impact of the European Green Deal on business operations-preliminary benchmarking. *Sustainability*, *15*(10), 7780. https://doi.org/10.3390/su15107780

Landis, F., & Heindl, P. (2019). Renewable energy targets in the context of the EU ETS: Whom do they benefit exactly? *The Energy Journal*, *40*. https://doi.org/10.5547/01956574.40.6.FLAN

Lane, D., Chatrchyan, A. M., Tobin, D., Thorn, K., Allred, S. B., & Radhakrishna, R. B. (2018). Climate change and agriculture in New York and Pennsylvania: risk perceptions, vulnerability and adaptation among farmers. *Renewable Agriculture and Food Systems*, *33*, 197–205. https://doi.org/10.1017/S1742170517000710

Lane, D., Murdock, E. A., Genskow, K., Betz, C. R., & Chatrchyan, A. M. (2019). Climate change and dairy in New York and Wisconsin: Risk perceptions, vulnerability, and adaptation among farmers and advisors. *Sustainability*, *11*(13), 3599 https://doi.org/10.3390/SU11133599

Larionova, M. V. (2021). The EU's policies for the Green Deal internationalization. *International Organisations Research Journal*, *16*(3), 124–160. https://doi.org/10.17323/1996-7845-2021-03-06

Larkin, S., Perruso, L., Marra, M. C., Roberts, R. K., English, B. C., Larson, J. A., Cochran, R. L., & Martin, S. W. (2005). Factors affecting perceived improvements in environmental quality from precision farming. *Journal of Agricultural and Applied Economics*, *37*, 577–588. https://doi.org/10.1017/S1074070800027097

Larsen, R. K., & Powell, N. (2013). Policy coherence for sustainable agricultural development: Uncovering prospects and pretence within the Swedish policy for global development. *PSN: Rural & Agricultural Development (Topic)*, *31*, 757–776. https://doi.org/10.1111/dpr.12034

Latynskiy, E., Berger, T., & Troost, C. (2014). Assessment of policies for low-carbon agriculture by means of multi-agent simulation. Corpus ID: 26248557.

Lawson, H. A. (2004). The logic of collaboration in education and the human services. *Journal of Interprofessional Care*, *18*, 225–237. https://doi.org/10.1080/13561820410001731278

Leonard, M., Pisani-Ferry, J., Shapiro, J. J., Tagliapietra, S., & Wolf, G. (2021). The geopolitics of the European Green Deal. *International Organisations Research Journal*, *16*(2), 204–235. https://doi.org/10.17323/1996-7845-2021-02-10

Li, J., & Sun, C. (2018). Towards a low carbon economy by removing fossil fuel subsidies? *China Economic Review*, *50*, 17–33. https://doi.org/10.1016/J.CHIECO.2018.03.006

Liang, X., He, J., Zhang, F., Shen, Q., Wu, J., Young, I. M., O'donnell, A. G., Wang, L., Wang, E., Hill, J., & Chen, D. (2020). Healthy soils for sustainable food production and environmental quality. *Frontiers of Agricultural Science and Engineering*, *7*(3), 347–355. https://doi.org/10.15302/j-fase-2020339

Lilliestam, J., Patt, A., & Bersalli, G. (2020). The effect of carbon pricing on technological change for full energy decarbonization: A review of empirical ex-post evidence. *Wiley Interdisciplinary Reviews: Climate Change*, *12*, e681. https://doi.org/10.1002/wcc.681

Linde, S. (2018). Political communication and public support for climate mitigation policies: A country-comparative perspective. *Climate Policy*, *18*, 543–555.

Linder, S. H. (1999). Coming to terms with the public-private partnership. *American Behavioral Scientist*, *43*, 35–51. https://doi.org/10.1177/00027649921955146

Linsner, S., Steinbrink, E., Kuntke, F., Franken, J., & Reuter, C. (2022). Supporting users in data disclosure scenarios in agriculture through transparency. *Behaviour & Information Technology*, *41*, 2151–2173. https://doi.org/10.1080/0144929X.2022.2068070

Littler, D., Leverick, F., & Wilson, D. F. (1998). Collaboration in new technology based product markets. *International Journal of Technology Management*, *15*, 139–159. https://doi.org/10.1504/IJTM.1998.002600

Liu, L., Zhu, Y., & Guo, S. (2020). The evolutionary game analysis of multiple stakeholders in the low-carbon agricultural innovation diffusion. *Complex*, *2020*, 1–12. https://doi.org/10.1155/2020/6309545

Lobell, D., Baldos, U. L., & Hertel, T. W. (2013). Climate adaptation as mitigation: The case of agricultural investments. *Environmental Research Letters*, *8*(1), 015012. https://doi.org/10.1088/1748-9326/8/1/015012

Lorenzoni, I., Nicholson-Cole, S. A., & Whitmarsh, L. (2007). Barriers perceived to engaging with climate change among the UK public and their policy implications. *Global Environmental Change-human and Policy Dimensions*, *17*, 445–459.

Lu, Y., Norse, D., & Powlson, D. S. (2020). Agriculture Green development in China and the UK: Common objectives and converging policy pathways. *Frontiers of Agricultural Science and Engineering*, *7*(1), 98–105. https://doi.org/10.15302/J-FASE-2019298

MacLeod, C. J., Brandt, A. J., Collins, K., & Dicks, L. V. (2021). Giving stakeholders a voice in governance: Biodiversity priorities for New Zealand's agriculture. *People and Nature*, *4*, 330–350. https://doi.org/10.1002/pan3.10285

Manjula, M., & Rengalakshmi, R. (2021). Making research collaborations: Learning from processes of transdisciplinary engagement in agricultural research. *Review of Development and Change*, *26*, 25–39. https://doi.org/10.1177/09722661211007589

Markantoni, M., & Woolvin, M. (2015). The role of rural communities in the transition to a low-carbon Scotland: A review. *Local Environment*, *20*, 202–219. https://doi.org/10.1080/13549839.2013.834880

Marotzke, J., & Forster, P. M. (2015). Forcing, feedback and internal variability in global temperature trends. *Nature*, *517*, 565–570, https://doi.org/10.1038/nature14117.

Massetti, E., & Mendelsohn, R. O. (2018). Measuring climate adaptation: Methods and evidence. *Review of Environmental Economics and Policy*, *12*, 324–341. https://doi.org/10.1093/reep/rey007

Mastrorilli, M., & Zucaro, R. (2019). Sustainable water management. In: Farooq, M. & Pisante, M. (Eds.), *Innovations in sustainable agriculture* (pp. 133–166). Cham: Springer. https://doi.org/10.1007/978-3-030-23169-9_6

Melović, B., Ćirović, D., Backovic-Vulić, T., Dudić, B., & Gubíniová, K. (2020). Attracting Green consumers as a basis for creating sustainable marketing strategy on the organic market-relevance for sustainable agriculture business development. *Foods*, *9*, 1552. https://doi.org/10.3390/foods9111552

Mi, Z., Gao, Y., & Liao, H. (2023). Carbon neutrality and socio-economic development. *Journal of Chinese Economic and Business Studies*, *21*, 1–3. https://doi.org/10.1080/14765284.2023.2182013

Middelberg, S. L. (2013). Sustainable agriculture: A review of challenges facing the South African agricultural sector. *Journal of Human Ecology*, *42*, 163–169. https://doi.org/10.1080/09709274.2013.11906590

Millot, A., Doudard, R., Gallic, T. L., Briens, F., Assoumou, E., & Maïzi, N. (2018). France 2072: Lifestyles at the core of carbon neutrality challenges. Part of the Lecture Notes in Energy book series (LNEN, volume 64). https://doi.org/10.1007/978-3-319-74424-7_11

Molin, J. P., Bazame, H. C., Maldaner, L. F., Corrêdo, L. D., Martello, M., & Canata, T.F. (2020). Precision agriculture and the digital contributions for site-specific management of the fields. *Revista Ciencia Agronomica*, 51, 1–10. https://doi.org/10.5935/1806-6690.20200088

Monasterolo, I., & Raberto, M. (2019). The impact of phasing out fossil fuel subsidies on the low-carbon transition. *Energy Policy*, *124*, 355–370 https://doi.org/10.1016/J.ENPOL.2018.08.051

Morton, L. W., Roesch-McNally, G., & Wilke, A. K. (2017). Upper Midwest farmer perceptions: Too much uncertainty about impacts of climate change to justify changing current agricultural practices. *Journal of Soil and Water Conservation*, 72, 215–225. https://doi.org/10.2489/jswc.72.3.215

Moschitz, H., Müller, A. W., Kretzschmar, U., Haller, L., Porras, M. D., Pfeifer, C., Oehen, B., Willer, H., & Stolz, H. (2021). How can the EU Farm to Fork strategy deliver on its organic promises? Some critical reflections. *EuroChoices*, *20*(1), 30–36. https://doi.org/10.1111/1746-692X.12294

Moser, P. (2019). Economics of research and innovation in agriculture. NBER Working Paper Series. https://doi.org/10.3386/w27080

Mudombi, S., Nhamo, G., & Muchie, M. (2014). Socio-economic determinants of climate change awareness among communal farmers in two districts of Zimbabwe. *Africa Insight*, *44*, 1–15.

Murty, D., Kirschbaum, M. U., McMurtrie, R., & Mcgilvray, H. (2002). Does conversion of forest to agricultural land change soil carbon and nitrogen? A review of the literature. *Global Change Biology*, *8*, 105–123. https://doi.org/10.1046/j.1354-1013.2001.00459.x

Muscat, A., Olde, E. M., Kovacic, Z., Boer, I. D., & Ripoll-Bosch, R. (2021). Food, energy or biomaterials? Policy coherence across agro-food and bioeconomy policy domains in the EU. *Environmental Science & Policy*, *123*, 21–30. https://doi.org/10.1016/J.ENVSCI.2021.05.001

Mutekwa, V. (2009). Climate change impacts and adaptation in the agricultural sector: the case of smallholder farmers in Zimbabwe. *Journal of Sustainable Development in Africa*, *11*, 237–256.

Naqvi, B., Mirza, N., Rizvi, S. K., Porada-Rochoń, M., & Itani, R. (2021). Is there a green fund premium? Evidence from twenty seven emerging markets. *Global Finance Journal*, *50*, 100656. https://doi.org/10.1016/j.gfj.2021.100656

Neef, A., & Neubert, D. (2011). Stakeholder participation in agricultural research projects: A conceptual framework for reflection and decision-making. *Agriculture and Human Values*, *28*, 179–194. https://doi.org/10.1007/S10460-010-9272-Z

Newton, P., Civita, N., Frankel-Goldwater, L., Bartel, K., & Johns, C. (2020). What is regenerative agriculture? A review of scholar and practitioner definitions based on processes and outcomes. *Frontiers in Sustainable Food Systems*, *4*, 194. https://doi.org/10.3389/fsufs.2020.577723

Niggli, U. A. (2014). Sustainability of organic food production: Challenges and innovations. *Proceedings of the Nutrition Society*, *74*, 83–88. https://doi.org/10.1017/S0029665114001438

Nilsen, H. R., Strømsnes, K., & Schmidt, U. (2018). A broad alliance of civil society organizations on climate change mitigation: Political strength or legitimizing support? *Journal of Civil Society*, *14*, 20–40.

Nozharov, S., & Nikolova, N. (2020). Shadow economy and populism – Risk and uncertainty factors for establishing low-carbon economy of Balkan countries (Case Study for Bulgaria). *Economic Studies Journal*, 121–144. Available at SSRN: https://ssrn.com/abstract=3647530

Ohlendorf, N., Jakob, M., Minx, J. C., Schröder, C., & Steckel, J. C. (2020). Distributional impacts of carbon pricing: A meta-analysis. *Environmental and Resource Economics*, *78*, 1–42. https://doi.org/10.1007/s10640-020-00521-1

Olejnik, T. P., & Sobiecka, E. (2017). Utilitarian technological solutions to reduce CO_2 emission in the aspect of sustainable development. *Problems of Sustainable Development*, *12*(2), 173–179.

Ossewaarde, M., & Ossewaarde-Lowtoo, R. (2020). The EU's Green Deal: A third alternative to Green growth and degrowth? *Sustainability*, *12*(23), 9825. https://doi.org/10.3390/su12239825

Ouhna, L. (2019). *Trust in relationships with agri-food distribution, new insights on trust in business-to-business relationships (Advances in Business Marketing and Purchasing, Vol. 26)* (pp. 103–119). Bingley: Emerald Publishing Limited. https://doi.org/10.1108/S1069-096420190000026009

Owen, G. (2020). What makes climate change adaptation effective? A systematic review of the literature. *Global Environmental Change-human and Policy Dimensions*, *62*, 102071. https://doi.org/10.1016/j.gloenvcha.2020.102071

Owusu, P. A., & Asumadu-Sarkodie, S. (2016). A review of renewable energy sources, sustainability issues and climate change mitigation. *Cogent Engineering*, *3*(1). Article ID: 1167990. https://doi.org/10.1080/23311916.2016.1167990

Parker, D. H. (2022). Targeted nitrogen and irrigation education in California's nitrate impacted regions. Corpus ID: 251163426.

Paustian, K. (2013). Bridging the data gap: Engaging developing country farmers in greenhouse gas accounting. *Environmental Research Letters*, *8*, 021001. https://doi.org/10.1088/1748-9326/8/2/021001

Pe'er, G., Bonn, A., Bruelheide, H., Dieker, P., Eisenhauer, N., Feindt, P.H., Hagedorn, G., Hansjürgens, B., Herzon, I., Lomba, Â., Marquard, E., Moreira, F., Nitsch, H., Oppermann, R., Perino, A., Röder, N., Schleyer, C., Schindler, S., Wolf, C., Zinngrebe, Y., & Lakner, S. (2020). Action needed for the EU Common Agricultural Policy to address sustainability challenges. *People and Nature*, *2*, 305–316. https://doi.org/10.1002/pan3.10080

Pham, L.D. (2016). Is it risky to go green? A volatility analysis of the green bond market. *Journal of Sustainable Finance & Investment*, *6*, 263–291. https://doi.org/10.1080/20430795.2016.1237244

Pianta, M., & Lucchese, M. (2020). Rethinking the European Green Deal. *Review of Radical Political Economics*, *52*, 633–641. https://doi.org/10.1177/0486613420938207

Prager, K., & Freese, J. (2009). Stakeholder involvement in agri-environmental policy making– learning from a local- and a state-level approach in Germany. *Journal of Environmental Management*, *90*(2), 1154–1167. https://doi.org/10.1016/j.jenvman.2008.05.005

Priyanto, M. W., & Mulyo, J. H. (2021). *Comparison of awareness and perception of climate change between Proklim and Non-Proklim farmers in Sleman District*. E3S Web of Conferences. https://doi.org/10.1051/E3SCONF/202123204007

Qin, L., Kırıkkaleli, D., Hou, Y., Miao, X., & Tufail, M. A. (2021). Carbon neutrality target for G7 economies: Examining the role of environmental policy, green innovation and composite risk index. *Journal of Environmental Management*, *295*, 113119. https://doi.org/10.1016/j.jenvman.2021.113119

Raghuvanshi, R., & Ansari, M. A. (2017). A study of farmers' awareness about climate change and adaptation practices in India. https://doi.org/10.11648/J.IJAAS.20170306.13

Raišienė, A. G., Podviezko, A., Skulskis, V., & Baranauskaitė, L. (2019). Interest-balanced agricultural policy-making: Key participative and collaborative capacities in the opinion of NGOs' experts. *Economics & Sociology*, *12*(3), 301–318. https://doi.org/10.14254/2071-789x.2019/12-3/20

Rausch, S., Metcalf, G. E., & Reilly, J. M. (2011). Distributional impacts of carbon pricing: A general equilibrium approach with micro-data for households. *Energy Economics*, *33*(1), S20–S33. https://doi.org/10.1016/J.ENECO.2011.07.023

Riaz, M., & Din, D. M. (2023). Collaboration as 21st century learning skill at undergraduate level. *sjesr*, *6*(1), 93–99. https://doi.org/10.36902/sjesr-vol6-iss1-2023(93-99)

Ricart, S., Castelletti, A., & Gandolfi, C. (2022). On farmers' perceptions of climate change and its nexus with climate data and adaptive capacity. A comprehensive review. *Environmental Research Letters*, *17*, 083002. https://doi.org/10.1088/1748-9326/ac810f

Richter, B. D., Mathews, R. E., Harrison, D. L., & Wigington, R. (2003). Ecologically sustainable water management: Managing river flows for ecological integrity. *Ecological Applications*, *13*, 206–224. https://doi.org/10.1890/1051-0761(2003)013[0206:ESWMMR]2.0.CO;2

Rissman, J., Bataille, C., Masanet, E. R., Aden, N. T., Morrow, W. R., Zhou, N., Elliott, N., Dell, R. W., Heeren, N., Huckestein, B. D., Cresko, J., Miller, S. A., Roy, J., Fennell, P. S., Cremmins, B., Blank, T. K., Hone, D., Williams, E. D., de la Rue du Can, S., Sisson, B., Williams, M., Katzenberger, J. W., Burtraw, D., Sethi, G., Ping, H., Danielson, D., Lu, H., Lorber, T., Dinkel, J., & Helseth, J. (2020). Technologies and policies to decarbonize global industry: Review and assessment of mitigation drivers through 2070. *Applied Energy*, *266*, 114848. https://doi.org/10.1016/j.apenergy.2020.114848

Rogel, L., Elzen, M., Hohne, N., Fransen, T., Fekete, H., Winkler, H., Schaeffer, R., Sha, F., Riahi, K., & Meinshausen, M. (2016). Paris Agreement climate proposals need a boost to keep warming well below 2°C. https://www.ncbi.nlm.nih.gov/pubmed/27357792

Rumble, J. N. (2015). Transparency in agriculture and natural resources: Defining transparent communication. *EDIS*, *2015*, 3. https://doi.org/10.32473/edis-wc225-2015

Russo, T. A., Alfredo, K., & Fisher, J. (2014). Sustainable water management in urban, agricultural, and natural systems. *Water*, *6*, 3934–3956. https://doi.org/10.3390/W6123934

Sani, S. (2016). Farmers' perception, impact and adaptation strategies to climate change among smallholder farmers in Sub-Saharan Africa: A systematic review. *Journal of Resources Development and Management*, *26*, 1–8.

Savaresi, A. (2016). The Paris Agreement: A new beginning? *Journal of Energy & Natural Resources Law*, *34*, 16–26. https://doi.org/10.1080/02646811.2016.1133983

Semieniuk, G., Campiglio, E., Mercure, J.-F., Volz, U., & Edwards, N. R. (2020). *"Low-carbon transition risks for finance"*, SOAS Department of Economics Working Paper No. 233. London: SOAS University of London.

Seo, S. N. (2017). Beyond the Paris Agreement: Climate change policy negotiations and future directions. *Regional Science Policy and Practice*, *9*, 121–140. https://doi.org/10.1111/RSP3.12090

Setegn, S. (2015). Water resources management for sustainable environmental public health, pp. 275–287. https://doi.org/10.1007/978-3-319-12194-9_15

Settle, Q. D., Rumble, J. N., McCarty, K. J., & Ruth, T. K. (2017). Public knowledge and trust of agricultural and natural resources organizations. *Journal of Applied Communications*, *101*, 8. https://doi.org/10.4148/1051-0834.1007

Shamsudin, M. N., Amir, H. M., & Radam, A. (2010). Economic benefits of sustainable agricultural production: The case of integrated pest management in cabbage production. *EnvironmentAsia*, *3*, 168–174.

Sharmina, M., Edelenbosch, O.Y., Wilson, C., Freeman, R., Gernaat, D.E., Gilbert, P., Larkin, A., Littleton, E. W., Traut, M., van Vuuren, D. P., Vaughan, N. E., Wood, F. R., & Le Quéré, C. (2020). Decarbonising the critical sectors of aviation, shipping, road freight and industry to limit warming to 1.5–2°C. *Climate Policy*, *21*, 455–474. https://doi.org/10.1080/14693062.2020.1831430

Sheng, R., Zhou, R., Zhang, Y., & Wang, Z. (2021). Green investment changes in China: A shift-share analysis. *International Journal of Environmental Research and Public Health*, *18*, 6658. https://doi.org/10.3390/ijerph18126658

Shukla, R., Sharma, S., & Thumar, V. (2016). Role and importance of public private partnerships in agricultural value chain and infrastructure. *International Journal of Commerce and Business Management*, 9, 113–118. https://doi.org/10.15740/HAS/IJCBM/9.1/113-118

Singh, A., Dhyani, B. P., Kumar, V., Singh, A. K., & Navsare, I. (2021). Potential carbon sequestration methods of agriculture: A review. *Journal of Pharmacognosy and Phytochemistry*, 10(2), 562–565 https://www.phytojournal.com/archives/2021/vol10issue2/PartG/10-2-32-592.pdf

Singh, S., & Ru, J. (2022). Accessibility, affordability, and efficiency of clean energy: a review and research agenda. *Environmental Science and Pollution Research*, 29, 18333–18347. https://doi.org/10.1007/s11356-022-18565-9

Sishodia, R. P., Ray, R. L., & Singh, S. (2020). Applications of remote sensing in precision agriculture: A review. *Remote Sensing*, 12, 3136. https://doi.org/10.3390/rs12193136

Škiudas, K., Kubilius, A., & Urbanavičius, J. (2020). *Climate change caused by human activity: Causes, consequences, possibilities*. Vilnius: European Public Policy Institute. ISSN 2669-1744

Smith, T. W., Kim, J., & Son, J. (2017). Public attitudes toward climate change and other environmental issues across countries. *International Journal of Sociology*, 47, 62–80. https://doi.org/10.1080/00207659.2017.1264837

Sollen-Norrlin, M., Ghaley, B. B., & Rintoul, N. L. (2020). Agroforestry benefits and challenges for adoption in Europe and beyond. *Sustainability*, 12(17), 7001. https://doi.org/10.3390/su12177001

Spielman, D. J., & Grebmer, K. V. (2006). Public-private partnerships in international agricultural research: An analysis of constraints. *The Journal of Technology Transfer*, 31, 291–300. https://doi.org/10.1007/S10961-005-6112-1

Spillias, S., Kareiva, P., Ruckelshaus, M. H., & McDonald-Madden, E. (2020). Renewable energy targets may undermine their sustainability. *Nature Climate Change*, 10, 974–976. https://doi.org/10.1038/s41558-020-00939-x

Srivastava, R. K. (2020). Influence of sustainable agricultural practices on healthy food cultivation. In: K. Gothandam, S. Ranjan, N. Dasgupta, & E. Lichtfouse (Eds.), *Environmental biotechnology Vol. 2. Environmental chemistry for a sustainable world Vol. 45*. Cham: Springer. https://doi.org/10.1007/978-3-030-38196-7_5

Stevanović, M., Popp, A., Lotze-Campen, H., Dietrich, J. P., Müller, C., Bonsch, M., Schmitz, C., Bodirsky, B. L., Humpenöder, F., & Weindl, I. (2016). The impact of high-end climate change on agricultural welfare. *Science Advances*, 2(8), e1501452. https://doi.org/10.1126/sciadv.1501452

Storm, S. (2020). The EU's Green Deal: Bismarck's 'what is possible' versus thunberg's 'what is imperative'. *Sustainability & Economics eJournal*. https://doi.org/10.36687/inetwp117

Syahruddin, N., & Kalchschmidt, M. (2012). Sustainable supply chain management in the agricultural sector: A literature review. *International Journal of Engineering Management and Economics*, 3, 237–258. https://doi.org/10.1504/IJEME.2012.049894

Szekely, I., & Jijakli, M. H. (2022). Bioponics as a promising approach to sustainable agriculture: A review of the main methods for producing organic nutrient solution for hydroponics. *Water*, 14(23), 3975. https://doi.org/10.3390/w14233975

Tahat, M. M., Alananbeh, K. M., Othman, Y. A., & Leskovar, D. I. (2020). Soil health and sustainable agriculture. *Sustainability*, 12(12), 4859 https://doi.org/10.3390/su12124859

Thompson, I. K. (2001). Collaboration in technical communication: A qualitative content analysis of journal articles, 1990–1999. *IEEE Transactions on Professional Communication*, 44, 161–173. https://doi.org/10.1109/47.946462

Torchia, M., Calabrò, A., & Morner, M. (2015). Public-private partnerships in the health care sector: A systematic review of the literature. *Public Management Review*, 17, 236–261. https://doi.org/10.1080/14719037.2013.792380

Trenberth, K. E., & Fasullo, J. T. (2013). An apparent hiatus in global warming? *Earth's Future, 1*, 19–32, https://doi.org/10.1002/2013EF000165

Tryndina, N. S., An, J., Varyash, I. Y., Litvishko, O., Khomyakova, L. I., Barykin, S. A., & Kalinina, O. V. (2022). Renewable energy incentives on the road to sustainable development during climate change: A review. *Frontiers in Environmental Science, 10*, 1715. https://doi.org/10.3389/fenvs.2022.1016803

Urwin, K., & Jordan, A. J. (2008). Does public policy support or undermine climate change adaptation? Exploring policy interplay across different scales of governance. *Global Environmental Change: Human and Policy Dimensions, 18*, 180–191.

Usuga, M., Jaimes, W. A., & Suarez, O. E. (2012). Coordination on agrifood supply chain. *World Academy of Science, Engineering and Technology, International Journal of Social, Behavioral, Educational, Economic, Business and Industrial Engineering, 6*, 2849–2853. https://publications.waset.org/6552/pdf

Van Dijk, J., Leopold, A., Flerlage, H., van Wezel, A. P., Seiler, T. B., Enrici, M. H., & Bloor, M.C. (2021). The EU Green Deal's ambition for a toxic-free environment: Filling the gap for science-based policymaking. *Integrated Environmental Assessment and Management, 17*, 1105–1113. https://doi.org/10.1002/ieam.4429

Van Hoof, S. (2023). Climate Change Mitigation in Agriculture: Barriers to the Adoption of Carbon Farming Policies in the EU. Sustainability 15, 10452. https://doi.org/10.3390/su151310452

Velasco-Muñoz, J. F., Aznar-Sánchez, J. A., López-Felices, B., & Balacco, G. (2022). Adopting sustainable water management practices in agriculture based on stakeholder preferences. *Agricultural Economics (Zemědělská ekonomika), 68*(9), 317–326. https://doi.org/10.17221/203/2022-agricecon

Velten, S., Jager, N.W., & Newig, J. (2021). Success of collaboration for sustainable agriculture: A case study meta-analysis. *Environment, Development and Sustainability, 23*, 14619–14641. https://doi.org/10.1007/s10668-021-01261-y

Velten, S., Leventon, J., Jager, N. W., & Newig, J. (2015). What is sustainable agriculture? A systematic review. *Sustainability, 7*, 7833–7865. https://doi.org/10.3390/SU706783

Waldman, K. B., Todd, P. M., Omar, S., Blekking, J., Giroux, S. A., Attari, S. Z., Baylis, K., & Evans, T. P. (2020). Agricultural decision making and climate uncertainty in developing countries. *Environmental Research Letters, 15*, 113004. https://doi.org/10.1088/1748-9326/abb909

Walker, P. H., & Elberson, K. L. (2005). Collaboration: Leadership in a global technological environment. *Online Journal of Issues in Nursing, 10*(1), 6. https://doi.org/10.3912/ojin.vol10no01man05

Wall, D. H., Nielsen, U. N., & Six, J. (2015). Soil biodiversity and human health. *Nature, 528*, 69–76. https://doi.org/10.1038/nature15744

Wang, F., Harindintwali, J. D., Yuan, Z., Wang, M., Wang, F., Li, S., Yin, Z., Huang, L., Fu, Y., Li, L., Chang, S. X., Zhang, L., Rinklebe, J., Yuan, Z., Zhu, Q., Xiang, L., Tsang, D. C., Xu, L., Jiang, X., Liu, J., Wei, N., Kästner, M., Zou, Y., Ok, Y. S., Shen, J., Peng, D., Zhang, W., Barceló, D., Zhou, Y., Bai, Z., Li, B., Zhang, B., Wei, K., Cao, H., Tan, Z., Zhao, L., He, X., Zheng, J., Bolan, N., Liu, X., Huang, C., Dietmann, S., Luo, M., Sun, N., Gong, J., Gong, Y., Brahushi, F., Zhang, T., Xiao, C., Li, X., Chen, W., Jiao, N., Lehmann, J., Zhu, Y., Jin, H., Schaeffer, A., Tiedje, J., & Chen, J. (2021). Technologies and perspectives for achieving carbon neutrality. *The Innovation, 2*. https://doi.org/10.1016/j.xinn.2021.100180

Wang, Y. (2023). The most promising new energy source-wind power. *Highlights in Science, Engineering and Technology, 33*, 20–23. https://doi.org/10.54097/hset.v33i.5240

Weisenburger, D. D. (1993). Human health effects of agrichemical use. *Human Pathology, 24*(6), 571–576. https://doi.org/10.1016/0046-8177(93)90234-8

Wezel, A., Bellon, S., Doré, T., Francis, C., Vallod, D., & David, C. (2009). Agroecology as a science, a movement and a practice. A review. *Agronomy for Sustainable Development, 29*, 503–515. https://doi.org/10.1051/agro/2009004

Wezel, A., Grard, B., & Gkisakis, V. (Eds.) (2023). *Agroecology in Europe. Country Reports Series, Vol. 1*. Lyon: ISARA; Corbais: Agroecology Europe. https://www.agroecology-europe.org/wp-content/uploads/2023/03/Mapping-book_1st-volume_FV.pdf

Wheeler, S. M. (2017). A carbon-neutral California: Social ecology and prospects for 2050 GHG reduction. *Urban Planning, 2*(4), 5–18. https://doi.org/10.17645/UP.V2I4.1077

Wilberforce, T., Baroutaji, A., El Hassan, Z., Thompson, J., Soudan, B., & Olabi, A. G. (2019). Prospects and challenges of concentrated solar photovoltaics and enhanced geothermal energy technologies. *The Science of the Total Environment, 659*, 851–861. https://doi.org/10.1016/J.SCITOTENV.2018.12.257

Wildauer, R., Leitch, S., & Kapeller, J. (2020). How to boost the European Green Deal's scale and ambition. Corpus ID: 226513145.

Willer, H., & Lernoud, J. (2018). The world of organic agriculture (Chinese) – Statistics & Emerging Trends 2018. Corpus ID: 169639005.

Willer, H., Schlatter, B., & Trávníček, J. (Eds.) (2023). *The world of organic agriculture. Statistics and emerging trends 2023*. Bonn: Research Institute of Organic Agriculture FiBL, Frick, and IFOAM - Organics International. Online Version 2 of February 23, 2023.

Williams, J. H., Jones, R. A., Haley, B., Kwok, G., Hargreaves, J. J., Farbes, J., & Torn, M. S. (2021). Carbon-neutral pathways for the United States. *AGU Advances, 2*, e2020AV000284. https://doi.org/10.1029/2020AV000284

Williams, L. J., Jaya, I. K., Hall, A., Cosijn, M., Rosmilawati, S., Suadnya, W., & Sudika, I. W. (2020). Unmasking partnerships for agricultural innovation: The realities of a research-private sector partnership in Lombok, Indonesia. *Innovation and Development, 12*, 417–436. https://doi.org/10.1080/2157930X.2020.1857949

Wirén-Lehr, S. V. (2001). Sustainability in agriculture – An evaluation of principal goal-oriented concepts to close the gap between theory and practice. *Agriculture, Ecosystems & Environment, 84*, 115–129. https://doi.org/10.1016/S0167-8809(00)00197-3

Wreford, A., Ignaciuk, A., & Gruère, G. P. (2017). Overcoming barriers to the adoption of climate-friendly practices in agriculture. https://doi.org/10.1787/97767DE8-EN

Xu, B., & Chen, J. (2021). How to achieve a low-carbon transition in the heavy industry? A nonlinear perspective. *Renewable & Sustainable Energy Reviews, 140*, 110708. https://doi.org/10.1016/J.RSER.2021.110708

Yang, Q., Zhu, Y., & Wang, F. (2021). Social media participation, low-carbon agricultural practices, and economic performance of banana farmers in Southern China. *Frontiers in Psychology, 12*, 790808. https://doi.org/10.3389/fpsyg.2021.790808

Yu, Q. (2011). *The cooperation in agri-food supply chain*, pp. 674–677. Harbin: MSIE. https://doi.org/10.1109/MSIE.2011.5707498

Zaraté, P., Alemany, M. M., Alvarez, A. E., Sakka, A., & Camilleri, G. (2020). Group decision support for agriculture planning by a combination of mathematical model and collaborative tool. ArXiv, abs/2006.08151. https://arxiv.org/ftp/arxiv/papers/2006/2006.08151.pdf

Zechter, R. H., Kossoy, A., Oppermann, K., Ramstein, C., Klein, N., Wong, L., Lam, L., Zhang, J., Quant, M., Neelis, M., Nierop, S. C., Ward, J., Kansy, T., Evans, S., & Child, A. (2017). State and Trends of Carbon Pricing 2019. https://doi.org/10.1596/978-1-4648-1435-8

Zhang, Y., Pan, C., & Liao, H. (2021). Carbon neutrality policies and technologies: A scientometric analysis of social science disciplines. *Frontiers in Environmental Science, 9*, 761736. https://doi.org/10.3389/fenvs.2021.761736

Zhao, D., & Zhou, H. (2021). Livelihoods, technological constraints, and low-carbon agricultural technology preferences of farmers: Analytical frameworks of technology adoption and farmer livelihoods. *International Journal of Environmental Research and Public Health, 18*(24), 13364. https://doi.org/10.3390/ijerph182413364

Zhao, Q. (2021). A review of pathways to carbon neutrality from renewable energy and carbon capture. *EDP Sciences, 245*, 01018. https://doi.org/10.1051/E3SCONF/202124501018

Zhao, W. (2022). China's goal of achieving carbon neutrality before 2060: Experts explain how. *National Science Review*, *8*(9), nwac115. https://doi.org/10.1093/nsr/nwac115

Zheng, H., Ma, J., Yao, Z., & Hu, F. (2022). How does social embeddedness affect farmers' adoption behavior of low-carbon agricultural technology? Evidence from Jiangsu Province, China. *Frontiers in Environmental Science*, *10*, 909803. https://doi.org/10.3389/fenvs.2022.909803

Zhu, Y., Lee, S., & Zhang, J. (2016). Performance evaluation on supplier collaboration of agricultural supply chain. *International Journal of Information Systems and Change Management*, *8*, 23–36. https://doi.org/10.1504/IJISCM.2016.077948

Zhu, Y., Wang, J., & Liu, L. (2021). Research on distribution and inventory cooperation of agricultural means supply chain. *Discrete Dynamics in Nature and Society*, *2021*, 1–12. https://doi.org/10.1155/2021/4428725

Zia, A., Alzahrani, M., Alomari, A., & Alghamdi, F. A. (2022). Investigating the drivers of sustainable consumption and their impact on online purchase intentions for agricultural products. *Sustainability*, *14*(11), 6563 https://doi.org/10.3390/su14116563

Zou, B. (2023). *Electrification of transport: Advantages and prospects*. Highlights in Science, Engineering and Technology. Vol. 43 (2023): 5th International Conference on Materials Science, Machinery and Energy Engineering (MSMEE 2023) https://doi.org/10.54097/hset.v43i.7418

2 GHG Emission and Energy Use in the EU Agriculture

Tomas Balezentis

2.1 INTRODUCTION

The agricultural sector has an important role for the whole society in that it serves as the source of food. In addition, the rural population is highly dependent on the development of the agricultural sector through employment. Besides the desirable effects associated with food security, the agricultural sector is also responsible for undesirable environmental effects that are caused by (unsustainable) farming practices (Clark & Tilman, 2017). Thus, there have been methodologies proposed to assess the sustainability of agricultural systems (Sabiha et al., 2016).

The case of agriculture is rather complex as the environment interacts with agriculture in a two-way manner. First, agricultural activities may affect the multi-functionality of the environment (Tilman, 1999; German et al., 2017). Such changes may be relevant for the whole society, rather than only being limited to the rural population. Second, the environmental change may affect the agricultural systems as well (Aydinalp & Cresser, 2008). In this case, the fluctuation in yields may increase due to the climate change, among other undesirable outcomes. Therefore, the environmental effects of the agricultural activity should be mitigated in order to prevent the degradation of the ecosystems that are important to different sectors and society in general. To tackle these effects, multiple measures are needed (Johnson et al., 2007; Lynch, 2009; Schieffer & Dillon, 2015; Zanin et al., 2022).

The environmental effects of economic activities (including farming) can be measured through the lens of greenhouse gas (GHG) emissions. The GHG emission comprises emissions of different gases expressed in the CO_2 equivalents. Therefore, looking at the GHG emission allows to comprehensively assess the contribution to the climate change exerted by the agricultural activities.

In general, the GHG emissions related to agricultural activities can be grouped into direct and indirect ones (Panchasara et al., 2021; Fuentes-Ponce et al., 2022). Direct emissions are due to farming activities (land management, manure management), whereas indirect emissions relate to the inputs used in the agriculture (energy, fertilizers). Dyer and Desjardins (2007) note that the production of agricultural inputs requires energy and, thus, considers them as indirect emissions related to agricultural activities.

DOI: 10.1201/9781003460589-3

The GHG emission from agricultural sector can be analysed from different perspectives. One can focus on either the energy-related or total GHG emission. The energy use can be analyzed as an important determinant of the GHG emission in the agricultural sector. Also, the energy use can be analyzed at different levels of aggregation (countries, sub-sectors, and farm groups).

The case of the European Union (EU) rewards particular attention. The EU Green Deal strategy calls for decarbonization in agriculture and other sectors (Matthews, 2023). There is a need to adapt the policy measures to ensure that the climate change mitigation objectives are suitably implemented. At the same time, the sustainable development goals require provision of affordable food as a component of the food security. Thus, the economic and environmental dimensions related to agricultural activities must be considered in a comprehensive manner.

This chapter discusses the patterns of the GHG emission and energy use targets in the EU agriculture. The country-level and disaggregate data from Eurostat and Farm Accountancy Data Network (FADN) databases are used.

2.2 GHG EMISSION IN THE EU AND ITS AGRICULTURE

According to the data from the United Nations Framework Convention on Climate Change (UNFCCC, 2023) database, the EU saw a reduction in the GHG emission (including land use and land use change and forestry) from 4.65 billion t CO_2 eq. in 1990 down to 3.24 billion t CO_2 eq. in 2021, which represents a 30% reduction (−1% per year). This result suggests a generally positive direction toward the mitigation of the climate change at the EU level.

The items related to the agricultural sector include the energy transformation (use of fuels and other energy sources for agricultural operations) and agricultural processes themselves (manure management, enteric fermentation). The dynamics in the total GHG emission in the EU and the two items related to the agricultural sector are presented in Figure 2.1. The data from the UNFCCC (2023) suggest that the

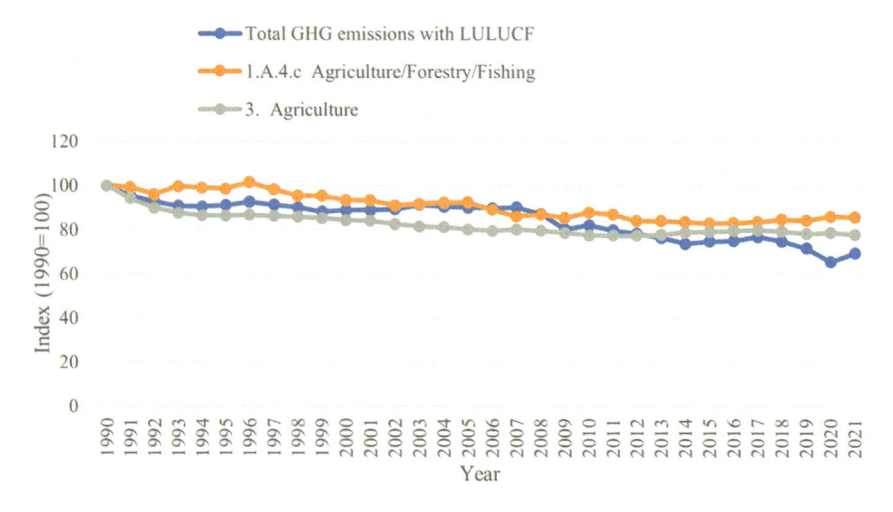

FIGURE 2.1 Indices for the GHG emissions in the EU (1990 = 100), 1990–2021. (Created by authors.)

GHG Emission and Energy Use in the EU Agriculture

energy-related agricultural GHG emission (item 1.A.4.c) declined by almost 15% in the EU during 1990–2021. The same trend was also noted for the agricultural emissions under item 3, where a decline over the same period amounted to 12%.

The period of 1990–2021 can be broken down into two sub-periods exhibiting different trends in the dynamics of the GHG emissions related to agricultural activities and the total GHG emission in the EU. The total GHG emission tended to decline until 1999 and then started to slightly rebound and remained in the area of a flat trend until 2008 when a substantial decline began. These fluctuations can be attributed to the major economic crises and economic cycles in general.

The agricultural emission (item 3) and energy-related agricultural emission (item 1.A.4.c) showed related trends. Indeed, the energy-related emission did not decline as fast as those directly related to agricultural processes. Anyway, it followed a very similar path after 2008. In general, one can observe that the agricultural sector of the EU demonstrated a positive change in the GHG emissions (i.e., a decline), yet it was lower than that observed for the total GHG emission.

As the agriculture-related items and the total GHG emission showed different rates of growth, the structure of the total GHG emission changed over time. The share of the GHG emission directly linked to agricultural activities (item 3) increased from 10.4% in 1990 up to 11.7% in 2021. The same trend was noted for the energy-related agricultural emission (item 1.A.4.c). The dynamics in the structure of the total GHG emission for the EU during 1990–2021 is given in Figure 2.2. The comparison between the trends in the shares of the energy-related and direct agricultural emissions in the total GHG emission of the EU are shown in Figure 2.3.

The energy-related GHG emission increased their share in the total GHG emissions to a lower extent if compared to the share of the direct agricultural GHG emission. The magnitude of these shares is also different with the direct agricultural emission comprising roughly ten times higher share than the energy-related agricultural emission. As for the period of 1990–2021, the energy-related agricultural GHG emission exhibited an increase from 2% up to 2.5% of the total GHG emission.

The possibilities to tackle the direct agricultural GHG emission are mostly related to the changes in animal farming. Specifically, the use of modern manure

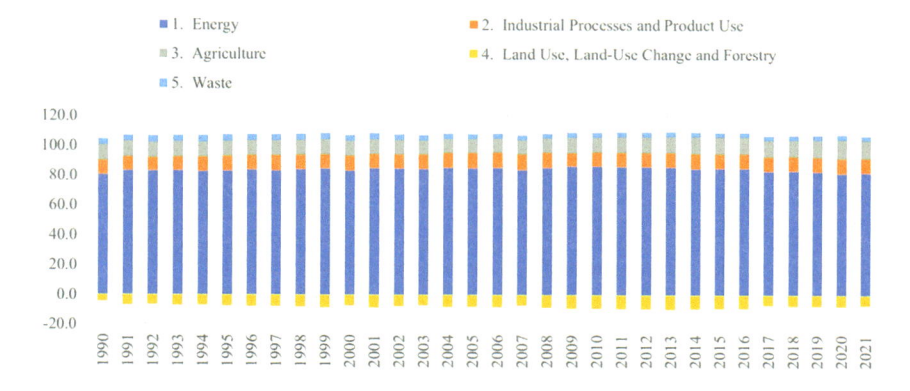

FIGURE 2.2 The structure of the total GHG emission in the EU (%), 1990–2021. (Created by authors.)

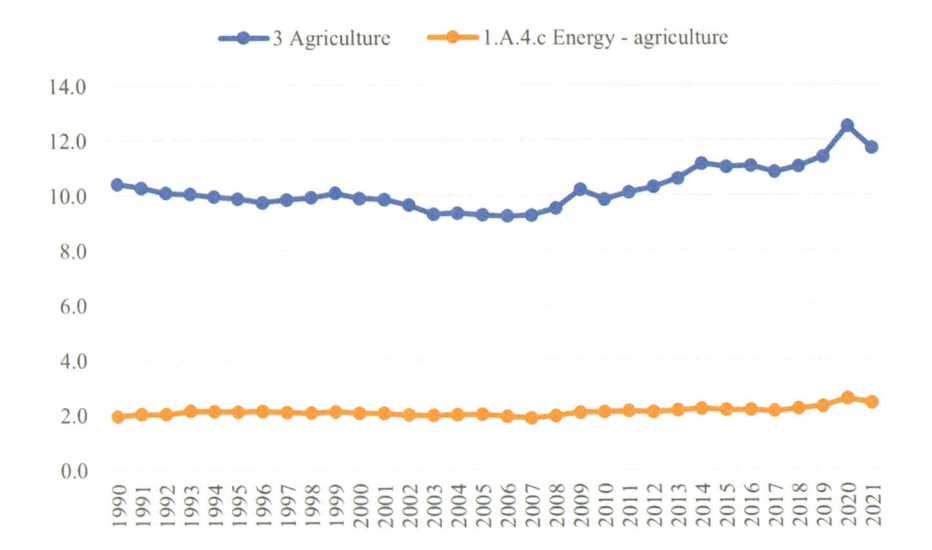

FIGURE 2.3 The shares of the energy-related and direct agricultural GHG emissions in the total GHG emission in the EU (%), 1990–2021. (Created by authors.)

management solutions and implementation of the circular economy principles may ensure that the emissions are not released without bringing any positive results. The changes in the scale of animal farming are a less viable solution given the food security objectives.

The energy-related GHG emission can be broken down into three components: (i) stationary energy sources, (ii) off-road vehicles, and (iii) fishing. All of these sub-categories showed a decline in the absolute levels. The steepest decline was noted for fishing (−2.8% per year). The lowest change was posted for the machinery operations (−0.1% per year). The stationary energy sources showed a decline of 1.1% per year. Thus, the use of the agricultural machinery requires further attention to ensure that the associated GHG emission is further reduced.

The structure of the energy-related GHG emission in the EU over 1990–2021 is presented in Figure 2.4. The share of energy consumed in fishing tended to decline steadily from 9.8% down to 5.5% (compared to item 1.A.4.c). The share of stationary energy sources declined, yet there was a rebound over 2010–2013. The latter share went down from 38.8% in 1990 to 32.8% in 2021. As for the machinery operations, the associated share increased from 51.4% up to 61.7% over 1990–2021. These changes once again suggest that the use of the agricultural machinery requires further attention in the sense of the energy use and the resulting GHG emission in the machinery of the agricultural sector. The efficiency of the fuel consumption and introduction of cleaner energy sources are topics in this regard.

The direct agricultural emission encompasses nine items that are relevant for the case of the EU. These include enteric fermentation, manure management, and agricultural soils which cover more than 96% of the direct agricultural emission in the EU. The structure of the direct agricultural emission did not vary with time: the

GHG Emission and Energy Use in the EU Agriculture 81

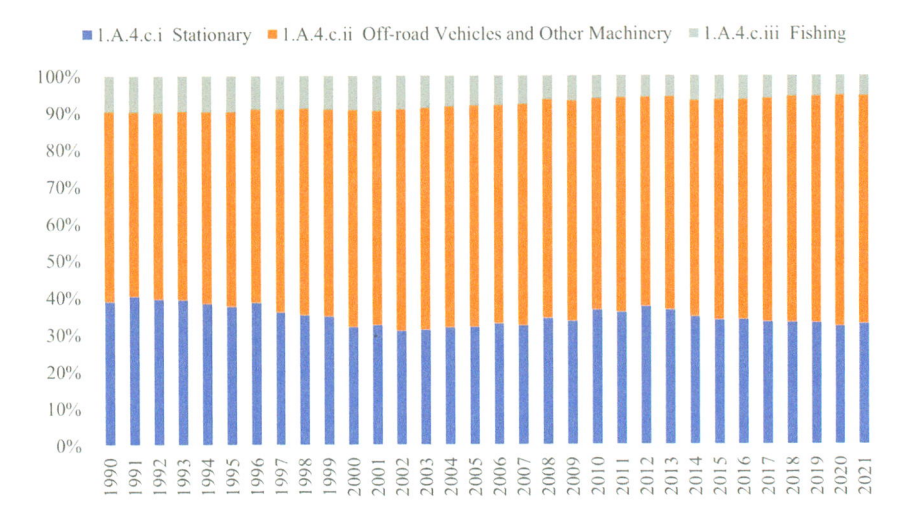

FIGURE 2.4 The structure of the energy-related GHG emission in the EU, 1990–2021. (Created by authors.)

shares of the three items mentioned before stood at 48.9%, 16.5%, and 30.6% in 1990 and at 48.2%, 16.6%, and 31.2% in 2021. Thus, animal farming can be related to some 65% of the direct agricultural emission as of 2021. The reduction in the emission related to animal farming can be ensured by changing the farming practices and limiting the scale of animal farming. Even though a switch to crop products in the diet may be considered welcome from the health viewpoint, it is less likely given the existing cultural standards (tastes).

2.3 ENERGY CONSUMPTION AND GHG EMISSION IN THE EU AGRICULTURE

The data at the country level allow to analyze the process of convergence with respect to the cleaner energy use and climate-neutral agriculture. Eurostat (2023) provides data for the energy use in various sectors, including agriculture. The energy balance provides information on the sources of energy. This information can be used to track the changes in the energy use and its mix. Further on, the energy-related GHG emissions are also provided by the Eurostat. Thus, one can calculate the carbon factor which indicates the success of the efforts to ensure cleaner energy mix. When using the data from Eurostat, we refer to the EU in the sense of the 27 countries (from 2020).

The final energy consumption in the EU agriculture increased from 24.7 million tonnes of oil equivalent (toe) up to 28.3 million toe during 2012–2021 (Figure 2.5). Therefore, the changes in the energy consumption are not beneficial for the environmental concerns and GHG emission in particular. This can also be related to the results related to item 1.A.4.c in the GHG inventory depicted in Figure 2.1.

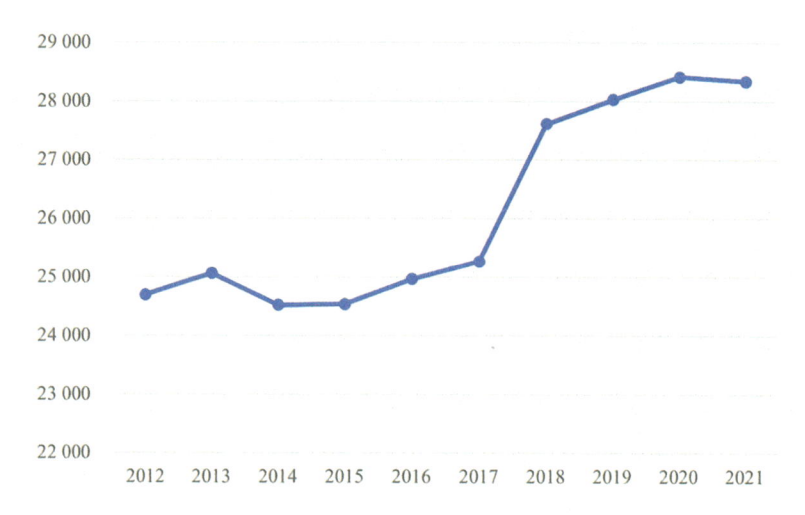

FIGURE 2.5 Final energy consumption in the EU agriculture (thousand tonnes of oil equivalent), 2012–2021. (Created by authors.)

The structure of the energy mix remained rather stable in the case of the EU agriculture concerning the fossil and renewable energy sources. The major changes can be observed within the fossil energy category. Energy use from solid fossil fuels declined steadily at the rate of 5.6% per year and the relevant input declined from 1.1 million toe down to 0.7 million toe within 2012–2021. The use of the natural gas as the source of energy in the EU agriculture also declined from 4.4 million toe in 2012 down to 3.8 million toe in 2021 (at 2% per year). These were replaced by oil and petroleum products (an increase from 12.6 million toe up to 15.8 million toe or 3.2% per year). The resulting share of the fossil fuel in the total final energy consumption dropped from 74% down to 72% throughout 2012–2021. The use of renewables increased from 2.1 million toe up to 3.2 million toe. The resulting average annual growth rate stood at 4.7%. The electricity and heat energy used in the EU agriculture also showed positive annual growth rates of 1.1% and 1.9%, respectively. The consumption of power went up from 4.1 million toe up to 4.4 million toe, whereas the use of the heat energy saw an increase from 284 thousand toe up to 338 thousand toe during 2012–2021. Figure 2.6 displays the structure of the energy mix in the EU agriculture.

The share of renewables went up from 8.5% to 11.4% during 2012–2021. However, one can also consider the electricity and heating energy as potentially renewable energy sources. Indeed, electrification is a major means for promoting renewables. The electrification of agricultural production can be considered as an important issue as the share of renewables and electricity and heating energy increases up to 26.4% and 28.3% for 2012 and 2021, respectively, at the EU level.

The discussed changes in the energy consumption in the EU agriculture suggest that further gains in energy efficiency are required to push the energy demand down. Also, the share of the renewables is still stagnant. The application of modern farming practices, including the novel machinery that is electrified, would bring further possibilities to improve the energy mix. The EU member states support the agricultural

GHG Emission and Energy Use in the EU Agriculture

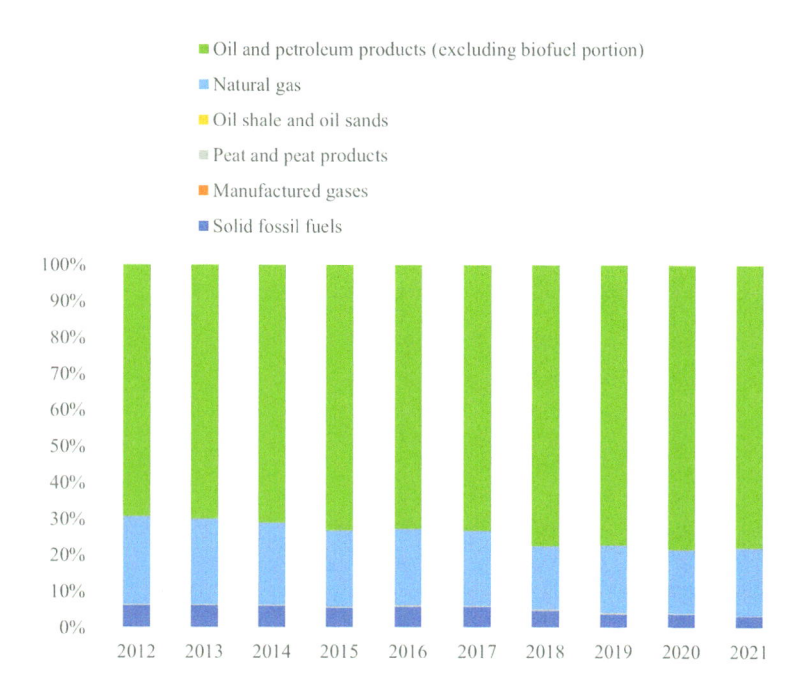

FIGURE 2.6 Energy-mix for the final consumption in the EU agriculture, 2012–2021. (Created by authors.)

sector by allowing for reductions in the gasoil excises. The abandonment of such practices would also increase the incentives for the green transition in the sector.

The volume of energy use and energy mix vary across the EU member states. Table 2.1 presents the data on the final energy consumption across the EU member states for the agricultural sector. The results indicate that six countries posted a decline in the energy use, whereas the rest showed an increase or values close to null.

The country-wise structure of the energy consumers in the agricultural sector changed with the highest gains for Germany (an increase by 7.3 p.p.) and multiple countries showing a slight decline. The steepest decline is observed for Poland (−1.7 p.p.) and France (−1.5 p.p.). The changes in the structure of the energy consumption may also be related to the related changes in the energy intensity and carbon factor.

As regards the change in the energy consumption, the stochastic trends are applied. Hence, the stochastic rates of change are obtained for each country. The resulting stochastic growth rate for the EU, thus, cannot be considered as the average value of the stochastic growth rates for the other countries.

The steepest decline in the final energy consumption in the agricultural sector was noticed for Denmark (−2% per year). As a result, the share of Denmark's energy use (compared to the total EU energy use) declined from 2.7% down to 2% during 2012–2021. Sweden, Slovakia, and Estonia showed rates of decline 0.8%, 0.9%, and 1.7%, respectively. The negligible rates of −0.2% per annum were noted for Bulgaria and Finland. Thus, both the Nordic countries and the new EU member states appeared among the countries that showed negative growth rates. Still, there are such countries

TABLE 2.1
The Final Energy Consumption in the Agricultural Sector across the EU Member States, 2012 and 2021

	Energy Use (thousand toe)			Share in Energy Use (%)		
Country	2012	2021	Growth (% per year)	2012	2021	Change in share (p.p.)
Germany	1,317.36	3,591.75	14.0	5.3	12.7	7.3
Malta	3.96	6.80	10.5	0.0	0.0	0.0
Hungary	398.23	665.92	4.6	1.6	2.3	0.7
Latvia	142.23	191.61	4.3	0.6	0.7	0.1
Portugal	322.26	434.15	3.2	1.3	1.5	0.2
Cyprus	36.93	45.82	2.7	0.1	0.2	0.0
Romania	497.77	566.58	2.5	2.0	2.0	0.0
Belgium	761.01	870.13	1.9	3.1	3.1	0.0
EU27_2020	24,692.20	28,343.64	1.9			
Croatia	207.34	233.86	1.4	0.8	0.8	0.0
Lithuania	108.52	121.75	1.4	0.4	0.4	0.0
Netherlands	3,702.72	3,971.12	1.2	15.0	14.0	−1.0
Poland	3 666.48	3,736.96	1.1	14.8	13.2	−1.7
Czechia	563.46	644.85	1.1	2.3	2.3	0.0
Italy	2,625.29	2816.91	0.9	10.6	9.9	−0.7
Luxembourg	24.98	27.19	0.7	0.1	0.1	0.0
Greece	284.45	298.15	0.4	1.2	1.1	−0.1
Ireland	250.56	232.30	0.3	1.0	0.8	−0.2
Slovenia	71.01	73.57	0.2	0.3	0.3	0.0
France	4,014.19	4,193.88	0.2	16.3	14.8	−1.5
Spain	2,675.25	2,781.19	0.1	10.8	9.8	−1.0
Austria	524.10	547.31	0.0	2.1	1.9	−0.2
Bulgaria	198.71	198.62	−0.2	0.8	0.7	−0.1
Finland	772.61	741.60	−0.2	3.1	2.6	−0.5
Sweden	599.27	566.33	−0.8	2.4	2.0	−0.4
Slovakia	143.48	131.74	−0.9	0.6	0.5	−0.1
Estonia	110.04	88.11	−1.7	0.4	0.3	−0.1
Denmark	670.00	565.44	−2.0	2.7	2.0	−0.7

Source: Created by authors.

as Austria, Spain, France, Slovenia, and Ireland where the growth in the final energy consumption did not exceed 0.3% per year.

The highest rate of growth in the final energy use in agriculture is observed for Germany (14% per year). As a result, the share of the final energy consumed in German agriculture went up from 5.3% up to 12.7% compared to the total consumption in the EU. The average annual growth rates exceeding the average value for the EU (1.9%) were noted for Malta (10.5%), Hungary (4.6%), Latvia (4.3%), Portugal (3.2%), Cyprus (2.7%), Romania (2.5%), and Belgium (1.9%).

The largest energy consumers are the Netherlands, Poland, France, and Germany. It was already mentioned that Germany saw a serious increase in the agricultural energy use and, accordingly, the share in the total consumption. The other large energy-consuming countries showed a decline in their shares.

The country-specific data for energy-related GHG emission (item 1.A.4.c) and GHG emission directly related to agricultural operations (item 3) are given in Tables 2.2 and 2.3, respectively. As shown in Table 2.2, the largest energy consumers are naturally the largest emitters as regards the energy-related GHG emission. A notable case is Germany where an increase in the energy-related GHG emission

TABLE 2.2
Energy-Related GHG Emission per Country (thousand tonnes CO_2 equivalent), 2012–2021

Country	2012	2021	Share 2012	Share 2021	Change (p.p.)	Rate of Growth (% per year)
Malta	19.72	22.87	0.0	0.0	0.0	5.8
Romania	1,255.37	1,703.92	1.6	2.2	0.5	5.1
Hungary	896.81	1,571.47	1.2	2.0	0.8	4.6
Belgium	2,052.12	2,857.67	2.7	3.7	1.0	3.7
Latvia	413.05	522.83	0.5	0.7	0.1	3.3
Portugal	1,143.66	1,362.79	1.5	1.7	0.3	1.8
Cyprus	80	85.21	0.1	0.1	0.0	1.6
Lithuania	235.33	255.91	0.3	0.3	0.0	0.9
Poland	11,560.78	11,436.34	15.2	14.7	−0.5	0.8
Croatia	710.73	750.32	0.9	1.0	0.0	0.7
Italy	7,520.27	7,777.07	9.9	10.0	0.1	0.6
EU-27	76,306.98	77,894.33				0.3
Austria	941.82	967	1.2	1.2	0.0	0.2
Ireland	750.63	672.62	1.0	0.9	−0.1	0.2
Spain	11,892.95	12,419.54	15.6	15.9	0.4	0.1
Germany	5936.3	6,333.46	7.8	8.1	0.4	0.1
Slovenia	232.88	232.78	0.3	0.3	0.0	0.0
Czechia	1,228.12	1,227.68	1.6	1.6	0.0	0.0
Netherlands	11,078.9	10,615.38	14.5	13.6	−0.9	−0.2
Finland	1,591.78	1,405.69	2.1	1.8	−0.3	−0.3
France	11,448.21	11,411.84	15.0	14.7	−0.4	−0.6
Bulgaria	532.42	499.31	0.7	0.6	−0.1	−1.0
Luxembourg	27.22	22.75	0.0	0.0	0.0	−1.0
Greece	942.08	609.78	1.2	0.8	−0.5	−1.0
Estonia	271.86	213.83	0.4	0.3	−0.1	−1.4
Denmark	1,608.32	1,395.8	2.1	1.8	−0.3	−2.2
Slovakia	397.05	376.53	0.5	0.5	0.0	−2.5
Sweden	1,538.61	1,143.94	2.0	1.5	−0.5	−3.6

Source: Created by authors.

TABLE 2.3
GHG Emission Directly Linked to the Agricultural Operations per Country (thousand tonnes CO_2 equivalent), 2012–2021

Country	2012	2021	Share 2012	Share 2021	Change (p.p.)	Rate of Growth (% per year)
Hungary	5840.89	7,201.98	1.6	1.9	0.3	2.1
Cyprus	555.21	626.2	0.1	0.2	0.0	1.9
Bulgaria	4,985	6,106.34	1.3	1.6	0.3	1.6
Ireland	19,706.42	22,953.53	5.2	6.1	0.8	1.6
Latvia	1,962.72	2,252.96	0.5	0.6	0.1	1.3
Estonia	1,409.72	1,583.94	0.4	0.4	0.0	1.1
Spain	31,623.81	34,369.39	8.4	9.1	0.7	1.1
Luxembourg	635.83	697.35	0.2	0.2	0.0	1.0
Portugal	6,793.96	7,258.18	1.8	1.9	0.1	0.9
Poland	31,826.63	34,035.26	8.5	9.0	0.5	0.7
Romania	17,933.31	19,169.3	4.8	5.1	0.3	0.6
Slovenia	1,724.58	1,813.47	0.5	0.5	0.0	0.6
Malta	84.23	87.71	0.0	0.0	0.0	0.5
Czechia	7,572.9	7,844.54	2.0	2.1	0.1	0.4
Denmark	12,061.25	12,074.39	3.2	3.2	0.0	0.1
Netherlands	17,704.18	17,974.7	4.7	4.7	0.0	0.1
Sweden	6,613.48	6,673.59	1.8	1.8	0.0	0.1
Finland	6,238.74	6,303.11	1.7	1.7	0.0	0.1
EU27	375,535.13	378,430.47				0.1
Lithuania	4287.37	4,327.72	1.1	1.1	0.0	0.1
Italy	33,199.96	32,717.22	8.8	8.6	−0.2	0.1
Austria	7,212.42	7,221.16	1.9	1.9	0.0	−0.1
Belgium	9,579.49	9,414.19	2.6	2.5	−0.1	−0.2
Slovakia	2,429.13	2,430.57	0.6	0.6	0.0	−0.3
France	71,600.63	66,213.86	19.1	17.5	−1.6	−0.8
Germany	60,052.38	56,332.89	16.0	14.9	−1.1	−0.9
Croatia	3,061.41	2,700.94	0.8	0.7	−0.1	−0.9
Greece	8,839.48	8,045.99	2.4	2.1	−0.2	−0.9

Source: Created by authors.

remains meagre even though this country showed a serious increase in the final energy use in agriculture (Table 2.1). This can partially be explained by the shift toward cleaner energy mix.

The highest rates of growth in the energy-related GHG emission are noted for Malta, Romania, Hungary, Belgium, and Latvia where the average annual stochastic growth rate exceeded 3%. Portugal, Cyprus, Lithuania, Poland, Croatia, and Italy also showed average growth rates that exceeded the EU average. Indeed, countries below the EU average showed negligible growth rates. This was caused by the presence of large emitters among countries with slightly negative growth rates. In general, the country-wise structure of the energy-related GHG emission in the EU agriculture remained rather stable. The highest rates of change in shares of the energy-related GHG emission were observed for small emitters only (one can consider the Netherlands as the only exception where a decline in the share of the energy-related GHG emission from 14.5% to 13.6% was noticed).

The dynamics in (the distribution of) the GHG emission directly related to the agricultural operations across the EU countries is provided in Table 2.3. As one can note, the ranking of the countries with respect to the average annual growth rate is different from the case of the energy-related GHG emission in Table 2.2. Indeed, the direct GHG emission is mostly linked to animal farming which explains the differences between the two types of rankings. The countries with intensive operation of the vegetable and fruit farming are often among those relying on the energy input.

The stochastic rates of growth in the direct GHG emission range in between 2.1% for Hungary and −0.9% for Greece. The countries with negative rates of growth (or even the slightly positive ones) may be considered as those with a declining share of livestock output in the total output structure. The two largest emitters, France and Germany, showed the steepest decline in their shares within this item. The share of France shrunk from 19.1 in 2012 down to 17.5 in 2021. Similarly, the share of Germany went from 16% down to 14.9%. Poland, Romania, Spain, and Ireland can be given as examples of countries with contributions to the direct agricultural GHG emission exceeding 5% and showing positive growth rates.

2.4 USE OF RENEWABLES IN EU AGRICULTURE

The share of renewables in the total energy mix increased marginally at the EU level. However, the countries show serious differences in the share of renewables and the dynamics thereof (Table 2.4). Austria, Germany, Sweden, Finland, and Czechia show the average shares of renewable resources in the total energy mix exceeding 20% over 2012–2021. Then, Slovakia, Poland, Latvia, Lithuania, and Greece exceed the average share for the EU, namely, 10.5%.

The highest increase in the share of renewables is noted for Sweden (from 16.6% in 2012 up to 35.6% in 2021), Slovakia (from 12.3% up to 20.7%), France (from 3.8% up to 9.8%), and the Netherlands (from 3% up to 10.1%). The data in Table 2.4 suggest that countries with low shares of renewables are rather stagnant in this regard. On the other hand, Germany showed the second-highest average share of renewables, yet it declined from 38.5% in 2012 down to 21.3% in 2021.

Renewables can be utilized for energy transformation while generating power and heat energy. Thus, the share of electricity and heat is added to that of the renewables in Table 2.5. These figures suggest the limits for the penetration of renewables in the medium run. Greece presents an exceptional case with the average share of renewables and power in the final consumption of 87.4% for 2012–2021. This can be attributed to the output mix in Greece where fruits and vegetables are among the most important outputs in the sense of the contribution to the total output value. Austria and Germany show the values close to 50%. Then, Sweden, Cyprus, Finland, and Denmark exceed the value of 40%. Czechia, Slovakia, and Lithuania are below the EU average with the mean values of 35%–30.8%. Croatia and Slovenia remain at

TABLE 2.4

The Share of the Renewable Energy Sources in the Final Energy Consumption in Agriculture, 2012–2021

Country	2012	2021	Average	Change (p.p. per year)
Austria	33.4	32.3	33.1	−0.3
Germany	38.5	21.3	30.5	−2.8
Sweden	16.6	35.6	28.8	2.3
Finland	21.8	23.4	22.0	−0.2
Czechia	14.7	21.2	20.5	0.5
Slovakia	12.3	20.7	17.4	1.5
Poland	13.8	15.5	13.8	−0.1
Latvia	8.4	11.2	13.1	0.6
Lithuania	10.9	14.7	12.8	0.5
Greece	10.8	9.1	11.0	−0.1
Luxembourg	16.3	2.3	10.8	−1.0
EU27_2020	8.5	11.4	10.5	0.3
Denmark	8.0	10.0	8.8	0.2
Hungary	7.0	7.2	7.4	0.2
France	3.8	9.8	6.6	0.8
Netherlands	3.0	10.1	6.4	0.8
Cyprus	7.4	7.0	6.4	0.0
Belgium	4.6	4.0	5.3	−0.2
Bulgaria	2.0	5.2	4.5	−0.1
Estonia	1.1	6.2	3.9	0.3
Slovenia	3.8	3.9	3.8	−0.1
Spain	2.6	2.6	2.8	0.0
Malta	0.5	2.9	2.1	0.4
Romania	1.1	1.3	2.0	−0.3
Italy	0.7	2.0	1.6	0.1
Croatia	0.0	0.7	1.3	0.1
Portugal	1.7	1.4	1.3	0.0
Ireland	0.0	0.0	0.0	0.0

Source: Created by authors.

the bottom of the ranking with mean values of 9.5% and 4.7%, respectively. Thus, it is necessary to promote electrification in countries with similar output mix in order to encourage the spread of renewables in the agricultural sector.

As regards the change in the share of the renewables and electricity, Germany shows a remarkable decline of −6.1 p.p. per year. The steepest increase is observed for such countries as Sweden (from 30.3% up to 54.9%), Slovakia (30.1% up to 36%), and the Netherlands (from 23.2% up to 36.2%). Note that Slovakia and the Netherlands appear as countries with relatively low penetration of renewables. These countries can be used as model cases for implementing the electrification of the agricultural sector.

TABLE 2.5
The Share of Renewables, Power and Heat Energy in the Final Consumption in Agriculture, 2012–2021

Country	2012	2021	Average	Change (p.p. per year)
Greece	93.2	83.9	87.4	−0.9
Austria	54.1	55.8	54.3	0.1
Germany	75.7	32.7	50.5	−6.1
Sweden	30.3	54.9	46.0	2.5
Cyprus	38.7	43.4	41.6	0.2
Finland	41.0	44.8	41.1	−0.1
Denmark	36.6	44.6	40.0	0.8
Czechia	28.9	38.6	35.0	0.8
Slovakia	30.1	36.0	34.4	1.4
Lithuania	30.1	32.1	32.2	0.4
Netherlands	23.2	36.2	30.8	1.6
EU27_2020	26.4	28.3	28.3	0.1
Latvia	21.4	24.0	26.5	0.6
Malta	22.2	21.4	25.5	−1.1
France	21.8	25.8	24.1	0.7
Portugal	26.4	24.3	23.3	−0.1
Belgium	23.0	20.2	23.1	−0.2
Luxembourg	29.3	12.5	22.3	−1.3
Bulgaria	18.7	25.1	21.7	0.7
Ireland	19.1	21.4	21.7	0.0
Italy	20.4	22.5	20.4	0.2
Spain	15.4	20.0	20.1	0.3
Hungary	23.8	19.8	20.1	0.0
Estonia	19.0	23.4	18.6	0.0
Poland	18.1	17.7	18.0	−0.2
Romania	21.4	10.8	17.8	−1.8
Croatia	6.9	8.9	9.5	0.2
Slovenia	3.8	7.2	4.7	0.3

Source: Created by authors.

Carbon factor is calculated as the ratio of the energy-related GHG emission to the final energy use in agriculture. The carbon factor tracks the dynamics in the energy mix as the higher values are associated with the increasing reliance on fossil fuels. The trend in the carbon factor for the EU agriculture is depicted in Figure 2.7. In general, the carbon factor showed a declining trend over 2012–2021. Specifically, it went from 3.09 kg CO_2 eq./kg oil eq. in 2012 down to 2.75 kg CO_2 eq./kg oil eq. in 2021. This indicates that the energy mix has become cleaner in the EU agriculture over the period covered.

The trajectory of the carbon factor suggests the presence of a structural break in 2018 when it sharply declined. Note that Figure 2.5 indicates a steep increase in the

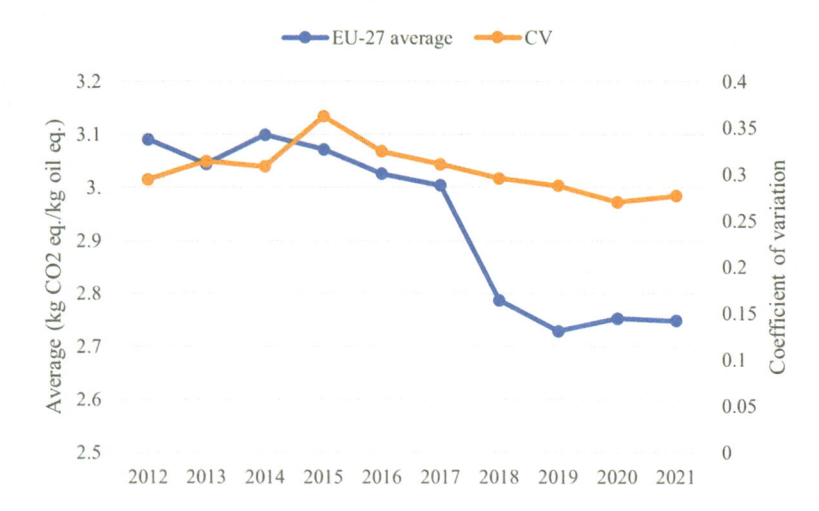

FIGURE 2.7 Carbon factor for the EU agriculture and its coefficient of variation across the member states, 2012–2021. (Created by authors.)

energy consumption during the same time period. This indicates that the declining use of the "dirty" energy sources allowed to achieve a decline in the carbon factor in the presence of increasing energy use.

The coefficient of variation allows one to assess the convergence among the EU countries with respect to the carbon factor related to the energy use in the agricultural sector. The trend in Figure 2.7 indicates that the coefficient of variation increased during 2012–2015 and then kept declining. However, there was no serious decline observed as the coefficient of variation went from 0.29 in 2012 down to 0.28 in 2021. This finding implies that the EU countries may require further attention to their energy mix, yet a complete convergence may be impossible due to different climatic conditions and output structure in the agricultural sector.

The country-specific carbon factors are presented in Table 2.6. The highest values of the carbon factor were noted in Spain, Malta, Germany, Portugal, Croatia, Slovenia, Poland, Ireland, and Slovakia. The mean carbon factors in these countries exceeded the EU average value of 2.94 kg CO_2 eq./kg oil eq. Austria and Luxembourg exhibited the lowest carbon factors. The changes in the energy-related GHG emission in the agricultural sector are chiefly related to the case of Germany where a steep decline in the carbon factor occurred. Romania showed an increase of the carbon factor of 2.5% per year leading to a growth from 2.52 kg CO_2 eq./kg oil eq. up to 3.01 kg CO_2 eq./kg oil eq. Sweden posted a decline of 2.8% per year that resulted in a decline from 2.57 kg CO_2 eq./kg oil eq. down to 2.02 kg CO_2 eq./kg oil eq.

The ratio of the energy expenditure in the agricultural economic accounts to the final energy consumption in the energy balance gives an estimate of the energy unit price. The weighted average for the EU-28 and the coefficient of variation are presented in Figure 2.8. The trend of the weighted average suggests that the energy price declined during 2012–2021. This indicates the increasing availability of energy resources over 2012–2021. However, the trend may have been reversed in 2022 due to the geopolitical situation that has increased the volatility in the energy markets

TABLE 2.6
Carbon Factors in the Agricultural Sector across the EU Member States (kg CO_2 eq./kg oil eq.), 2012–2021

Country	2012	2021	Average	Rate of Growth (% per year)
Spain	4.45	4.47	4.70	0.0
Malta	4.98	3.36	4.22	−4.7
Germany	4.51	1.76	3.43	−13.9
Portugal	3.55	3.14	3.34	−1.4
Croatia	3.43	3.21	3.27	−0.8
Slovenia	3.28	3.16	3.24	−0.3
Poland	3.15	3.06	3.12	−0.3
Ireland	3.00	2.90	2.94	−0.1
Slovakia	2.77	2.86	2.94	−1.6
EU–27	3.09	2.75	2.94	−1.6
Belgium	2.70	3.28	2.90	1.8
Italy	2.86	2.76	2.86	−0.3
France	2.85	2.72	2.75	−0.8
Netherlands	2.99	2.67	2.75	−1.5
Latvia	2.90	2.73	2.72	−1.0
Romania	2.52	3.01	2.71	2.5
Bulgaria	2.68	2.51	2.63	−0.8
Estonia	2.47	2.43	2.51	0.3
Denmark	2.40	2.47	2.37	−0.3
Hungary	2.25	2.36	2.36	−0.1
Greece	3.31	2.05	2.13	−1.4
Lithuania	2.17	2.10	2.13	−0.5
Sweden	2.57	2.02	2.09	−2.8
Finland	2.06	1.90	2.03	−0.1
Czechia	2.18	1.90	2.00	−1.1
Cyprus	2.17	1.86	1.97	−1.1
Austria	1.80	1.77	1.80	0.3
Luxembourg	1.09	0.84	0.91	−1.7

Source: Created by authors.

among other outcomes. The reduced energy price may also encourage energy consumption and, in turn, energy-related GHG emission. Note that the PPS is measured at the constant prices of 2015. Note that these trends may be explained by the favorable regime of the gas oil excises and other exemptions that are applicable to the agricultural sector.

The trend in the coefficient of variation suggests an increasing polarization of the EU countries with respect to the energy unit price. The coefficient of variation increased steadily over 2012–2021. Therefore, energy prices tended to increase their variation across the EU member states. This may be a result of different energy

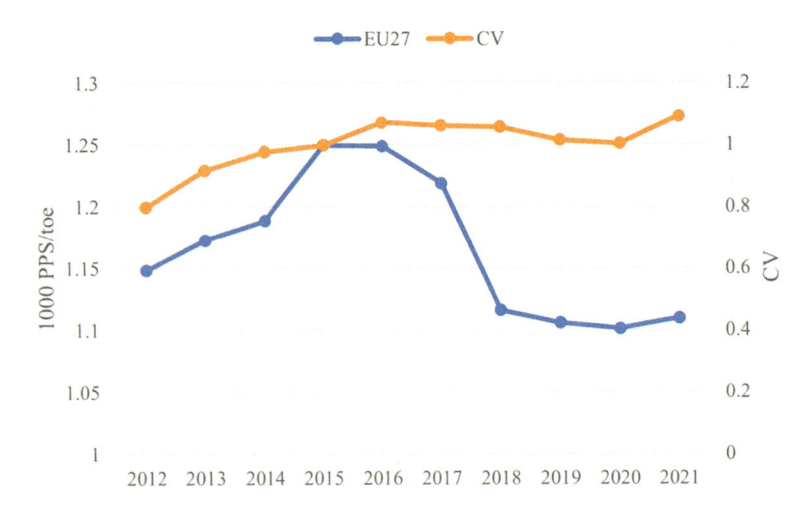

FIGURE 2.8 The energy unit price and its coefficient of variation in the EU Agriculture, 2012–2021. (Created by authors.)

sources and endowments available across the countries. Also, different countries are associated with different patterns of the energy demand.

The stochastic rates of growth reported in Table 2.7 suggest that such countries as Belgium and Germany showed the steepest decline in the energy prices in the agricultural sector (−15.2% and −14.8% per year on average, respectively). These countries exceeded the EU average in 2012, yet managed to reduce the energy prices so that they appeared below the EU average in 2021. The highest rates of growth are observed for Romania and Croatia (2.4% per year and 8.3% per year, respectively). The energy price in Croatia still remained below the EU average. Thus, high energy prices are observed in the new EU member states where renewables are not easily accessible.

The correlation between the share of renewable energy sources and the energy unit price (considering the average values for 2012 to 2021) is −0.2 indicating a weak negative relationship. This suggests that the increasing share of renewables may be associated with a decline in the energy price. This can be reasonable given the high dependency on the imported fossil energy. Similar results are obtained by adding the power to the renewables when calculating the share in the final energy consumption in the EU agriculture. Figure 2.9 provides a graphical presentation of the relationship between these two variables. The resulting coefficient of determination is rather low (0.04) which calls for further analysis of the determinants of the energy price and, especially, the effects of the renewables.

The period of 2021–2022 saw an increasing volatility in the energy markets due to geopolitical shifts. It may be assumed that the trade dynamics were already affected in 2021 due to the upcoming events in 2022, e.g., the war in Ukraine. Also, the pandemic of COVID-19 still played a role on the supply chain performance which affected the input prices. The resulting price dynamics have had a serious effect on agricultural activities as some farming types are highly reliant on the energy input.

TABLE 2.7
The Energy Unit Prices in the Agricultural Sectors of the EU Countries (1,000 PPS/toe), 2012–2021

Country	2012	2021	Average	Rate of Growth (% per year)
Belgium	1.34	0.35	0.63	−15.2
Germany	2.48	0.82	1.87	−14.8
Malta	2.77	1.64	2.87	−10.2
Latvia	1.87	1.22	1.52	−6.0
Hungary	2.61	1.81	1.94	−3.4
Portugal	1.43	1.15	1.28	−2.0
Denmark	0.50	0.43	0.49	−1.8
Finland	0.48	0.36	0.43	−1.7
Slovakia	3.39	2.77	2.63	−1.7
Greece	4.07	3.64	4.23	−1.5
Netherlands	0.43	0.40	0.42	−1.4
EU27	1.15	1.11	1.17	−0.9
Estonia	0.94	0.97	0.77	−0.7
Austria	0.71	0.67	0.69	−0.4
Slovenia	1.79	1.75	1.75	−0.1
Ireland	1.40	1.61	1.67	0.0
Luxembourg	0.53	0.56	0.55	0.0
France	0.83	0.82	0.82	0.1
Italy	1.11	1.10	1.09	0.2
Czechia	1.36	1.45	1.35	0.3
Bulgaria	5.20	6.14	6.20	0.8
Sweden	0.54	0.58	0.54	1.0
Spain	0.78	0.84	0.85	1.0
Cyprus	1.04	1.11	1.05	1.2
Poland	1.34	1.60	1.58	1.3
Lithuania	3.24	3.57	3.71	1.6
Romania	5.57	8.92	7.93	2.4
Croatia	0.69	1.10	0.92	8.3

Source: Created by authors.

In such cases, the production may cease. Table 2.8 presents the energy price indices for the EU countries in the economic accounts for agriculture.

The EU-27 countries saw the average increase in the energy prices of 53% over 2021–2022 in nominal terms and 44% in real terms. Compared to 2015, the energy prices went up by 72% and 48% in the nominal and real terms, respectively. This suggests a serious increase in the production costs of the agricultural produce in the EU.

The changes in energy prices vary across the countries due to different types of energy used and types of energy markets the countries are integrated into. For instance, Malta faced no serious price growth as it is a small economy relying on

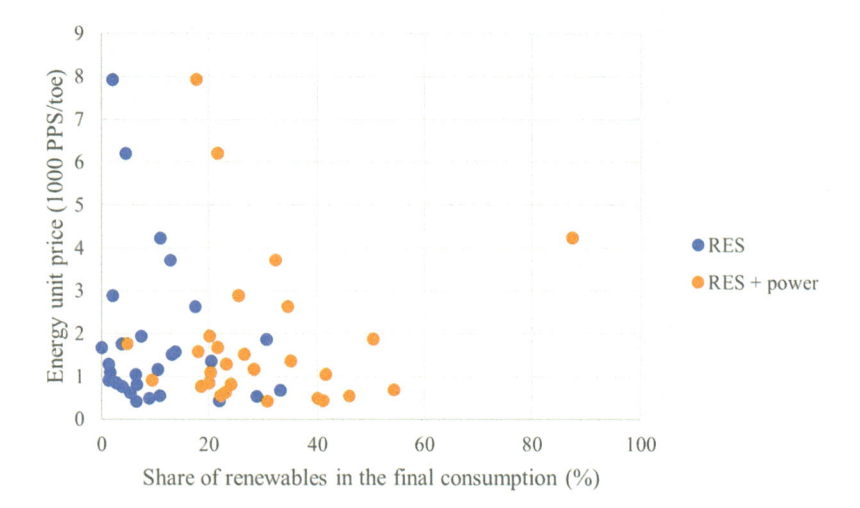

FIGURE 2.9 The share of the renewables in the final energy consumption and the energy unit price (average values for 2012 to 2021) across the EU member states. (Created by authors.)

energy imports (mostly, liquified natural gas) from the competitive markets. As for the other countries, they relied on long-term contracts with such suppliers as Russia and faced increasing energy prices during the war with Ukraine. Slovenia and Bulgaria showed rates of growth in the energy price of 28–29% in the nominal terms (or 15–20% in the real terms) over 2021–2022.

The highest price growth was noted for the Netherlands (104% or 93% in the nominal or real terms, respectively). As for Belgium, a 71% increase in nominal terms or a 60% increase in real terms was noted. Belgium is highly dependent on gas imports via pipeline from the Netherlands and the latter country stopped production in the Groningen gas field. Thus, these countries faced serious challenges in the energy market due to the Europe-wide developments related to the sanctions against Russia and internal adjustments due to the closure of the gas field.

The results show that the EU countries are sensitive to disruptions in the energy markets and the prices may become highly volatile in periods of uncertainty. This may change the competitiveness of the agrifood products and impact the energy use in general. Such changes, in turn, may affect the energy-related GHG emission. Further research may embark on relating these multiple factors in a quantitative framework.

2.5 CONCLUSIONS

The results suggest that EU countries managed to decrease the GHG emission from fuel combustion and production processes. The energy-related agricultural emission remained more stable and require further attention in ensuring that the agricultural operations do not cause environmental pressures. Emissions directly linked to agricultural processes comprise a higher share in the total GHG emission and are related to the scale and type of farming. Thus, adjustment in this type of emission requires consideration of multiple objectives.

TABLE 2.8
The Price Indices (2015 = 100) for the Energy Input in the Agricultural Sector, 2021 and 2022

Country	Nominal 2021	Nominal 2022	Growth (%)	Real 2021	Real 2022	Growth (%)
EU-27	113	172	53	102	148	44
Belgium	193	329	71	172	276	60
Bulgaria	108	139	29	82	94	15
Czechia	106	149	41	90	116	29
Denmark	138	242	75	127	215	69
Germany	110	153	40	98	129	33
Estonia	127	231	82	105	167	59
Ireland	113	161	42	107	139	30
Greece	117	164	40	117	151	29
Spain	116	176	52	107	157	47
France	120	168	39	112	152	36
Croatia	62	105	69	57	92	61
Italy	115	172	50	108	157	45
Cyprus	94	122	30	90	112	24
Latvia	112	185	65	94	140	49
Lithuania	107	209	96	87	147	69
Luxembourg	119	185	55	103	150	46
Hungary	120	180	50	91	126	38
Malta	91	91	0	80	77	−5
Netherlands	119	242	104	106	205	93
Austria	110	147	34	99	126	28
Poland	115	203	76	99	154	56
Portugal	108	153	42	98	133	36
Romania	110	155	41	82	105	27
Slovenia	108	138	28	97	117	20
Slovakia	112	174	55	102	147	44
Finland	120	175	46	111	154	39
Sweden	126	168	34	110	138	26

Source: Created by authors.

The final energy consumption in the EU agriculture increased over 2012–2022. This calls for actions toward not only cleaner energy mix but also toward energy conservation. The reduction of the support measures for the energy inputs (e.g., reduced excises for gas oil) may increase the efficiency of the fuel use in agriculture. This warrants further research.

The results show that the EU countries appeared to be rather sensitive to the geopolitical developments in 2022. As the energy prices increased in all the EU countries save Malta, one can highlight the need for further integration into competitive market of the energy sources. Also, the increase in the energy efficiency may reduce the undesirable impacts of the price spikes in the energy markets on the agricultural sector.

In this research, we did not consider the (changes in) land use which is yet another item in the GHG emission inventory that relates to the agricultural sector. Future research may embark on the analysis of the latter item to identify the shifts in agricultural sector leading to changes in the GHG emission. In addition, the energy demand functions may be established to assess the impacts of various factors on the demand for energy in agriculture. The rebound effect needs to be assessed in the EU agriculture. Finally, emission change related to energy efficiency and energy mix should be analyzed. Relating farm data and aggregate data may also provide insights into the sources of the GHG emission change in the EU agriculture.

REFERENCES

Aydinalp, C., & Cresser, M. S. (2008). The effects of global climate change on agriculture. *American-Eurasian Journal of Agricultural & Environmental Sciences*, 3(5), 672–676.

Clark, M., & Tilman, D. (2017). Comparative analysis of environmental impacts of agricultural production systems, agricultural input efficiency, and food choice. *Environmental Research Letters*, 12(6), 064016.

Dyer, J. A., & Desjardins, R. L. (2007). Energy based GHG emissions from Canadian agriculture. *Journal of the Energy Institute*, 80(2), 93–95.

Eurostat (2023). Database. https://ec.europa.eu/eurostat/data/database

Fuentes-Ponce, M. H., Gutiérrez-Díaz, J., Flores-Macías, A., González-Ortega, E., Mendoza, A. P., Sánchez, L. M. R., Novotny, I., & Espíndola, I. P. M. (2022). Direct and indirect greenhouse gas emissions under conventional, organic, and conservation agriculture. *Agriculture, Ecosystems & Environment*, 340, 108148.

German, R. N., Thompson, C. E., & Benton, T. G. (2017). Relationships among multiple aspects of agriculture's environmental impact and productivity: A meta-analysis to guide sustainable agriculture. *Biological Reviews*, 92(2), 716–738.

Johnson, J. M. F., Franzluebbers, A. J., Weyers, S. L., & Reicosky, D. C. (2007). Agricultural opportunities to mitigate greenhouse gas emissions. *Environmental Pollution*, 150(1), 107–124.

Lynch, D. (2009). Environmental impacts of organic agriculture: A Canadian perspective. *Canadian Journal of Plant Science*, 89(4), 621–628.

Matthews, A. (2023). Agricultural emissions: A case of limited potential or limited ambition? In Tim Rayner, Kacper Szulecki, Andrew J. Jordan, & Sebastian Oberthür (Eds.), *Handbook on European Union Climate Change Policy and Politics* (p. 275). Glos: Edward Elgar Publishing Limited.

Panchasara, H., Samrat, N. H., & Islam, N. (2021). Greenhouse gas emissions trends and mitigation measures in Australian agriculture sector – a review. *Agriculture*, 11(2), 85.

Sabiha, N. E., Salim, R., Rahman, S., & Rola-Rubzen, M. F. (2016). Measuring environmental sustainability in agriculture: A composite environmental impact index approach. *Journal of Environmental Management*, 166, 84–93.

Schieffer, J., & Dillon, C. (2015). The economic and environmental impacts of precision agriculture and interactions with agro-environmental policy. *Precision Agriculture*, 16, 46–61.

Tilman, D. (1999). Global environmental impacts of agricultural expansion: The need for sustainable and efficient practices. *Proceedings of the National Academy of Sciences*, 96(11), 5995–6000.

UNFCCC (2023). Greenhouse gas inventory data – Detailed data by party (unfccc.int). https://di.unfccc.int/time_series

Zanin, A. R. A., Neves, D. C., Teodoro, L. P. R., da Silva Júnior, C. A., da Silva, S. P., Teodoro, P. E., & Baio, F. H. R. (2022). Reduction of pesticide application via real-time precision spraying. *Scientific Reports*, 12(1), 5638.

3 Sustainable Agriculture

Dalia Streimikiene

3.1 AGRICULTURE AND SUSTAINABLE DEVELOPMENT

An astounding 24% of the world's Gross Domestic Product (GDP) is derived from the agricultural sector, proving its importance to the global economy. It employs a staggering 1.3 billion individuals or 22% of the worldwide population. Numerous developing nations have made agricultural production expansion a significant component of their development programs (Musvoto et al., 2015). Agriculture is the most prevalent land use globally, with 1.2–1.5 billion hectares under cultivation and more than 3.5 billion hectares used for grazing. In addition, four billion hectares of forest are utilized by humans for various purposes. In order to meet projected population growth and rising food demand per capita, it is imperative that agricultural production increases persist (Elawama, 2016).

Since the publication of the Brundtland Report in 1987, the importance of sustainable agriculture has increased due to human needs and challenges such as climate change, biodiversity loss, land degradation, and water pollution, as well as rising production costs, rural poverty, and a decline in the number of farms. This adheres to the broader notion of sustainable development (Velten et al., 2015).

Prioritizing agricultural productivity over sustainability in developing countries has led to the depletion of natural resources and soil and water degradation. This disproportionately affects impoverished and developing nations significantly dependent on agriculture and natural resources (Bathaei & Štreimikienė, 2023).

Sustainable agriculture is an "integrated system of plant and animal production practices with site-specific application that will, over the long term: (a) satisfy human food and fiber needs; (b) enhance environmental quality; (c) make efficient use of non-renewable resources and on-farm resources and integrate appropriate natural biological cycles and controls; (d) sustain the economic viability of farm operations; and (e) enhance the quality of life for farmers and their communities" (Hildén et al., 2012).

"For a farm to be sustainable, it must produce sufficient quantities of high-quality food, safeguard its resources, and be profitable and environmentally friendly. A sustainable farm relies as much as possible on beneficial natural processes and renewable resources derived from the farm itself, rather than on purchased materials such as fertilizers" (Reganold et al., 1990).

Sustainable Agriculture encompasses "management procedures that work with natural processes to conserve all resources, minimize waste and environmental impact, prevent problems and promote agroecosystem resilience, self-regulation, evolution, and sustained production for the nourishment and fulfillment of all" (Phatak, 1992).

DOI: 10.1201/9781003460589-4

Agricultural sustainability guarantees food security and eradicates starvation in a swiftly expanding population. To feed 9–10 billion people by 2050, global food production must increase by 60%–110%, according to projections. Therefore, the current emphasis on agricultural productivity enhancement must be shifted to agricultural sustainability. Sustainable methods of agrarian production seek not only to increase productivity but also to reduce negative environmental impacts (Ben Ayed & Hanana, 2021).

Agriculture is a vital and stable industry that provides basic materials to the food and animal feed industries. However, limited natural resources (such as production land, water, and soil) and a growing global population necessitate the development of agriculture that is both sustainable and efficient (Prasad et al., 2017). Sustainable development is a multidisciplinary concept that has attracted the attention of researchers and scientific institutions around the globe. Agriculture is important to the economy, society, and environment. It includes food processing, land cultivation, livestock propagation, animal husbandry, etc. Given that there are 7.41 billion people on Earth and 6.38 billion hectares of land, sustainable agricultural development is of the utmost importance. Sustainability is a crucial issue that requires the integration of social, technological, and natural sciences to meet societal requirements (More et al., 2022; Wasieleski et al., 2021) because approximately 1.3 billion people worldwide engage in agriculture and food production. Therefore, sustainability is a multifaceted concept that is essential for human survival and societal growth.

The global population was estimated to reach 9.7 billion by 2050 and 10.9 billion by 2100, up from 7.8 billion in 2020. On 4.75 billion hectares, 883.5 million people were employed in agriculture in 2019, emphasizing sustainable soil productivity through optimal nutrients and minimal heavy metals and chemical residues (Kalinowska et al., 2022). Organic agriculture is a viable alternative (Muller et al., 2017).

Despite the importance of sustainability in agriculture, the assessment of sustainable agricultural development is ongoing. Sustainable development evaluation should be based on indicators that quantify the economic, social, and environmental impact of agricultural outputs (Rigby et al., 2001). Various obstacles that have lasting effects on agriculture impede sustainable development. To assess the social and environmental impacts of the current agrifood supply systems, it is essential to analyze the relationship between these issues and sustainable development (Vinuesa et al., 2020). Sustainable agriculture development resolves fundamental food production issues in an environmentally responsible manner (Rosset & Altieri, 1997).

The agricultural sector is plagued by environmental issues such as water pollution, excessive pesticide use, hydrological changes that increase the risk of inundation and damage to livestock, and greenhouse gas (GHG) emissions that cause air pollution and environmental degradation (Raimi et al., 2021). Due to the impact of population growth on global agricultural systems, sustainable development principles are required. Agriculture is essential for attaining global food security and economic, social, and environmental objectives (Patel et al., 2020).

Harmonized policies and measures to reduce carbon dioxide and toxic substance emissions, mitigate climate change, promote technological development in the agricultural sector, and encourage renewable energy solutions in agriculture are required for sustainable development (Polasky et al., 2019). The concept of sustainable development is fundamental to agriculture and the entire economy, contributing to the emergence of the circular economy and green economy paradigms. These production and consumption models emphasize the elimination of waste, the recycling of materials and products, the protection of the natural environment, and the promotion of eco-friendly development (Schröder et al., 2020).

1983 saw the introduction of the concept of sustainable development by the Brundtland Commission. It seeks to satisfy present requirements without compromising the ability of future generations to fulfill their own (Borowy, 2013). Sustainable development focuses on minimizing impacts on the environment and natural resources while meeting socioeconomic requirements through modifications to culture, production systems, energy supply, agriculture, and the distribution of healthier products (Tomislav, 2018). Sustainable development requires monitoring environmental, social, and economic factors for long-term growth (Costanza et al., 2014). The concept of sustainable development has become a primary focus for national governments in designing and implementing development strategies, and it is extensively debated in the European Union in various contexts, such as intelligent development, innovation, and the employment field (Carayannis & Rakhmatullin, 2014).

Sustainable development, conservation of basic resources, and technological advancement require using modern farming techniques and considering environmental concerns (Altieri, 2011). An eco-agricultural approach to farming seeks to balance environmental and economic resources to ensure sustainability of agricultural activities (Bastan et al., 2016).

Natural factors, such as soil and water, directly and indirectly affect agricultural business activities. Insufficiency of these factors can significantly alter dynamics and outcomes. The Sustainable Development Committee of the United Nations proposes systemic measures for social, economic, and environmental divisions to accomplish sustainable development. See Figure 3.1 for a synthesis of ecological, cultural, social, and economic factors (Díaz et al., 2015).

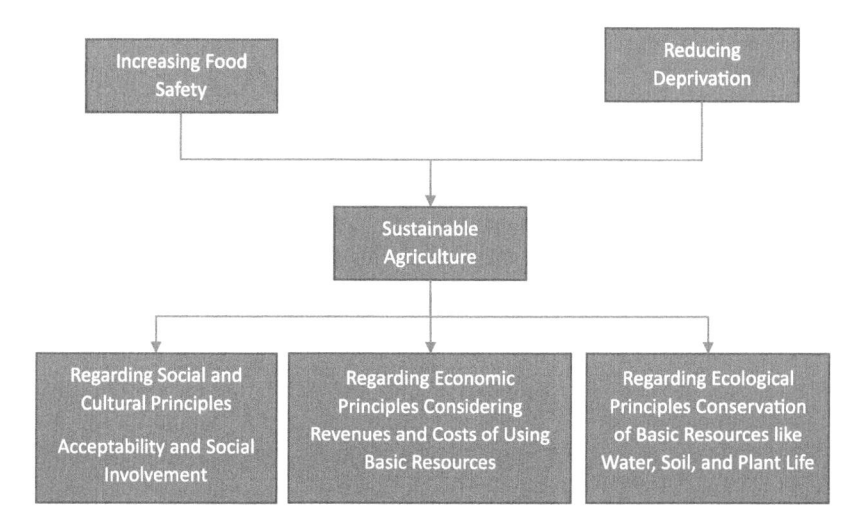

FIGURE 3.1 The main issues of sustainable agriculture. (Created by authors.)

Developing a mechanism for a sustainable agricultural enterprise requires a focus on competent production management and implementing competitive policies. However, many agricultural enterprises must consider all relevant factors during strategic planning and ongoing operations (Cheney et al., 2014).

Food security in the agricultural sector is a global priority governed by the principles of sustainable development (Lyulyov et al., 2019). Food security is now viewed as assuring access to safe and nutritious food that is also environmentally sustainable, transferring the emphasis from simply gratifying hunger to a more comprehensive approach (Mugambiwa & Tirivangasi, 2017).

Instead of relying on the starvation index as the primary indicator of poverty in a country, the author (Santeramo, 2015) suggests estimating the relationship between food security, income level, and the country's political climate. A collection of scientists is investigating the institutional environment's impact on sustainable agriculture development. The most important indicators are food security and lifelong physical, social, emotional, and cognitive development (Sachs et al., 2019).

The sustainable development of the agricultural sector requires implementing pertinent political decisions that impact the integrity of the institutional environment. According to scientists, democracy is the most effective political system for promoting sustainable development in all economic sectors, including agriculture (Lyulyov et al., 2019).

The implementation of sustainable development principles in the agrarian sector of the economy influences the growth of food security in three primary ways: the level of democracy, the rights and freedom of expression, and the effectiveness of political institutions. Harris argued that democratic institutions reduce poverty by promoting equitable and open competition that encourages adherence to the principles of sustainable agricultural and industrial development (Harris, 2014). Therefore, countries with high levels of destitution and starvation and low food security indices are more likely to experience widespread social unrest than nations with above-average indices as such, one of the primary responsibilities of a nation's political institutions is to promote food security through the establishment of sustainable agricultural infrastructure, the provision of free market access, the subsidization of agriculture, and the expeditious regulation of food prices. The established institutions must be responsive to the community's requirements and flexible to new threats (Headey & Ecker, 2013).

Rossignoli and Balestri (2017) discovered that a country's political governance has a significant impact on its food security level, a critical indicator of sustainable agricultural development. Sustainable agrarian development practices are more effective in comprehensive democracies, increasing food security and decreasing poverty (Rossignoli & Balestri, 2018).

Blaydes and Kayser (2011) provide empirical evidence that democratic and hybrid political administrations are more effective at modifying a country's calorie consumption (Blaydes & Kayser, 2011). Swinnen and Vandevelde (2018) state that structural changes in economic development alter the advantages and benefits of stakeholder groups, resulting in a rise in the political openness of the government field (Swinnen & Vandevelde, 2018).

The government must establish an effective political system and immediately modify the political and economic equilibrium (Liu, 2019). Regarding the relationship between sustainable development in agriculture and democracy, there are two opposing viewpoints. Comparatively speaking, democracy positively influences the sustainable economic development of nations. From the perspective of conflict, democracy hinders the sustainable economic development of nations (Wang et al., 2018).

Fu Jia et al. (2012) used the Agriculture Effect Policy System Dynamics method to analyze the negative effects of organic farming's sluggish development from 2009 to 2050. They also discussed the unstable energy structure and ways to mitigate its negative impacts to propose improvement policies (Li et al., 2012). In another study conducted in Slovenia, Rozman et al. (2012) utilized system dynamics to examine the growth of organic agriculture. They presented several enhancement policies, such as farm industry subsidies (Rozman et al., 2012).

3.2 SUSTAINABLE AGRICULTURE DEVELOPMENT CONCEPTS

Since its introduction in the Brundtland Report in 1987 (Ruggerio, 2021), the concept of sustainable development has become a standard for scientific research on the environment and has acquired a paradigmatic status for development. Since the Earth Summit in Rio de Janeiro, the concept has become hegemonic and has been incorporated into international treaties, national constitutions, and laws in many countries worldwide. It has also been applied to issues concerning business, agricultural production, industry, and urban development, and it has become the conceptual basis for theoretical approaches such as green economy and circular economy. It has even become part of the common sense of a significant portion of the global population and political slogans for environmental protection (Ruggerio, 2021).

As observed in the academic and scientific literature (Ruggerio, 2021), the concepts of sustainable development and sustainability are frequently used synonymously. Nevertheless, various schools of thought assert that sustainable development is contradictory due to the impossibility of sustaining infinite economic growth on a finite planet and highlighting the contradictions in its goals (Munda, 2016). This position draws attention to the incalculable problem – not only epistemological but also social, political, economic, cultural, and environmental – of basing local and global environmental policies and actions on a contradictory or poorly defined concept (Ruggerio, 2021). The definition of sustainable agriculture is provided in Table 3.1.

3.2.1 Sustainable Agriculture Dimensions

Sustainable agriculture is a subject of great interest and lively debate in many world segments. The disputes stem largely from differing viewpoints on sustainable agriculture (Alrøe & Noe, 2016). Sustainable agriculture is defined as a system that, "over the long term, enhances environmental quality and the resource base on which agriculture depends; provides for basic human food and fiber needs; is economically viable; and enhances the quality of life for farmers and society as a whole" (Ikerd,

TABLE 3.1
The Definition of Sustainable Agriculture

Authors/Publication and Year	Meaning and Understanding of Sustainable Development
WCED (1987)	Sustainable development is a development that satisfies the requirements of the present without compromising future generations' capacity to satisfy their own needs.
Pearce et al., 1998)	Sustainable development entails a socioeconomic system that ensures the sustainability of objectives in the form of real income achievement and advancement of educational standards, health care, and overall quality of life.
Harwood (1990)	Sustainable development is an unbounded system of development that focuses on attaining greater benefits for humans and more efficient resource use in harmony with the environment necessary for all humans and all other species.
IUCN, UNDP and WWF (1991)	Sustainable development is the process of enhancing human life quality within the carrying capacity of sustainable ecosystems.
Meadows (1998)	Sustainable development is a social construction derived from the long-term evolution of a highly complex system – the integration of human population and economic development with the Earth's ecosystems and biochemical processes.
PAP/RAC (1999)	Sustainable development is development determined by an ecosystem's carrying capacity.
Vander-Merwe and Van-der-Merwe (1999)	Sustainable development is a program that modifies the economic development process to assure a minimum standard of living while safeguarding ecosystems and other communities.
Beck and Wilms (2004)	Sustainable development is a potent global contradiction to the culture and lifestyle of the contemporary West.
Vare and Scott (2007)	Sustainable development is a process of change in which resources are increased, the orientation of investments is determined, the development of technology is focused, and the work of various institutions is synchronized, thereby increasing the likelihood of meeting human needs and desires.
Sterling (2010)	Sustainable development is a reconciliation of the economy and the environment on a new development path that will facilitate the long-term sustainable development of humanity.
Marin et al. (2012)	Sustainable development affords the opportunity for indefinite interaction between society, ecosystems, and other forms of life, without depleting vital resources.
Duran et al. (2015)	Sustainable development is a development that protects the environment, because a sustainable environment facilitates sustainable development.

1993; Scott et al., 2018). From this statement, numerous definitions emerged, but the concept surrounding agricultural sustainability remains the same. Also, sustainable agriculture is defined as a commitment to satisfy human food and fiber needs and to enhance the quality of life for farmers and society as a whole, now and in the future.

Consequently, there is that no brief, universally acceptable definition of sustainable agriculture has yet emerged (Ramcilovic-Suominen & Pülzl, 2018). This is because sustainable agriculture is viewed more often as a management philosophy rather than a method of operation. As such, acceptance or rejection of any definition is linked to one's value system. Agriculture has changed dramatically, especially since the end of World War II (Raven & Wagner, 2021). Food and fiber productivity increased due to new technologies, mechanization, increased chemical use, specialization, and government policies favoring maximizing production. Agriculture is highly susceptible to climate variability and its related effects (Tesfaye & Seifu, 2016). Food security and maintenance of sustainable ecological balance are major challenges for thinkers, researchers, conservationists, and policymakers. Sustainable agriculture should be taken as an eco-system approach, where soil–water–plants–environment–living beings live in harmony with a well-balanced equilibrium of food chains and their related energy balances (Qin et al., 2021). The goal is to address environmental issues of natural resource management to sustain significant increases in farm productivity through the efficient use of land and other resources, provide better economic returns to individuals, and contribute to the quality of life and economic development (Valentine et al., 2012).

To ensure sustainable agriculture and productivity, it is essential to employ innovative technologies such as modern irrigation systems, improved varieties, enhanced soil quality, and resource conservation technologies (Shah & Wu, 2019). Even though these changes have had many positive effects and reduced many hazards in agriculture, there have been substantial costs. The depletion of topsoil, the contamination of groundwater, the decline of family farms, the prolonged neglect of farm laborers' living and working conditions, the escalation of production costs, and the deterioration of economic and social conditions in rural communities are prominent examples. Sustainable agriculture arose as part of an increasing critique of the detrimental environmental effects of unquestioned modern agricultural practices (Tal, 2018). Even though the concept of sustainable agriculture is still relatively novel, the issue of sustainable agriculture is gaining increasing support and acceptability in conventional agriculture. In addition to addressing numerous environmental and social concerns, sustainable agriculture provides innovative and economically viable opportunities for producers, consumers, and policymakers (Ahmadzai et al., 2021). There are three dimensions of sustainable agriculture: social, economic, and environmental.

3.2.1.1 Social Dimension

Social sustainability pertains to individuals and distinguishes between two primary categories (Terrier & Marfaing, 2015). First, social sustainability is important at the level of the agricultural community. This has to do with the welfare of farmers and their families. Lebacq et al. (2013) classified the indicators identified in the literature into three primary groups: education, working conditions (measured by working time, burden including discomfort, and workforce), and quality of life (measured by isolation and social involvement) (Lebacq et al., 2013). Van Cauwenbergh et al. (2007) only considered the quality of life as a social theme. Still, they separated it into physical well-being (working conditions and health) and psychological well-being (education, gender equality, family access to infrastructures and services, and the farmer's sense of autonomy). Other aspects of well-being, such as the physical health of workers (Van Cauwenbergh et al., 2007), can also be considered, although this can also be seen as a result of working conditions. Second, social sustainability is significant at the societal level. This is "related to society's demands, depending on its values and concerns" (Desiderio et al., 2022).

Lebacq et al. (2013) classified the indicators found in the literature into three main categories: multifunctionality (including quality of rural areas, contribution to employment, and ecosystem services), acceptable agricultural practices (including environmental impacts and animal welfare), and product quality (including food safety and quality processes) (Lebacq et al., 2013). Van Calker et al. (2007) examined the contribution to the rural economy, which is less stringent than the contribution to employment but could also be included in Lebacq et al.'s (2013) quality of rural areas study (Lebacq et al., 2013; Van Calker et al., 2007). Van Cauwenbergh et al. (2007) had equity in addition to heritage, cultural, spiritual, and aesthetic values. Additionally, the theme of succession is sometimes incorporated into the social sustainability dimension (Van Cauwenbergh et al., 2007). For instance, Gómez-Limón and Sanchez-Fernández (2010) assessed intergenerational continuity in agriculture, and Dillon et al. (2009) analyzed demographic viability. Unlike most environmental and economic indicators, many social indicators are qualitative. As they are frequently subjective, they are challenging to quantify. Often, indicators of the agricultural community are derived from the self-evaluation of farmers through surveys or interviews (Dillon et al., 2010; Gómez-Limón & Sanchez-Fernandez, 2010).

Beginning in the 1960s, the concept of sustainability emerged in response to the public's environmental concerns. The most frequently cited definition of sustainable development states that "humanity can make development sustainable to ensure that it meets the needs of the present without compromising future generations' ability to meet their own needs" (Dempsey et al., 2011). Even though this definition focuses on people, some authors contend that the emphasis should be on environmental sustainability (Kopnina, 2014). "Economic and social aspects are only relevant in this approach to the extent that the ecologization of social development must also be compatible from an economic and social perspective" (Janker et al., 2019; Littig & Griessler, 2005). Some scholars assert that the historically embedded dualism between social and natural science and the use of heterogeneous terminologies and methods (Janker et al., 2019) explain why social science has played a smaller role in conceptualizations of sustainability. This dualism has decreased gradually in sustainability science over the past two decades as social science concepts have emerged in this context (Kuhlicke et al., 2011).

Current research on the social dimension of sustainability examines multiple fields, levels, and conceptual approaches, including development studies, political studies (Valdes-Vasquez & Klotz, 2013; Vifell & Soneryd, 2012), project development (Missimer et al., 2017), and business and management studies (Longoni & Cagliano, 2015), amongst others. Scholars in food-related studies examine participatory approaches and social learning among farmers and rural communities (Pilgeram, 2011) or consumers (Testa et al., 2016). The scope of these studies ranges from housing issues to communities (Magis, 2010), urban studies (Magee et al., 2013; Weingaertner & Moberg, 2014), and globalization and sustainability. Some papers approach the social dimension of sustainability through a theoretical lens, addressing the term's meaning and reviewing the state of the art (Vallance et al., 2011). In contrast, others propose measuring or assessing the social dimension of sustainability using indicators (Hicks et al., 2016; Wolsko et al., 2016) or, more recently, value chain- or supply chain-based approaches (Mani et al., 2016; Petti et al., 2016).

Although the incorporation of the social dimension into sustainability definitions, conceptions, and assessments has advanced, the emergence of an increasing number of social sustainability conceptions has led to a problem on which the vast majority of authors agree: the understanding of the social dimension of sustainability lacks cohesion (Eizenberg & Jabareen, 2017).

3.2.1.2 Economic Dimension

Economic sustainability is defined as the economic viability of agricultural systems, i.e. their ability to be profitable in order "to provide prosperity to the farming community" (Lebacq et al., 2013). According to Van Cauwenbergh et al. (2007), agriculture should "provide prosperity to the farming community." In this context, economic sustainability is typically regarded as financial viability, or the ability of an agricultural system to endure in a changing economic environment over the long term. Variability in output and input prices, yields, output destinations, and public support and regulation may impact the economic context. The term "long term" can refer to the farmer's entire professional career or to multiple generations. This pertains to transferability, or the capacity of a property to be passed on to a successor. The primary indicators of economic viability are profitability, liquidity, stability, and productivity (Van Cauwenbergh et al., 2007).

Profitability is determined by comparing revenue and cost as a difference or a ratio or by income variables such as agricultural income as proxies. Stability is typically measured by the proportion and growth of equity capital, whereas liquidity refers to the availability of currency to satisfy immediate and short-term obligations. Productivity measures the capacity of production factors to generate output. It is commonly measured as a partial productivity indicator, which is the ratio of output to one input, as well as by measures that account for the possibility of input or output substitution, such as total factor productivity (TFP) and technical efficiency (Latruffe et al., 2016). Indicators of profitability and productivity are predominantly quantitative and conveyed in monetary terms or as ratios; reference scales are employed infrequently. Although the measurement of economic sustainability typically only extends beyond these economic indicators, a broader range of indicators has been proposed to capture other economic properties of agricultural systems associated with sustainability (Dantsis et al., 2010).

Sustainable development requires the integration of economic, social, and environmental dimensions (ESCAP & Scientific, 2015). According to Sulewski et al., this approach is crucial because the long-term development of one subsystem is partially dependent on the development of the others. According to them, economic and ecological components should be regarded together, including their interaction, because they comprise a single system in the end (Sulewski et al., 2018). Bardy et al. offer a similar perspective (Bardy et al., 2015). Sadok et al. highlight a similar issue, noting that to assess sustainability realistically, we require "the integration of diverse information regarding economic, social, and environmental objectives; and the handling of conflicting aspects of these objectives as a function of the views and opinions of the individuals involved in the assessment" (Sadok et al., 2008).

According to a study by Pani, the economic aspect of sustainability has been neglected and ignored; for example, "Enhancing production and productivity to safeguard agro-ecology" has been replaced with "Minimising production costs and environmental factors." In addition, the high production costs, low returns, and high environmental costs violate the standard for agricultural sustainability (Pani et al., 2020).

The main indicators of economic viability are profitability, liquidity, stability, and productivity. Profitability is determined by comparing revenue and cost, either as a difference or as a ratio, or by income variables such as farm income as approximations (Rashid, 2021). Stability is typically measured by the proportion and growth of equity capital, whereas liquidity refers to the availability of currency to satisfy immediate and short-term obligations. Productivity measures the capacity of production factors to generate output (Nachum, 1999). It is commonly measured as a partial productivity indicator, which is the ratio of output to one input, and by measures that account for the possibility of input or output substitution, such as TFP and technical efficiency (Sheng et al., 2020).

Indicators of profitability and productivity are predominantly quantitative and conveyed in monetary terms or as ratios; reference scales are employed infrequently. Although the measurement of economic sustainability typically only extends beyond these economic indicators, a broader set of indicators has been proposed to capture additional economic properties of agricultural systems that are associated with sustainability (Garibaldi et al., 2017). Some studies use "autonomy" (or dependence) to measure financial sustainability. Autonomy is one of every system's fundamental characteristics: freedom (De Clercq & Brieger, 2021; Senderowicz, 2020). Therefore, independence may also be considered a social indicator. It can be viewed in terms of inputs, indicating that farms that depend less on external inputs (such as feed or fertilizers) are less susceptible to input availability and price fluctuations. Additionally, autonomy is regarded in terms of financing or terms of the burden of obligations. Income diversification (whether farm or household income) is another facet of independence. Farm income can be diversified by instituting non-agricultural activities such as direct sales, on-farm processing, or agritourism. In contrast, household income can be diversified by farmers or their families holding off-farm employment (this is known as income diversification). Subsidy dependence is an additional aspect of autonomy: if farms rely on public support, any policy reform that reduces subsidies could jeopardize farm sustainability (Gomiero, 2018; Khanna et al., 2022).

3.2.1.3 Environmental Dimension

Due to the growing concern for sustainability and environmental issues over the past two decades, numerous initiatives with diverse indicators have been proposed (Govindan, 2018). Lebacq et al. (2013) classified ecological indicators from the scientific literature into eleven environmental themes/topics that either focus on observable physical features of the environment or human activities with significant ecological impact (Lebacq et al., 2013). These include soil quality, biodiversity, nutrients, pesticides, nonrenewable resources (such as energy and water), land management, GHG emissions, and acidification-causing compounds. Generally, three categories of environmental topics can be distinguished:

Themes associated with local or global impacts that have implications for the functional units used to convey the indicators; Themes according to the action chain, namely, the ultimate objective (e.g., human health), the process to attain the objective (e.g., environmental function balance), and the means (e.g., ecological compartment protection); Themes based on frameworks oriented toward system properties and frameworks oriented toward objectives to be achieved.

Despite the various ways sustainable agriculture is conceptualized, its many dimensions, including economic, environmental, and social concerns, are frequently emphasized (Lee, 2005). Pramanik (2016) used Analytical Hierarchy Process (AHP) and Geographic Information System (GIS) methodologies to conduct a suitability study for agricultural land use in the Darjeeling district (Pramanik, 2016). Bozdağ et al. (2016) conducted a land suitability analysis for Cihanbeyli (Turkey) County using AHP and GIS (Bozdağ et al., 2016).

Performance must be measured and compared to industry standards to determine how well an agricultural system operates and how sustainable it may be. Due to pervasive interest in sustainability, numerous indicators have been devised and are frequently incorporated into sustainability frameworks (Bell & Morse, 2012). Moreover, instead of supporting stakeholder learning and steering their activities during the transition to sustainability, many indicators developed for sustainability studies have been devised and implemented for evaluation and assessment purposes (Torres-Delgado & Saarinen, 2014). The stakeholder-driven and inclusive prioritization of adaptation alternatives (Khatri-Chhetri et al., 2019) improves the capacity and motivation of stakeholders to adopt technologies and practices in agriculture and related sectors.

There are numerous examples in the literature of the use of agroecological indicators to evaluate the environmental impacts of agricultural systems (Aleksandrova et al., 2016; Meuwissen et al., 2019; Migliorini et al., 2018). Several typologies have been presented in the literature (Latruffe et al., 2016; Sharifi, 2020) to organize these indicators. These are based on the causal chain between agricultural practices and their effects and highlight the indirect link between practices and outcomes due to the influence of external factors such as soil characteristics or weather (Reed et al., 2021; Seman et al., 2019). These typologies vary according to the description of each level of the causal chain. Based on these analyses, we use a typology defining several types of indicators: (i) means-based indicators assessing technical means and inputs used on the farm, e.g., the livestock stocking rate; (ii) system-state indicators about the state of the farming system, e.g., the amount of post-harvest soil nitrate; (iii) emission indicators about the farm's polluting emissions into the environment and the potential impact of these emissions, e.g., there is a dichotomy between means-based and effect-based indicators in terms of environmental relevance and measurability. Environmental relevance answers the query, "Does the indicator reflect environmental impacts?" (Bockstaller et al., 2015).

On the one hand, means-based indicators are simple to implement regarding data availability and calculation, but they need better environmental impact prediction (Himeur et al., 2022). In contrast, effect-based indicators have a high ecological relevance due to their direct relationship with the objectives and context specificity. Still, they are challenging to implement from a methodological or practical standpoint. Moreover, data collection is frequently more costly. Consequently, assessment tools employing such indicators are more difficult to implement because they are more complex, time-consuming, or require data that are not readily accessible (Di Maio & Rem, 2015). Typically, effect-based indicators encompass spatial scales larger than the farm, such as regional or watershed scales. Indicators based on means are more appropriate for farm use because they are simple to implement and attuned to production practices (Einarsson, 2020). In contrast, effect-based indicators do not permit cause–effect relationship monitoring, making it difficult to formulate specific advice for producers (Coteur et al., 2020). However, some authors emphasize that producers are then free to choose their methods for reducing environmental impact. For example, the cultivator can choose which measures to implement to reduce nitrate concentration in groundwater (Richard et al., 2020).

3.2.2 Obstacles Preventing Sustainable Agriculture Development

By incorporating new technologies and modernizing production methods, agricultural intensification seeks to maximize crop productivity in space and time (Silveira et al., 2018). It is characterized by specialized agri-food production in crop or livestock systems and low genetic and landscape diversity (Mantino & Vanni, 2018). Agricultural intensification is characterized by the high use of external inputs, particularly energy, and agrochemicals, which harms the environment via soil degradation, progressive depletion of soil organic matter (SOM), a decline of soil quality and agrobiodiversity, GHG emissions (Di Bene et al., 2022), and nutrient losses from agricultural soils that cause water pollution and eutrophication (Khan & Mohammad, 2014; Withers et al., 2014).

To achieve a balance between economically viable and socially equitable food production and environmental objectives, scientists concur that agricultural sustainability must undergo fundamental changes (Morton et al., 2017; Tomlinson, 2013). This is especially evident in the Mediterranean Basin, where highly specialized agricultural systems are predominantly based on cereal-based intensive cropping systems under rain-fed or irrigated conditions as monoculture, or short-rotations such as wheat-summer irrigated crops, or mixed succession with bare fallow (Di Bene et al., 2022), resulting in high incidences of pests and diseases, loss of soil fertility, and biodiversity, among other things (Asante et al., 2017).

Therefore, implementing suitable crop diversification strategies and alternative management practices in typical intensive systems is essential for fostering the redesign of agricultural systems (Altieri et al., 2017; Di Bene et al., 2022). This transition is critical for achieving the objectives of assuring the availability of resources (e.g., nutrients, water, and land) by increasing reliance on ecosystem services that minimize the use of external inputs (Brussaard et al., 2010; Gaba et al., 2015) and promote healthy agroecosystems (French et al., 2021).

Enhancing temporal and spatial crop diversity in arable cropping systems can increase crop productivity and resource use efficiency by delivering multiple ecosystem services via crop rotations, integration of cover crops as agro-ecological service crops, green manure, and species mixtures such as multiple cropping and/or intercropping that can include legumes, leys, grassland, and minor crops of local interest, with overall socio-economic benefits (Blickensdörfer et al., 2022; Redlich, 2020). Coupling agricultural diversification (AD) with more diverse management strategies by adopting cover crops for green manure or fodder, conservation agriculture (i.e., tillage, crop diversification, and residue management), organic farming, and fertilization management also increase crop yields, profitability, and cropping system resilience (Sanderson et al., 2013). However, the short-term economic costs may outweigh the environmental and ecological benefits, so financial instruments may be required to promote the adoption of combined AD strategies instead of single-crop diversification systems (Makate et al., 2016).

Despite a large scientific consensus on the potential agro-ecological and socio-economic benefits of crop diversification, agronomic solutions for crop diversification strategies are frequently hampered and not always affordable due to a variety of technical, organizational, and institutional barriers linked to the overall functioning of the dominant agro-food chains. In this context, novel commodities may be off the market or need more technical knowledge and production-related skills, particularly during the initial phases of implementation (Dora et al., 2021; Quisumbing et al., 2015).

Little is known about the advantages of crop rotations and the costs of machinery or new labor organization, while market uncertainty remains high. These and other forces of simplification influence farmers' decisions regarding the use of their agricultural land and entrepreneurial resources, indicating a scenario of specialization of various farm products in multiple regions (Moraine et al., 2014). Therefore, research and policy are crucial in promoting more sustainable agri-food production practices and ensuring environmental improvement (Notarnicola et al., 2017). At the EU level, the Farm to Fork and Biodiversity Strategies launched within the Green Deal intended to promote a more sustainable and resilient form of food production systems with a neutral or positive environmental impact (Mowlds, 2020). In this context, food legumes and legume-inclusive production systems can play a vital role by providing multiple services consistent with sustainability principles (Stagnari et al., 2017).

Multiple obstacles impede the adoption of sustainable practices. The challenges of organic farming adoption, while not unique, are emblematic of the difficulties across sustainable philosophies; therefore, this discussion focuses on organic production (Youngberg & DeMuth, 2013). Organic food and cultivation is one of the few food system development areas. In North America, consumer demand typically exceeds supply, and many wholesalers, manufacturers, and retailers cannot procure adequate supplies to meet customer demands. Despite this demand and the substantial price premiums, conventional producers are reluctant to convert to organic production. This has prompted numerous analysts to question why adoption is so slow and how to accelerate it (Popkin & Reardon, 2018).

Farmers cite a variety of motives for adopting organic farming, but the question of adoption transcends individual characteristics and motivations (Läpple, 2013). Earlier studies attempted to identify the factors of organic adopters using the innovation-diffusion model, either implicitly or explicitly (see discussion of its limitations below) (Ruzzante et al., 2021; Ward et al., 2018). Other studies have identified a variety of farmer motivations and structural elements (sometimes referred to as exogenous factors) associated with transition (Greiner, 2015; Hansson et al., 2013), such as personal values and convictions about the environment, social equity and health, conversion of a neighbor, pressure from family members, health problems of family members and animals, avoiding chemicals because of an accident, anticipating consumer demand, economic opportunities, overcoming reluctance, and adapting to new technologies (Haque et al., 2020).

3.2.3 Challenges of Agricultural Sustainability

There are difficulties in the adoption of sustainable agriculture theory. Much of the early research on predicting the adoption of conservation strategies centered on three models: farmer demographics, farm structure, and the diffusion of information process. Particularly, farm structural dimensions and profit maximization frameworks have received considerable attention. In addition, traditional diffusion frameworks suggest that influencing attitudes will result in behavior changes and that effective communication will result in rational adoption (Adnan et al., 2017).

The classic innovation-diffusion models of extension, used to convey adoption messages for many of these agri-environmental program designs (Juhasz, 2014), have been useful for comprehending the adoption of practices or technologies with an obvious commercial benefit. Unfortunately, they have not effectively explained the adoption of conservation practices and systems, particularly for "preventive" rules and procedures with historically low adoption rates (Lowenberg-DeBoer & Erickson, 2019). In addition, voluntary adoption is typically insufficient for practices or systems with many off-site benefits that equal or outweigh on-site benefits unless additional incentives are available or farmers have already developed strong stewardship attitudes (Biardeau et al., 2016).

3.2.3.1 Barriers to Organic Adoption and Environmental Systems

There are typically distinctions in environmental attitudes between organic adopters and non-adopters, a phenomenon not consistently reported for adopting other environmental systems (Barnes et al., 2019). However, other factors play a significant role. Although attitudes motivate many organic pioneers, they may be less significant for later adopters. To comprehend organic adoption at the farm level, we reviewed dozens of peer-reviewed reports on barriers to adopting environmental practices and systems (Chen & Lin, 2018; Teng & Wang, 2015) published over the past 25 years.

3.2.3.2 Financial Anxiety (Investments, Markets, and Revenue Stream)

When a significant change is being contemplated, monetary concerns arise. Most studies indicate that organic farming is more profitable than conventional farming (Daghagh Yazd et al., 2019), but more than this fact is needed to convince many farmers to convert. Even substantially higher organic commodity prices will not necessarily attract traditional producers because a combination of other barriers (described below) is so substantial that price and income alone are insufficient incentives (Duesberg et al., 2013). The fixed costs of conversion and policy uncertainty may explain the reluctance to convert (Regan et al., 2015) in the face of large organic price premiums. According to Rogers, the trial ability of organic agriculture is quite low (Kernecker et al., 2021). The lack of nearby processing and marketing infrastructure exacerbates these monetary concerns (Serebryakova et al., 2017).

3.2.3.3 Labor Difficulties

Organic farms typically require more labor per unit of production area, but the character of the work is frequently more diverse than on conventional farms. Nonetheless, given the current farm labor scarcity in Canada and the need for more specialized knowledge, organic producers may be at an even greater disadvantage (Seufert & Ramankutty, 2017).

3.2.3.4 Problems Acquiring Information

As the inherent complexity of conservation technology increases, information becomes increasingly essential; organic cultivation is complex. Although details on organic and sustainable farming have dramatically increased over the past two decades, farmers cannot consistently rely on the dominant agricultural development institutions to provide relevant information (Reganold & Wachter, 2016). Although there is evidence that greater farmer participation in selecting research priorities correlates with higher adoption rates of the practices, such opportunities are common. Information is frequently deemed more valuable when the recipient has a relationship with the source (Bravo-Monroy et al., 2016).

3.2.3.5 Challenges in Obtaining Sufficient Instruments or Inputs

Organic producers are limited in their use of inputs and have historically found that some conventional agricultural equipment needs to be better suited to their conditions. As a result, the transition frequently entails the disposal of traditional equipment, which may result in capital losses (Coquil et al., 2018; Kreitzman et al., 2022).

3.2.3.6 Lack of Self-Assurance

The knowledge required to manage ecological processes typically results in increased management requirements in organic agriculture (Bedoussac et al., 2015; Seufert et al., 2017).

3.2.3.7 Fear Alteration

When the transition is considered, numerous farms are already experiencing financial or health-related strain. Frequently, the prospect of even greater stressors during conversion is a significant barrier (Stringer et al., 2020).

3.2.3.8 Concern about Alterations

Farmers are influenced by their social environment and peer acceptance (for some theoretical considerations on this, see (Hayden et al., 2018)). Agriculture has undergone numerous revolutions, including the domestication of animals and plants a few thousand years ago, the systematic use of crop rotations and other improvements in farming practices a few centuries ago, and the "green revolution" with systematic breeding and the widespread use of man-made fertilizers and pesticides a few decades ago (Walter et al., 2017). The exponentially expanding use of information and communication technology (ICT) in agriculture has sparked a fourth industrial revolution. Autonomous, mechanized vehicles have been developed for agricultural purposes, such as mechanical weeding, fertilizer application, and produce harvesting.

The development of unmanned aerial vehicles with autonomous flight control and lightweight and powerful hyperspectral snapshot cameras that can be used to calculate biomass development and fertilization status of crops paves the way for sophisticated farm management advice (Walter et al., 2017). Moreover, available decision-tree models currently enable producers to differentiate between plant diseases based on optical data (Rakhra et al., 2021). Virtual fence technologies permit cattle herd management using remote-sensing signals, sensors, and actuators affixed to livestock. Collectively, these technological advances comprise a technological revolution that will result in disruptive changes to agricultural practices (El Bilali et al., 2020).

This trend applies to farming not only in developed countries but also in developing countries, where ICT deployments (e.g., use of mobile phones, access to the Internet) are being adopted rapidly and could become game-changers in the future (e.g., seasonal drought forecasts, climate-smart agriculture [CSA]). Such profound changes in practice present both opportunities and challenges (De Bon et al., 2010). To avoid "lock-ins," it is crucial to identify them early on in this revolution: advocates and skeptics of technology must engage in an open dialogue about the future development of farming in the digital era (Boyer, 2020). The term "smart farming" should only be applied to farming in the digital era if technological aspects, crop and livestock systems diversity, and networking and institutions (i.e., markets and policies) are considered concurrently in the dialogue. Abundant Opportunities "Intelligent farming reduces agriculture's ecological imprint. In precision agriculture systems, the minimal or site-specific application of inputs, such as fertilizers and pesticides, will mitigate leaching issues and greenhouse gas emissions" (Walter et al., 2017).

Using current ICTs, it is possible to construct a sensor network that enables near-constant farm monitoring (Cario et al., 2017). Similarly, theoretical and practical frameworks to link the conditions of plants, animals, and soils with the requirements for production inputs, such as water, fertilizer, and medications, are attainable with current ICT worldwide (El-Ramady et al., 2014). Smart farming can increase the profitability of agriculture for the cultivator. Reduced resource inputs will save the farmer time and money, while improved spatially explicit data will reduce hazards. Based on dense weather and climate data network, optimal site-specific weather forecasts, yield projections, and probability maps for diseases and calamities will enable optimal crop cultivation (Saiz-Rubio & Rovira-Más, 2020). Additionally, site-specific information allows new insurance and business opportunities for the entire value chain, including technology and input suppliers, producers, processors, and retailers in both developing and developed societies. If automated sensors record all farming-related data, the time required for resource allocation prioritization and administrative oversight is reduced. Additionally, smart farming can increase consumer acceptance (Maddikunta et al., 2021).

By optimizing management practices, smart farming can increase agricultural profitability and product quality. Using techniques such as optimal fruiting densities in orchards and individualized feeding regimens for livestock, it is possible to produce healthier products with a higher market value. Using ICT to capture production and processing information can facilitate greater transparency along the value chain, allowing for direct communication between producers and consumers (El Bilali & Allahyari, 2018).

However, several obstacles must be overcome. Data ownership and use pose concerns regarding property rights, necessitating regulatory frameworks to assure data quality and foster stakeholder trust. Data misuse poses legal and ethical challenges that require oversight and regulation. Access to technology may need to be improved by adoption barriers, such as high costs and limited knowledge and skills, particularly in developing nations. This could limit the benefits of ICT to industrialized countries and main commodities, leading to intensification practices that need to be more sustainable (Magrach & Sanz, 2020).

The digitalization of agriculture may also affect agricultural employment opportunities and job profiles. The responsibility and accountability of new technologies must be defined, particularly in the event of mismanagement or errors with economic or environmental repercussions. Emerging challenge is balancing the role of producers' knowledge and experiences with new technologies (Deichmann et al., 2016).

Smart farming provides a path forward by diversifying the agricultural sector's technologies, production systems, and networks. The adoption of ICT can be supported by identifying the mechanisms that impede the sustainable use of technology and implementing the appropriate actions. The policy environment should provide a transparent legal framework that guarantees effective ownership and user rights (Gruskin et al., 2013).

ICT enables new forms of farm diversification and facilitates management advice, thereby reducing resistance issues and encouraging a greater diversity of production systems. However, producers continue to play an important role as scientists and watchdogs, monitoring unforeseen situations and devoting time to disease treatment and individualized livestock care. Acceptance from consumers, farmers, and decision makers is crucial for the success of intentional diversification, which is supported by transparent data transmission along the food production chain (Riccaboni et al., 2021).

In addition to facilitating information exchange, cooperation, and peer review among farmers, ICT also fosters self-organized networks and institutional innovation. Sharing equipment and applications, similar to platforms such as Airbnb and Uber, can facilitate the private exchange of agricultural operations. However, explicit policies and a transparent data management system are required to meet regulatory requirements and ensure data privacy and security (Tikkinen-Piri et al., 2018). Smart farming can only be implemented successfully with the proactive development of policies, legal frameworks, and market architecture. To realize the potential benefits of ICT and data management in agriculture (Devi, 2023), dialogue between proponents and detractors of technology and careful consideration of emergent ethical concerns will be essential.

3.3 CLIMATE SMART AGRICULTURE

The impacts of climate change on the environment, including climate patterns, biodiversity, and natural resources, are significant and global. Climate change is predominantly caused by human activities, resulting in global warming, glacier melting, rising sea levels, ocean acidification, and extreme weather events (Chan et al., 2018). As a susceptible system, agriculture is significantly impacted by climate change, resulting in a decline in food production and increased prices. Due to their agricultural-based economies, higher climates, frequent exposure to extreme weather events, and limited resources for adaptation, developing countries face more severe impacts (Nnadi et al., 2013).

The limits of biodiversity loss, human interference in the nitrogen cycle, and climate change have already been exceeded, resulting in environmental degradation and diminished Earth system resilience. Agriculture and climate change are intricately intertwined, with the agricultural sector contributing approximately 21% of global GHG emissions. Despite increasing agricultural production, the Green Revolution has had socio-ecological consequences, including pollution, biodiversity loss, land degradation, and erosion of traditional agrarian knowledge (Patel et al., 2020).

Climate change must be mitigated and adapted to if its effects are to be mitigated. CSA aims to increase food production, promote climate adaptation and resilience, and reduce GHG emissions. As a climate-smart strategy, traditional agriculture practices, rooted in local agricultural experiences over thousands of years, are gaining popularity. Conventional agroecosystems have characteristics such as high productivity, conservation of biodiversity, minimal energy inputs, and climate change mitigation. There is the potential for practices such as agroforestry, intercropping, crop rotation, cover cropping, organic decomposition, and integrated crop-animal husbandry to increase crop productivity and mitigate climate change (Singh & Singh, 2017).

However, the transition toward modern agriculture has resulted in abandoning traditional agricultural practices, which has caused environmental issues. Conventional agriculture is frequently restricted to small producers and is threatened by habitat loss, the introduction of new varieties, social stigma, and a shift in lifestyles. Adopting climate-smart traditional practices is essential for achieving sustainable food production in a changing climate. Integrating traditional and modern agriculture can reconcile the divide between indigenous and modern producers, improve human–nature relationships, and enhance the socio-ecological integrity of agroecosystems (Singh & Singh, 2017).

Agriculture and climate change have a reciprocal relationship. Significant long-term alterations in the mean condition of the climate and an increase in the frequency and severity of weather extremes constitute climate change. Human activities, such as industrialization and agricultural practices, are the primary contributors to the emission of GHGs. Agriculture significantly contributes to global warming, accounting for 10% and 12% of human-caused GHG emissions (Goss et al., 2020). The primary GHGs emitted by agriculture are carbon dioxide, methane, and nitrous oxide. Climate change poses a grievous danger to agroecosystems, as well as to food production and human health. Climate change directly affects agriculture by modifying agroecological conditions and indirectly by increasing demand for agricultural products (Singh & Singh, 2017). It involves some agrarian factors, including produce yields, invasive parasites, livestock production, and aquaculture. Compared to temperate and developed nations, developing and tropical countries are especially vulnerable to the effects of climate change on food production (Tubiello et al., 2013).

3.3.1 CLIMATE-SMART AGRICULTURE: PRINCIPLES AND OBJECTIVES

To resolve food security in the context of global warming, the Food and Agriculture Organization (FAO) developed the concept of CSA. Sustainable enhancement of agricultural productivity to support income growth, food security, and development; increasing adaptive capacity and resilience to shocks at various levels; and reducing GHG emissions and encouraging carbon sequestration (Lipper et al., 2014). CSA's primary objective is to accomplish sustainable food production while reducing GHG emissions and bolstering the resilience of agricultural systems. CSA entails transforming and reorienting agroecosystems to produce food and adapt to climate change (Heeb et al., 2019). It is an emerging strategy that seeks to increase food production, biodiversity, environmental quality, agroecosystem resilience, livelihoods, and economic development while confronting climate change impacts (Singh & Singh, 2017). The particular priorities of CSA may vary by location, with a greater emphasis on productivity and adaptability for small producers in developing nations. Due to its potential to improve agricultural productivity, agroecosystem resilience, and GHG emission reduction, CSA has garnered increasing attention, particularly in developing countries (Jost et al., 2015).

3.3.2 CONCEPT AND AGROECOLOGICAL CHARACTERISTICS OF TRADITIONAL AGRICULTURE

Traditional knowledge is a vast corpus of information with applications in numerous fields, including agriculture, climate, soils, hydrology, vegetation, animals, forests, and human health. It includes the ancient practices of husbandry and agriculture, which have played a significant role in human interaction with nature and the management of ecosystem services (Dhakal & Kattel, 2019). Traditional agriculture is the consequence of generations of local farming practices and has substantially contributed to the advancement of agricultural science. Farmers, particularly in developing regions, rely on local, traditional, or landrace varieties of both minor and major commodities (Qaim, 2020). These practices have sustained large populations for centuries and continue to provide sustenance for many regions around the globe. Despite the widespread adoption of modern agriculture practices, approximately 1.9–2.2 billion people still practice traditional agriculture. Smallholder farmers, who account for about 84% of the world's farms, are crucial in preserving conventional agricultural practices and possess the indigenous knowledge necessary to adapt to environmental changes (Díaz Bonilla et al., 2014). Numerous modest farms exist in China, India, Indonesia, Bangladesh, and Vietnam, among other nations. Smallholder farmers account for 80% of farms in Sub-Saharan Africa, and their traditionally cultivated fields tend to be more productive than those of large-scale farmers. In Mexico and the traditional agroecosystems of the Himalayas, where producers significantly rely on local resources and indigenous technology, conventional agricultural practices are also prevalent (Singh & Singh, 2017).

Traditional agricultural landscapes are distinguished by conserving sustainable farming practices and biodiversity. These landscapes are valued for their aesthetic, natural, cultural, historical, and socioeconomic qualities (Slámová & Belčáková, 2019).

The Western Ghats in India, the Satoyama landscapes in Japan, the Milpa cultivation systems in Mexico, the traditional village systems in Eastern Europe, and the terrace landscapes of Southwestern China are examples of regions where farming practices have remained relatively unchanged for an extended period. The Hani rice terraces in China's Yunnan Province are a well-known example of agricultural systems in mountainous regions that have been designated as World Cultural Heritage sites by UNESCO (Zhang et al., 2017).

Agrobiodiversity has decreased over the past few decades due to the prevalent adoption of monoculture agricultural practices. Approximately 80% of arable land is devoted to a handful of commodities, including maize, wheat, rice, and soy legume (Wingeyer et al., 2015). About 75% of the world's food crop diversity has been lost due to this transition during the twentieth century. Traditional farmers are recognized as essential stewards of natural resources and biodiversity because they preserve unique genotypes and crop varieties that are adapted to environmental stresses, such as climate change (Singh & Singh, 2017). Traditional agricultural systems in developing nations are distinguished by a high level of vegetational diversity and the use of indigenous knowledge (Nunes et al., 2015).

Energy-intensive components of modern agriculture include synthetic nitrogen fertilizers. Historically, peasant farmers have relied on local knowledge and locally accessible resources to develop sustainable agricultural systems. Integrated crop and livestock production reduce reliance on external inputs such as fossil fuels, fertilizers, and pesticides (D'Odorico et al., 2018). As they recycle agriculture and other residues, traditional agricultural systems have a bidirectional link between the agroecosystem and consumers, unlike modern agriculture. The function of agrobiodiversity in providing ecosystem services to agriculture and reducing the need for off-farm inputs is crucial (Allen et al., 2014). Composting and manure application enhance soil microbial and invertebrate communities, thereby enhancing nutrient cycling. Indigenous cultivators in Asia, Africa, and Latin America have created agricultural systems that can withstand extreme climatic conditions with minimal external inputs (Altieri et al., 2015).

3.3.3 ALTERNATIVE PRACTICES FOR CLIMATE CHANGE MITIGATION IN TRADITIONAL AGRICULTURE

Climate change mitigation refers to human-led efforts to reduce GHG emissions and enhance GHG absorbers to reduce the anthropogenic impact on the climate system. Indigenous communities have demonstrated their ability to observe weather and climate changes and have devised strategies for adaptation and mitigation (Meier et al., 2014). Traditional agroecosystems are acquiring recognition as viable alternatives to industrial agriculture, providing biodiversity conservation and climate-resilient food production. Indigenous agricultural systems are renowned for their diversity, adaptability, environmentally conscious methods, and productivity (Johns et al., 2013). Diverse vegetation, such as cereals and trees, plays a role in converting CO_2 into organic forms, thereby reducing global warming. In addition to reducing the risk of crop failure, parasites, and diseases, mixed cultivation diversifies the food supply (Malhi et al., 2021). It is estimated that traditional multiple cultivation systems alone provide 15%–20% of the world's food supply. Agroforestry, intercropping, crop rotation, cover cropping, conventional organic decomposition, and integrated crop-animal farming are prominent traditional agricultural practices advocated for a climate-smart agricultural approach (Singh & Singh, 2017).

3.3.4 Traditional Agriculture: A Sustainable Adaptation Strategy to Climate Change

Adaptation is essential for assessing climate change vulnerability and devising mitigation policies. Adapting to observed or anticipated climate change and extreme weather events to reduce exposure and increase resilience. Adaptation to climate change is crucial for global food security and environmental quality (Keenan, 2015). Farmers in various regions have diverse perceptions and adaptation strategies for coping with climate change, and their insights are valuable for designing policies and integrating scientific and traditional knowledge (Ajani et al., 2013). Indigenous producers have a lengthy history of adapting their agricultural methods to fluctuating climatic conditions. Combining traditional expertise with climate change policies makes adaptation and mitigation strategies cost-effective and sustainable. Climate change poses significant challenges to agrarian production without adaptation, but adaptation techniques can mitigate these effects (Singh & Singh, 2017).

Environmental conditions, geographic location, socioeconomic standing, cultural differences, and knowledge systems influence the perceptions of climate change and adaptation strategies. Farmers with a comprehensive understanding of climate change and its effects are better suited to adapt. Adaptation strategies that are sustainable to climate change include altering planting and harvesting dates, crop diversification, integrated crop-livestock farming, and cultivating drought-resistant and water-sensitive plants. Compared to monoculture, polyculture systems demonstrate greater yield stability and less productivity decline during drought (Ferdushi et al., 2019). Indigenous knowledge should be implemented into approaches and policies for climate change adaptation. Small farmers use techniques such as local drought-tolerant varieties, polyculture, agroforestry, water harvesting, and soil conservation to mitigate crop failure caused by climate change (Altieri & Nicholls, 2017). To forecast weather patterns, indigenous communities also rely on natural indicators such as variations in the behavior of flora and fauna. Examples include observing the behavior of grasshoppers and birds and the scattering of foliage by specific tree species to predict the advent of drought or rainfall (Radeny et al., 2019).

Traditional agricultural systems play a crucial role in fostering the sustainability of agroecosystems by conserving soil, harvesting water, and cultivating drought-tolerant crop varieties under conditions of water stress, limited resources, and low technology (Gobarah et al., 2015). Farmers in various regions, such as Ethiopia, South Africa, Zimbabwe, and Bangladesh, employ different adaptation strategies to deal with drought and climatic change. These strategies include modifying planting dates, operating drought-tolerant crop varieties, implementing water harvesting techniques, and modifying animal husbandry practices (Abdur Rashid Sarker et al., 2013). Traditional crops and vegetables, such as sweet potato, kachori, cassava, and diverse rice varieties, are valued for their drought resistance, contribution to sustainable food production, and adaptability to climate change. Floating agriculture is an indigenous agricultural method practiced in flood-prone regions of Bangladesh, where crops are cultivated on floating platforms constructed from locally accessible materials. Climate change poses a significant hazard to agrarian productivity worldwide, especially in regions such as Sub-Saharan Africa that depend heavily on rain-fed agriculture. Agriculture must endure a transformation to guarantee sufficient food supplies and resolve the challenges posed by climate change in light of projected population growth (Lipper et al., 2017).

To satisfy the rising food demand of a growing global population, two main scenarios are frequently considered: expanding production areas through land clearance and intensifying production on existing agricultural lands. However, expanding urban areas has resulted in land scarcity issues, especially in Sub-Saharan Africa (Byerlee et al., 2014). Sustainable intensification of smallholder agriculture is viewed as a viable strategy for meeting food needs. Under altering climate conditions, the concept of CSA has emerged to achieve agricultural security. CSA intends to increase productivity, improve resilience, decrease GHG emissions, and support national food security and development objectives (Steenwerth et al., 2014). It incorporates traditional and innovative practices promoting productivity, income generation, adaptation to climate change, and reduction of GHG emissions. Increasing agricultural productivity, enhancing adaptive capacity at various scales, and reducing emissions are the three pillars of CSA. While the potential of CSA is acknowledged, there are divergent opinions regarding how to expand CSA options. Prior agricultural investments focused on technology development without considering farming systems' institutional context and complexity, resulting in limited success (Reed & Stringer, 2015). There is a shift toward a more comprehensive systems-oriented approach that considers the broader context, including policy, market, and institutional factors that influence agricultural systems. This strategy recognizes that technology alone is insufficient and considers the enabling and constraining factors for effective and sustainable agrarian transformation (Ingram, 2018).

3.4 SUSTAINABILITY INDICATORS FOR AGRICULTURE

As the importance of sustainability has grown over the past several decades, organizations are placing a greater emphasis on how to measure sustainability. Assessing a program or initiative based on current best practices is one method for measuring the sustainability of highways. INVEST includes sustainability best practices that exceed minimum requirements. Using sustainability as a metric typically necessitates expanding the traditional business reporting framework to include social and environmental performance alongside economic performance (the Triple Bottom Line). These three essential principles must be assessed, but they must provide a single measurement system. Therefore, many organizations are devising organization- or industry-specific measurement instruments and best practices to assist them in achieving the appropriate equilibrium between Social, Environmental, and Economic principles (Sharifi & Murayama, 2013).

While achieving sustainability by harmonizing the triple-bottom-line principles is an optimal objective that may aid in decision making, it will only be possible for some undertakings. However, achieving balance becomes more feasible when these principles are applied to a portfolio of initiatives, an entire system planning program, or an operations and maintenance program (Too & Weaver, 2014).

Measuring sustainable agriculture typically involves evaluating various environmental, social, and economic indicators and metrics. Common measurement criteria used to evaluate sustainable agriculture are shown in Table 3.2.

TABLE 3.2
The Main Measurement Criteria Used to Evaluate Sustainable Agriculture

Criteria	Definition
Soil health	Indicators such as organic matter content, nutrient levels, soil erosion rates, soil compaction, and soil biodiversity can be evaluated to determine the health and quality of agricultural soils.
Water efficiency and conservation	Monitoring water consumption and irrigation practices to guarantee water efficiency and minimize water waste, taking into account irrigation techniques, water source management, and water recycling.
Biodiversity conservation	Evaluating the diversity and abundance of plant and animal species within agricultural ecosystems, such as the presence of beneficial insects, birds, and pollinators.
Use of agrochemicals	Tracking the use of pesticides, herbicides, and fertilizers in order to minimize their harmful effects on the environment and human health, while promoting integrated pest management and organic agricultural practices.
Energy conservation	Assessing the energy inputs and outputs of agricultural systems, including energy use for irrigation, machinery, processing, and transportation, with an emphasis on reducing energy consumption and promoting the use of renewable energy sources.
Footprint of carbon	Measuring greenhouse gas emissions from agricultural activities, such as methane from livestock, nitrous oxide from fertilizers, and carbon dioxide from the use of fossil fuels, and implementing strategies to reduce emissions and increase carbon sequestration.
Economic viability	Evaluating the profitability and financial stability of farming systems, taking into account input costs, market access, income diversification, and resource efficiency, in order to ensure the economic sustainability of agricultural operations over the long term.
Social prosperity	Assessing the social impacts of agricultural practices, such as labor conditions, community involvement, access to nutritious food, food security, and equitable distribution of resources and benefits to various stakeholders.
Climate change resilience	Considering factors such as crop diversification, water management, conservation practices, and the use of climate-resilient crop varieties, assess the capacity of agricultural systems to adapt to and mitigate the effects of climate change.
Certification and standards	Considering adherence to specific sustainability certifications and standards, such as organic certification, Fair Trade, Rainforest Alliance, or other well-known sustainability labels, which provide criteria for sustainable agriculture.

Source: Created by authors.

This can be accomplished via field surveys, data collection, remote sensing, interviews, and analysis of pertinent indicators. Various regions and contexts may necessitate customized measurements based on local agricultural practices, priorities, and sustainability objectives.

3.4.1 SOIL QUALITY

Soil health is essential for evaluating sustainable agriculture because it directly influences agricultural systems' productivity, resilience, and environmental viability. It refers to the soil's condition, functioning, and capacity to support plant growth, maintain ecosystem services, and resist ecological pressures (Ingram, 2018). Key components and indicators used to evaluate soil health include:

SOM is an essential component that provides vital nutrients, improves soil structure, increases water retention capacity, and stimulates microbial activity. Measuring organic matter levels assists in assessing soil fertility and the potential for nutrient cycling.

It is essential to evaluate the soil's availability and balance of essential nutrients (such as nitrogen, phosphorus, potassium, and micronutrients) to determine its ability to support plant growth. Soil analysis and testing can reveal nutrient deficiencies or excesses, guiding fertilizer application practices.

Soil erosion: Evaluating soil erosion rates assists in determining the degree to which soil particles are lost due to wind or water erosion. Excessive erosion can reduce soil fertility, deplete topsoil, and degrade the environment. Employing techniques such as erosion modeling, sedimentation measurements, and erosion control practices is possible.

Measuring soil compaction levels evaluates the degree to which soil particles are compacted, which can inhibit root development, water infiltration, and nutrient movement. Typically, soil penetrometers or bulk density measurements are used to assess compaction.

Assessing the diversity and abundance of soil organisms, such as bacteria, fungi, earthworms, and other beneficial organisms, reveals the soil ecosystem's biological activity and ecological equilibrium. Methods including DNA analysis, assessments of microbial biomass, and visual observations can provide insight into soil biodiversity.

pH and soil acidity: The pH of the soil influences the availability of nutrients and microbial activity. Monitoring pH levels assists in determining whether the acidity or alkalinity of the soil is optimal for plant growth. Typically, pH meters or soil testing tools are utilized for pH measurement.

Water-holding capacity: Measuring the soil's ability to retain and provide water to plants is essential for determining its capacity to sustain produce during arid periods. Water-holding capacity can be determined using soil moisture sensors or laboratory analysis techniques.

Assessing the stability of soil aggregates, clusters of soil particles held together, reveals the soil's ability to resist erosion, maintain structure, and permit root penetration. Aggregate stability can be measured using moist sieving and slake experiments.

Biological activity: Evaluating microbial activity, enzymatic functions, and other natural processes in the soil offers insight into nutrient cycling, organic matter decomposition, and the overall functioning of the soil ecosystem. Utilize enzyme assays, respiration analyses, and microbial biomass evaluations.

To ensure food safety, environmental protection, and human health, it is essential to evaluate soil for contaminants such as heavy metals, pesticides, or pollutants. Typically, soil sampling and laboratory analysis identify and quantify contaminants.

These indicators assist farmers, researchers, and policymakers in comprehending the current status of soil health, identifying areas for improvement, and implementing suitable management practices to improve soil fertility, structure, and resilience. Regular monitoring of soil health indicators enables adaptive management strategies and encourages sustainable agricultural practices that conserve soil resources and support long-term productivity.

3.4.2 Water Conservation

Regarding the efficient and responsible use of water resources in agricultural systems, water use efficiency is vital to measuring sustainable agriculture. It is the ratio of agricultural produce (such as crop yield or livestock production) per unit of water consumed or utilized. Optimizing water resources, minimizing water waste, and ensuring sustainable water management in agriculture are facilitated by maximizing water use efficiency (Lawson & Blatt, 2014). The following factors and indicators are utilized to evaluate water use efficiency:

Crop water productivity: Also known as water productivity or water-use efficiency at the crop level, crop water productivity measures the quantity of crop yield generated per unit of water ingested. Generally, it is expressed as yield per water unit, such as kilograms of crop produced per cubic meter of water ($kg/m3$). This indicator aids in evaluating the efficacy of water use in crop production systems and directs irrigation methods.

Irrigation efficiency refers to the efficacy with which irrigation systems deliver water to crops. It considers water application inefficiencies such as evaporation, discharge, and subsurface percolation. A high irrigation efficacy ensures that a greater proportion of supplied water reaches the root zones of plants, thereby minimizing water loss.

Sustainable water use in agriculture necessitates implementing efficient water management practices, such as precision irrigation, trickle irrigation, and mulching. These practices seek to optimize water delivery, reduce evaporation, and improve root zone water infiltration and distribution. Evaluating the adoption and efficacy of these practices contributes to evaluating water use efficiency.

The water footprint of agricultural products measures the total volume of water utilized throughout the production process, including direct water use, indirect water use (included in inputs), and virtual water trade. It includes both blue water (surface and groundwater) and green water (rainwater) components in its comprehensive evaluation of the water resources required to produce specific agricultural commodities.

Indicators of water stress help evaluate the equilibrium between water availability and demand in agricultural systems. They consider variables such as rainfall patterns, evapotranspiration rates, and crop water needs. Monitoring indicators such as the water stress index, crop water deficit, and water balance reveals the adequacy of the water supply and the need for water-saving measures.

Water recycling and reuse: Evaluating the extent of water recycling and reuse practices in agriculture is essential for enhancing water use efficiency. Reusing and recycling effluent, drainage water, and other water sources within the agricultural system reduces freshwater extraction and overall water demand.

Rainwater harvesting: Rainwater harvesting techniques, such as collecting and storing precipitation for agricultural use, contribute to water conservation and enhance water availability during arid periods. Evaluating the adoption and efficacy of rainwater harvesting systems assists in determining their influence on water use efficiency.

Integrated water management: Integrated water management approaches take into account the interactions between various water sources (e.g., surface water, groundwater), competing water demands (agriculture, industry, and domestic), and environmental factors. Assessing the integration of water management strategies promotes sustainable water use across sectors and reduces water resource conflicts.

Farmers and policymakers can optimize water resources, reduce water scarcity risks, and promote sustainable agricultural practices by measuring and enhancing water use efficiency. Improving water use efficiency is essential for assuring agricultural security, adapting to climate change, and protecting water ecosystems in the face of escalating water challenges.

3.4.3 Preservation of Biodiversity

Biodiversity conservation is essential to gauging sustainable agriculture because it acknowledges the significance of preserving and fostering diverse ecosystems, species, and genetic resources within agricultural landscapes. It entails practices and strategies designed to preserve and enhance biodiversity while assuring farm productivity and food security (Wezel et al., 2016). Here are a few important considerations for comprehending the significance of biodiversity conservation in assessing sustainable agriculture:

Ecosystem services: Biodiversity in agricultural landscapes provides essential ecosystem services for sustainable farming systems. This includes pollination, natural pest control, nutrient cycling, maintaining soil fertility, water regulation, and climate regulation. The contribution of biodiversity conservation to sustainable agriculture can be measured by assessing the presence and functionality of these ecosystem services.

Sustainable agriculture endeavors to preserve and enhance agricultural landscapes' natural habitats, such as wetlands, forests, hedgerows, and riparian zones. These habitats serve as biodiversity centers, foster the diversity of native species, and perform essential ecological functions. Measuring the extent of habitat preservation and the connectivity between various habitats aids in evaluating the conservation efforts within agricultural systems.

Agroforestry and landscape design: Agroforestry practices contribute to biodiversity conservation in agricultural landscapes by integrating trees or woody perennials with crops or livestock. Agroforestry systems support diverse plant and animal species, which provide diverse habitats, improve soil fertility, mitigate climate change, and foster biodiversity. Evaluating the adoption and effectiveness of agroforestry practices and landscape designs that promote biodiversity conservation is important for assessing sustainable agriculture.

The significance of sustaining genetic diversity in cereals, livestock, and other agricultural species is acknowledged by sustainable agriculture. Genetic diversity confers resistance to parasites, diseases, and a fluctuating environment. Assessing the conservation of traditional and locally adapted crop varieties, livestock strains, and untamed relatives helps measure the conservation of agricultural genetic resources.

Conservation agriculture promotes agricultural methods that minimize soil disturbance, preserve soil cover, and diversify crop rotations. These practices aid in the preservation of soil biodiversity, improvement of soil health, and reduction of soil erosion. Measuring biodiversity conservation in sustainable agriculture is facilitated by evaluating the adoption and impact of conservation agriculture practices.

Establishing protected areas and buffer zones in agricultural landscapes can help protect biodiversity. These areas provide refuge for indigenous species, sustain ecological processes, and prevent habitat fragmentation. Assessing the establishment and efficacy of protected areas and buffer zones in agricultural regions contributes to the measurement of biodiversity conservation.

Management of invasive species: In agricultural systems, invasive species threaten native biodiversity. Monitoring and managing the spread and impact of invasive species serves to preserve native species diversity and ecosystem function. Evaluating invasive species control and prevention measures contributes to evaluating biodiversity conservation efforts.

Participation of farmers in biodiversity conservation practices and facilitation of knowledge exchange among agricultural communities are crucial to the development of sustainable agriculture. Measuring the integration of biodiversity conservation in agriculture requires evaluating the participation of farmers in biodiversity conservation initiatives, their comprehension of the value of biodiversity, and their adoption of sustainable land management practices.

Farmers, policymakers, and researchers can promote resilient and environmentally friendly agricultural systems by recognizing biodiversity conservation as a crucial aspect of sustainable agriculture. Monitoring and evaluating biodiversity-related indicators assists in tracking progress, identifying obstacles, and informing decision-making processes to ensure the long-term viability of agricultural landscapes.

3.4.4 Use of Agrochemicals

Using agrochemicals is an essential factor to consider when measuring sustainable agriculture. Agrochemicals, such as synthetic pesticides, herbicides, fungicides, and fertilizers, play a crucial role in agricultural production by controlling pests, diseases, and nutrient deficiencies. However, their misuse or overuse can have adverse environmental and health effects (Singh et al., 2021). Here is a closer look at the role of agrochemical use in determining whether agriculture is sustainable:

Integrated Pest Management (IPM): Sustainable agriculture encourages the adoption of IPM strategies, which seek to minimize the use of chemical pesticides by integrating multiple pest control methods. Before resorting to chemical interventions, IPM emphasizes prevention, monitoring, and using the least hazardous pest control methods. Measuring the implementation and efficacy of IPM practices allows for evaluating the reduction in agrochemical use.

Pesticide toxicity and environmental impact: The selection and use of pesticides substantially affect the environmental impact of agricultural systems. Sustainable agriculture aims to reduce the use of highly toxic pesticides and promote less hazardous alternatives. Measuring the toxicity of used pesticides and evaluating their potential adverse effects on non-target organisms, such as beneficial insects, wildlife, and aquatic life, contributes to assessing the environmental impact of agrochemical use.

Applying fertilizer is essential for sustaining soil fertility and promoting crop growth. However, excessive fertilizer use can result in nutrient imbalances, water contamination, and GHG emissions. Sustainable agriculture emphasizes optimizing fertilizer management via techniques such as soil testing, precision application, and the use of organic fertilizers. Measuring nutrient application rates, fertilizer use efficiency, and implementing environmentally benign fertilizer management practices contributes to evaluating sustainable agrochemical use.

Sustainable agriculture recognizes the significance of maintaining nutrient cycling and soil health to reduce reliance on synthetic fertilizers. Cover cropping, crop rotation, and organic matter management increase nutrient availability, enhance soil structure, and sustain beneficial soil organisms. Sustainable nutrient management can be measured by evaluating the adoption of these practices and their effect on decreasing agrochemical inputs.

Agrochemical residues in agricultural products pose hazards to human health and the environment. Sustainable agriculture encourages the implementation of Good Agricultural Practices (GAPs) to reduce chemical residues in crops and potential health risks. Measuring chemical residue levels in harvested produce and evaluating compliance with GAPs aid in evaluating the reduction of agrochemical residues.

Risk assessment and monitoring: Sustainable agriculture emphasizes the need for agrochemical use risk assessment and monitoring. This includes assessing the possible effects on human health, fauna, and ecosystems. Measuring the implementation of risk assessment protocols, monitoring programs, and adopting precautionary measures helps evaluate agrochemical use commitment.

Improving farmers' knowledge and comprehension of agrochemical use and its potential consequences is essential for promoting sustainable practices. Training programs and educational initiatives on the appropriate handling, application, and storage of agrochemicals contribute to reducing associated risks. Measuring farmers' participation in training programs and their knowledge of safe agrochemical practices aids in assessing the integration of education and training in sustainable agriculture.

By considering agrochemical use as a quantifiable aspect of sustainable agriculture, it becomes possible to evaluate adopting practices that minimize adverse environmental impacts, safeguard human health, and maximize agricultural productivity.

Creating a balance between the need for agrochemicals and their responsible and prudent use contributes to developing sustainable agricultural systems that promote agricultural resilience and ecological health over the long term.

3.4.5 Energy Conservation and Energy Efficiency

When assessing agriculture's sustainability, energy efficiency is a crucial factor to consider. It entails optimizing agricultural systems' energy inputs and outputs to ensure energy is used efficiently, thereby minimizing waste and environmental impact (Thomson et al., 2017). Here is a thorough look at the role of energy efficiency in determining whether agriculture is sustainable:

Sustainable agriculture seeks to minimize energy consumption on farms by adopting energy-efficient practices. This includes maximizing the energy efficiency of irrigation, machinery, heating, ventilation, and lighting systems. Measuring and monitoring energy consumption on farms allows for evaluating energy use efficiency and identifying improvement opportunities.

Adoption of renewable energy: Incorporating renewable energy sources into agricultural operations is a crucial component of sustainable agriculture. Utilizing solar panels, wind turbines, biomass energy, and biogas systems to generate clean and renewable energy on farms is included. Measuring the adoption and utilization of renewable energy technologies contributes to the evaluation of the reduction of GHG emissions and reliance on fossil fuels.

Energy audits and assessments: Conducting energy audits and assessments on farms can provide valuable insight into energy consumption patterns and improvement opportunities. Measuring the implementation of energy audits and the adoption of energy-saving recommendations facilitates the evaluation of agricultural operations' commitment to energy efficiency.

Energy efficiency in agriculture also includes the transportation of agricultural inputs and products. Sustainable agriculture promotes the use of fuel-efficient vehicles, optimized transportation routes, and alternative modes of transportation such as rail and marine transportation. Measuring the adoption of sustainable transportation practices and evaluating fuel consumption and emissions from agricultural transportation contributes to the evaluation of energy efficiency in the agricultural supply chain as a whole.

Sustainable agriculture acknowledges the significance of energy efficiency throughout the entire agricultural value chain, including refining, packaging, and distribution. Measuring energy consumption at these stages and promoting energy-efficient practices contributes to the reduction of energy waste and environmental impact beyond the farm level.

Adoption of energy-efficient technologies and apparatus can substantially contribute to energy efficiency in the agricultural sector. This includes the use of energy-efficient machinery, precision agriculture technologies, and energy-efficient irrigation systems. Measuring the adoption of these technologies and evaluating their effect on reducing energy consumption aids in assessing agricultural practices' energy efficiency.

Education and awareness: Promoting education and awareness regarding energy efficiency in agriculture is essential for fostering sustainable practices. The improvement of farm energy efficiency is facilitated by training programs, seminars, and educational materials emphasizing energy-saving techniques and best practices. Measuring farmers' participation in energy efficiency programs and their knowledge and awareness of energy-saving measures assists in evaluating the integration of education and awareness in sustainable agriculture.

By incorporating energy efficiency as a measurable aspect of sustainable agriculture, it is possible to evaluate the adoption of practices that optimize energy use, reduce GHG emissions, and promote sustainable resource management. Increasing energy efficiency in agriculture mitigates climate change and improves the resilience and sustainability of agricultural systems.

3.4.6 Footprint of Carbon

It focuses on quantifying and reducing GHG emissions from agricultural activities. The carbon footprint of agricultural systems refers to the total amount of GHGs, namely, carbon dioxide (CO_2), methane (CH_4), and nitrous oxide (N_2O), emitted over the entire life cycle of agricultural practices and products (Gan et al., 2014). The function of carbon footprint in measuring sustainable agriculture is examined in greater detail below.

Measuring the carbon footprint of agricultural activities requires evaluating the quantity of GHG emissions emitted at various phases of the agricultural value chain. This encompasses, among other things, emissions from land use change, energy consumption, fertilizer use, enteric fermentation (methane emissions from livestock), manure management, and rice cultivation. By quantifying these emissions, the main sources of GHG emissions in agricultural systems can be identified.

Measuring the carbon footprint of agriculture facilitates the identification of opportunities for emission reductions and the implementation of mitigation strategies. Adopting practices that reduce emissions and encourage carbon sequestration, sustainable agriculture seeks to minimize the carbon footprint. This includes conservation agriculture, agroforestry, organic farming, enhanced nutrient management, efficient use of fertilizers, the adoption of renewable energy, and livestock management practices that reduce methane emissions. Measuring the adoption and effectiveness of these mitigation strategies helps evaluate the commitment to reducing the agricultural sector's carbon footprint.

Carbon sequestration: Sustainable agriculture also emphasizes enhancing carbon sequestration, which is the process of capturing and storing atmospheric carbon dioxide. Afforestation, reforestation, cover cropping, and conservation tillage can increase carbon sequestration in soils and vegetation, thereby mitigating agricultural emissions. Measuring the potential for carbon sequestration and implementing these practices contributes to determining the carbon balance of farming systems as a whole.

Life cycle assessment: To measure the carbon footprint of agricultural products, sustainable agriculture frequently conducts life cycle assessments (LCAs). LCAs evaluate the environmental impacts of a product's entire life cycle, from the extraction of basic materials to its final disposal. LCAs comprehensively comprehend the carbon emissions associated with specific agricultural products by considering the carbon footprint of farming inputs, production processes, transportation, and refuse management. This data assists consumers, policymakers, and producers make informed decisions regarding the carbon footprint of their agricultural practices and food choices.

Carbon markets and certification: Measuring the carbon footprint of agricultural activities enables farmers to participate in carbon markets and certification schemes that recompense them for their emission reduction and carbon sequestration efforts. Agriculture carbon offset initiatives, such as reforestation and methane capture from manure management, can generate carbon credits that can be sold to entities seeking to offset their own emissions. Measurement and verification of emission reductions are crucial for the credibility and transparency of these schemes.

Sustainable agriculture seeks to reduce GHG emissions, encourage carbon sequestration, and contribute to climate change mitigation efforts by measuring the carbon footprint of agricultural practices and products. It aids in the identification of opportunities for emission reductions, guides the implementation of mitigation strategies, and encourages the adoption of climate-friendly farming practices.

3.4.7 Economic Viability

Economic viability is essential to assessing sustainable agriculture because it evaluates agricultural practices' financial profitability and long-term economic sustainability. Sustainable agriculture seeks to balance environmental stewardship and economic viability, recognizing that agrarian activities must be economically profitable to ensure the producers' livelihoods and the farm sector's long-term success (Reganold & Wachter, 2016). Examining the role of economic viability in measuring sustainable agriculture.

Economic viability requires evaluating the costs and benefits associated with agricultural practices in terms of their cost-effectiveness. It considers input costs, such as seedlings, fertilizers, pesticides, labor, and equipment, and evaluates agricultural production returns. Sustainable agriculture seeks to maximize the use of inputs and reduce wasteful expenditures while maintaining or increasing productivity. Evaluate the cost-effectiveness of various practices and technologies to ensure that the benefits outweigh the costs.

Profitability: Sustainable agriculture's economic viability focuses on generating profits for producers. The profitability of agricultural production is determined by comparing the revenue generated with the expenses incurred. Long-term, sustainable agricultural practices should be economically viable, allowing farmers to generate sufficient income to support their livelihoods, invest in their properties, and adapt to shifting market conditions. It requires consideration of market prices, yield levels, production costs, and market access.

Sustainable agriculture takes market demand and consumer preferences into consideration. To be economically viable, producers must produce products that are in demand and can command appropriate market prices. Assessing market trends, consumer preferences for sustainably produced commodities, and the potential for niche markets or products with added value. By matching agricultural production to market demand, producers can enhance their economic viability and develop sustainable business models.

Diversification and resilience: In order for sustainable agriculture to be economically viable, it is often necessary to diversify income streams and reduce reliance on a single commodity or product. Diversification may entail the cultivation of multiple commodities, the incorporation of livestock, agroforestry, or the exploration of non-agricultural income sources. Diversification increases resiliency by distributing risks and decreasing sensitivity to market fluctuations and climate-related challenges. It enables farmers to adjust to fluctuating economic conditions and maintain economic viability over time.

Comparing the costs and benefits of various agricultural practices or investment decisions, cost-benefit analysis is frequently used to evaluate the economic viability of a project. This analysis assists producers and policymakers in making informed decisions by examining the monetary implications of employing sustainable practices. It considers the short-term and long-term costs, such as initial investment costs, operational expenses, and prospective returns or savings associated with enhanced productivity, reduced input use, or increased resource efficiency.

The economic viability of sustainable agriculture is influenced by supportive policies, access to financial resources, and incentives for environmentally friendly practices. Governments, organizations, and financial institutions support adopting sustainable agricultural practices by providing resources, loans, grants, or subsidies. Measurement of economic viability considers the availability of such support mechanisms and evaluates their efficacy in facilitating the transition to sustainable agriculture.

Sustainable agriculture seeks to assure agricultural systems' long-term profitability and economic resilience by evaluating economic viability. It requires optimizing costs, generating profits, responding to market demands, diversifying income sources, and making prudent investment decisions. Together with environmental sustainability and social welfare, economic viability is a key pillar of sustainable agriculture.

3.4.8 Social Prosperity

As it concentrates on agricultural practices' social dimensions and effects, social well-being is a fundamental metric for quantifying sustainable agriculture. The significance of promoting the well-being and livelihoods of farmers, agricultural laborers, rural communities, and society as a whole is acknowledged by sustainable agriculture (Latruffe et al., 2016). The role of social well-being in measuring sustainable agriculture is examined in greater detail below.

Sustainable agriculture seeks to improve the producers' and farmworkers' standard of living and income. It entails providing fair and acceptable employment opportunities, ensuring equitable benefit distribution, and fostering economic opportunities in rural areas. Measurement of social welfare takes into account the influence of agricultural practices on income generation, employment security, and poverty reduction. It seeks to develop agricultural systems that enable producers to maintain and enhance their standard of living.

Social well-being in sustainable agriculture places an emphasis on equity and social justice. It involves promoting agricultural systems that are inclusive and address gender equality, social inclusion, and the rights of marginalized groups. It takes into account the equitable allocation of resources, opportunities, and benefits among various stakeholders. Measuring social well-being evaluates the extent to which agricultural practices contribute to reducing social disparities, assuring access to resources, and empowering vulnerable or marginalized groups in rural communities.

Food security and nutrition: The goal of sustainable agriculture is to improve food security and nutrition. Measuring social well-being examines the availability, accessibility, and affordability of nutritious food for producers and consumers. It takes into account the effects of agricultural practices on food production, diversity, and quality, as well as the well-being of individuals and communities in terms of their food consumption patterns, dietary diversity, and nutritional status. Sustainable agriculture endeavors to ensure that agricultural systems contribute to the well-being of people by providing food that is secure, nutritious, and culturally acceptable.

In sustainable agriculture, social well-being takes into consideration the health and well-being of producers, agricultural employees, and communities. It examines the effects of agricultural practices on occupational health and safety, exposure to agrochemicals, and the physical and mental health of agricultural workers. It promotes practices that reduce health hazards, improve working conditions, and enhance agricultural communities' overall quality of life.

Community engagement and participation: The social well-being measurement takes into account the extent of community engagement and participation in agricultural decision-making processes. It acknowledges the significance of involving local communities, farmer organizations, and other stakeholders in the development of agricultural policies, practices, and resource management. The objective of social well-being is to promote social capital, cooperation, and concerted action within rural communities in order to address common challenges and promote sustainable development.

Sustainable agriculture values and respects local cultures, traditional knowledge systems, and the social cohesion of rural communities. Measurement of social well-being takes into account the cultural heritage and identity associated with agricultural practices. Examines the role of agriculture in sustaining social cohesion, preserving cultural traditions, and promoting a sense of belonging and pride in rural communities.

By emphasizing social welfare, sustainable agriculture acknowledges the interdependence of agriculture and society. It seeks to develop agricultural systems that contribute to individuals' and communities' well-being, livelihoods, and empowerment. Measuring social well-being helps assess the social impacts of farming practices and ensures that agricultural development is inclusive, equitable, and advantageous for all stakeholders.

3.4.9 Climate Change Resilience

Resilience to climate change is an essential indicator of sustainable agriculture. Agricultural systems can absorb disturbances caused by climate change, adapt to shifting conditions, continue functioning, and provide crucial services such as food production while minimizing negative environmental impacts (Agovino et al., 2019). In the context of measuring sustainable agriculture, the following is a more in-depth explanation of resilience:

Climate change poses numerous challenges to agricultural systems, including an increase in the frequency and intensity of extreme weather events (such as droughts, floods, and storms), altered precipitation patterns, rising temperatures, shifting pest and disease dynamics, and a shift in the growing season. These alterations can impair agricultural productivity, jeopardize food security, and result in environmental degradation.

Resilience is an indicator used to evaluate how well agricultural systems can withstand and recuperate from these climate-related challenges. It concentrates on enhancing the ability of agricultural systems to adapt, endure disturbances, and preserve or restore productivity and ecosystem services. Measuring resilience enables the identification of vulnerabilities, the development of effective adaptation strategies, and the improvement of the long-term sustainability of agricultural practices.

Components of Resilience: Resilience in agriculture includes the following components:

a. **Diversification**: Promoting crop diversification, intercropping, agroforestry, and livestock integration can increase system resilience by decreasing its susceptibility to pests, diseases, and extreme weather. Diverse systems are more likely to resist disruptions and offer a variety of goods and services.
b. **Adaptive Capacity**: Increasing farmers' adaptive capacity through knowledge transfer, training, access to information, and resources enables them to respond and adapt to altering climatic conditions. This can be accomplished by employing new technologies, implementing climate-smart practices, and enhancing water management techniques.
c. **Soil Health and Water Management**: Keeping soils healthy through sustainable practices such as conservation agriculture, organic matter management, and erosion control improves water infiltration, nutrient availability, and crop resilience in general. In addition, effective water management, including irrigation methods adapted to fluctuating water availability, is essential.
d. **Risk Management**: Integrating risk management strategies such as insurance programs, early warning systems, and enhanced forecasting models can assist farmers in anticipating and mitigating the effects of climate-related risks, thereby reducing their vulnerability and enhancing their resilience.
e. **Socioeconomic Factors**: Socioeconomic aspects must be considered when measuring resilience. This includes assessing the social and economic factors that influence producers' adaptive capacity, access to resources, markets, and financial support, as well as confronting issues of social equity and justice in relation to climate change's effects.

Measurement and Evaluation: In sustainable agriculture, measuring resilience requires a combination of qualitative and quantitative approaches. Multiple indicators

Sustainable Agriculture

must be evaluated, such as crop yield stability, income diversification, access to resources, adoption of climate-smart practices, and ecosystem services provided by the agricultural system. Using instruments such as resilience indices, vulnerability assessments, and participatory approaches, it is possible to assess resilience levels at various scales, ranging from individual farms to entire landscapes.

By comprehending and assessing resilience in sustainable agriculture, policymakers, researchers, and producers can identify effective strategies and practices for developing agricultural systems that are more climate resilient. This is essential to ensure long-term food security, environmental sustainability, and the prosperity of farming communities in the face of climate change.

3.4.10 CERTIFICATION AND STANDARDS

In measuring sustainable agriculture, certification and standards play a crucial role. They offer frameworks and guidelines that allow farmers, producers, and consumers to evaluate and verify the sustainability of agricultural practices (Latruffe et al., 2016). In the context of gauging sustainable agriculture, here is a comprehensive explanation of certification and standards:

Certification refers to a formal process in which an independent third party evaluates and verifies that a particular agricultural product or production system meets specific standards or criteria. In contrast, standards are predetermined guidelines or benchmarks that define what constitutes sustainable agriculture. They outline the necessary requirements and practices for achieving sustainable outcomes.

Purpose of Certification and Standards Certification and standards in sustainable agriculture serve primarily to:

a. Standards provide a distinct set of criteria and recommendations for sustainable agricultural practices. They encompass areas such as soil health, water management, biodiversity conservation, labor rights, and fair trade in outlining environmental, social, and economic considerations that must be taken into account.
b. Certification verifies that agricultural products or practices conform to the established criteria. It provides a credible and transparent method for assuring consumers, retailers, and other stakeholders that a product has been produced sustainably by meeting certain environmental and social benchmarks.
c. **Drive Continuous Improvement**: Certification and standards also encourage agricultural practices' continuous improvement. They serve as a guide for farmers and producers to employ more sustainable practices, encouraging innovation, efficiency, and the implementation of best practices.
d. **Increase Market Access and Value**: Sustainable certifications can differentiate agricultural products on the market. They provide a competitive advantage by satisfying consumer demand for sustainably produced products, expanding market access, increasing market prices, and enhancing producer reputation.

Different types of certification and standards exist for measuring sustainable agriculture. Among the most recognizable are:

a. Organic certification ensures that agricultural products are grown without synthetic pesticides, genetically modified organisms (GMOs), or synthetic fertilizers. It emphasizes ecological balance, the preservation of biodiversity, and the use of renewable resources.
b. **Equitable Trade Certification**: It ensures that farmers and laborers receive equitable prices and acceptable working conditions. It promotes social equity, community development, and sustainable environmental practices.
c. This certification emphasizes sustainable agricultural practices that conserve biodiversity, safeguard ecosystems, and promote the well-being of laborers and local communities.
d. **Roundtable on Sustainable Palm Oil (RSPO)**: RSPO certification ensures sustainable palm oil industry practices, promoting environmental responsibility, community involvement, and wildlife conservation.
e. **GlobalGAP (Good Agricultural Practices)**: GlobalGAP (Good Agricultural Practices) is a private sector standard that promotes safe and sustainable agricultural practices, including product safety, traceability, and environmental stewardship.

Certification bodies and auditors evaluate compliance through on-site inspections, documentation evaluations, and interviews with farmers and producers. They assess land management practices, chemical inputs, waste management, labor conditions, social responsibility, and regulatory compliance.

Challenges and Limitations: Certification and standards are valuable instruments but also subject to obstacles and constraints. Some examples include:

a. Certification processes can be complicated, time-consuming, and expensive, especially for small-scale producers. Compliance with the requirements and certification maintenance may incur additional costs and administrative burdens.
b. **Fragmentation and Multiple Standards**: Multiple certification schemes and standards can lead to market fragmentation and confusion. Harmonization and alignment efforts are required for certification landscape simplification and streamlining.
c. **Verification and Enforcement**: To ensure the validity of the certification, robust verification and enforcement mechanisms are necessary. Issues such as fraudulent claims, greenwashing, and a lack of transparency in the certification process must be addressed.
d. **Changing Standards**: Standards must change and adapt to new sustainability challenges and scientific advances. Regular evaluations and revisions are required to ensure their continued relevance and efficacy in addressing emerging environmental and social issues.

Certification and standards provide a structure for assessing and promoting sustainable agriculture. They empower farmers, producers, and consumers to make informed decisions supporting environmentally responsible, socially responsible, and economically viable agricultural practices.

3.5 MCDM-BASED AGRICULTURAL SUSTAINABILITY MEASUREMENTS

3.5.1 Agricultural Sustainability Indicators Frameworks

Agricultural sustainability can be assessed through three dimensions of sustainable development: economic, environmental, and social. Analysis of indicators of sustainable development in agriculture can serve as a tool for successful sustainable agriculture development policy implementation (Alaoui et al., 2022; Dabkiene et al., 2021).

Indicators show the state of the agriculture sector and allow the formulation of aims and monitoring of achieved progress (Gómez-Limón & Sanchez-Fernandez, 2010; Nadaraja et al., 2021). Effective analysis of sustainability indicators in agriculture facilitates decision making, management, and information dissemination to stakeholders, enabling the creation of shared scenarios for sustainable agriculture development (Yu & Mu, 2022; De Olde et al., 2017; Iocola et al., 2020).

Selecting indicators and frameworks for sustainable development in agriculture is very important because the conclusions depend on it. Many scientific studies have been carried out, which confirm that the selection of indicators must be carried out taking into account: economic, social, and environmental aspects; the simplicity of the indicators and the possibility of their practical application; the possibility of using the indicators taking into account the cultural and geographical differences of the country (countries); the ability to meet the expectations of stakeholders and make political decisions; meaningfulness to end users; economic efficiency in data collection and processing (Alaoui et al., 2022; Yu & Mu, 2022; Bathaei & Streimikiene, 2023; Streimikiene & Mikalauskiene, 2023).

The summarized indicators framework for sustainable agricultural development is provided in Table 3.3.

European Commission developed agri-environmental indicators for the integration of environmental concerns into the Common Agricultural Policy (CAP) in EU member states. In Table 3.4, the agri-environmental indicators framework was constructed based on EU agri-environmental indicators (European Commission, 2023; EUROSTAT, 2023). Due to the data limits, the framework of indicators for sustainability assessment of agriculture was developed from agri-environmental indicators based on available data in EUROSTAT.

Having frameworks of sustainability assessment in agriculture in order to assess and compare countries based on sustainable agriculture indicators multi-criteria decision making (MCDM) may be applied as these indicators have different measurement units. In addition, countries can show good results in some areas but

TABLE 3.3
Indicators of Sustainable Agricultural Development

Dimension	Factors
Economic indicators	Utilized agricultural area (UAA) per annual work unit (AWU) (SE025/SE010) in ha/AWU
	Net investment on fixed assets in EUR (SE521)
	Cash flows in EUR (SE526)
	Equity capital in EUR per hectare
	Total liabilities in EUR (SE485)
	Population connected to organized water supply (%)
	Population connected to sewage network (%)
	Population connected to gas network (%)
	Energy consumption per 1 resident (kWh/resident)
	Drinking water from water network consumption per 1 resident (m^3)
	Difference between populations connected to water supply and sewage systems (%)
	Rural water supply and sewage network length (km)
	Rural landfills and their area (−) and (ha)
	Annual investments in rural water supply and sewage networks (EUR)
	Number of rural wastewater treatment plants, collective and individual (−)
	Number of annually constructed rural wastewater treatment plants, collective and individual (−)
	Groundwater quality (%)
	Ammonia emissions from agriculture (% of total emission)
	Greenhouse gas emissions from agriculture (% of total emission)
	Estimated soil loss by water erosion (tone/hectare)
	Area of organic farming (%)
Social indicators	Employment in agriculture (in thousands of jobs)
	Domestic consumption of farm products (SE260)
	Income per person in EUR (SE430)
	Total number of work-related accidents (fatal and nonfatal) per year
	Migrations (residents)
	Life expectancy (years)
	Birthrate per 1,000 residents (−)
	Infant deaths per 1,000 residents (−)
	Live births per 1,000 residents (−)
	Deaths per 1,000 residents (−)
	Rural population age distribution (−)
	Unemployment rate (%)
	Accidents at work in individual farming (−)
	Farm labor force (full-time equivalents)
	Share of agriculture in government expenditure in %

(*Continued*)

TABLE 3.3 (*Continued*)
Indicators of Sustainable Agricultural Development

Dimension	Factors
Environmental indicators	Ammonia emissions from agriculture in kg/ha per year
	Share of permanent grassland in the UAA in %
	Harmonized risk indicator 1 for pesticides by category of active ingredients (Directive 2009/128/EC) per year
	Final energy consumption in agriculture/forestry per hectare of agricultural area in kilograms of oil equivalent per hectare (KgOE/ha)
	Consumption of inorganic fertilizers in tons of nitrogen (N)/year
	Community income per 1 resident (EUR)
	Community expenditures per 1 resident (EUR)
	Mean disposable agricultural and self-employment income ratio (%)
	Agricultural production per 1 ha (EUR)
	Agricultural products purchase per 1 ha (kg) or (L)
	Mean retirement (EUR)
	Governmental funds granted to agricultural research and development (EUR/resident)
	People at risk of poverty or social exclusion (%)
	Gross value added to agriculture (EUR)
	Energy productivity (EUR/kg oil equivalent)
	Agritourist facilities (−)
	Use of agritourist accommodation (%)

Source: Created by authors based (Bockstaller et al., 2008; Rozman et al., 2012; De Olde et al, 2017; Gómez-Limón et al., 2010; Nadaraja et al., 2021; Alaoui et al., 2022; Bathaei, Streimikiene, 2023; Clerino et al., 2023).

TABLE 3.4
Agricultural Sustainability Indicators

Agri-Environmental Indicator Factsheets	Most Recent Data (year)	Responsible
1. Area under organic farming	2020	Eurostat
2. Nitrogen fertilizer consumption	2020	Eurostat
3. Phosphorus fertilizer consumption		
4. Pesticides sales	2020	Eurostat
5. Energy use	2019	Eurostat
6. Livestock density index	2020	Eurostat
7. Ammonia emissions from agriculture	2020	EEA
8. Greenhouse gas emissions from agriculture	2020	EEA

Source: Created by the authors based (European Commission, 2023; Eurostat, 2023).

negative results in other areas of agricultural sustainability. Therefore, MCDM is necessary to evaluate countries based on all criteria of sustainability (Streimikiene & Mikalauskiene, 2023).

3.5.2 MULTI-CRITERIA DECISION AIDING MODELS FOR SUSTAINABILITY ASSESSMENT IN AGRICULTURE

MCDM is a methodological approach that can be employed to evaluate and rank various criteria in decision-making procedures. MCDM can assist in evaluating and ranking multiple agricultural sustainability indicators based on their relative importance and performance when it comes to measuring agrarian sustainability (Musvoto et al., 2015). To apply MCDM to agricultural sustainability, the following steps may be taken:

Identify pertinent sustainability criteria: Begin by identifying a set of criteria that captures the essential aspects of agricultural sustainability. These criteria must incorporate economic, environmental, and social considerations. Examples of potential criteria include soil health, water consumption, GHG emissions, biodiversity, labor conditions, profitability, and food security (Elawama, 2016).

After identifying the criteria, you must allocate weights to each criterion according to its relevance. The weights reflect the preferences of the decision maker. They can be determined through expert opinions, consultations with stakeholders, or mathematical techniques such as Analytic Hierarchy Process (AHP) or Analytic Network Process (ANP) (Velten et al., 2015).

Collect data and assess performance: To evaluate the performance of agricultural practices or systems, collect data pertinent to each criterion. This could involve gathering information on farm management practices, resource utilization efficacy, environmental impacts, social indicators, and economic viability. The data may be collected via surveys, field measurements, literature evaluations, or existing databases (Bathaei & Štreimikienė, 2023).

Normalize the data, as various sustainability indicators may use varying measurement units and scales. Normalization permits meaningful comparisons between distinct criteria. Min-max scaling, z-score normalization, and fuzzy logic-based normalization are frequent normalization techniques (Reganold & Wachter, 2016).

Evaluate performance against criteria: Using appropriate evaluation methods, compare the performance of agricultural practices or systems against each criterion. This may entail calculating performance scores, rankings, or categorical evaluations. The precise evaluation method will depend on the nature of the data and the assessment's objectives. Weighted summation, TOPSIS (Technique for Order of Preference by Similarity to Ideal Solution), and ELECTRE (ELimination Et Choix Traduisant la Réalité) are prevalent techniques (Napitupulu & Hasibuan, 2017).

Once the evaluations are complete, aggregate and interpret the results to produce an overall classification or performance score for each agricultural practice or system. This is accomplished by combining the weighted scores derived in the preceding phase. The aggregation method may entail simple weighted summation, weighted average, or more complex methods such as the VIKOR (VIekriterijumsko KOmpromisno Ranging) method (Mir et al., 2016).

Conduct sensitivity analysis and validation to determine the robustness of the results by adjusting the weights assigned to the criteria or the normalization techniques. In addition, validate the results with domain experts or stakeholders to ensure that they align with their expectations and knowledge (Suresh et al., 2021).

By incorporating multiple criteria and quantifying their performance, MCDM provides a standardized method for measuring agricultural sustainability. It can assist decision makers in prioritizing actions, policies, and investments that promote sustainable agricultural practices and maximize resource allocation (Ikram et al., 2021).

Using an integrated methodology that combines ISM and DEMATEL, Narwane et al. investigated thirty-eight obstacles to the sustainable development of biofuels. The results revealed four obstacles: a lack of government support, subsidies/incentives for generating competition, a lack of support for entrepreneurship, and the absence of biomass supply chain standards. These obstacles can be eliminated using various strategies (Kalinowska et al., 2022).

In order to evaluate and assess proposed adaptation scenarios to climate change in the Jarreh agricultural water resources system in southwest Iran, Zamani et al. developed a fuzzy-based decision support system. The results indicated an increase in annual mean temperature, decreased discharge into the reservoir, and decreased system dependability. According to performance standards, six adaptation scenarios were proposed and ranked. It was discovered that increasing irrigation efficiency and decreasing cultivated area are preferable (More et al., 2022; Zamani et al., 2020).

Kannan et al. proposed a hybrid MCDM strategy for identifying viable locations for establishing solar facilities in the eastern region of Iran. Two robustness functional measures are used to compare the performance of the methods. The results indicate that the construction cost, initial investment, devastation of ecosystems, solar radiation intensity, and distance to catchment reservoirs are crucial (Keenan, 2015).

To estimate the extent of wind energy development in Turkey, Bozdağ et al. (2016) created an instrument that combines GIS and MCDM techniques. It utilized a literature review and regulations to identify infeasible sites and pairwise comparisons to create a three-class suitability map (Bozdağ et al., 2016).

E. Reig-Martínez et al. (2011) examined the appraisal of sustainability at the farm level by integrating DEA and MCDM methodologies. Economic and environmental composite sustainability indicators are positively correlated, but social indicators are not. The variables farm size, cooperative membership, and technical education all have a positive effect on sustainability (Reig-Martínez et al., 2011).

Ecer proposed a structural method for evaluating the sustainability performance of extant onshore wind farms based on the triple bottom line of sustainability. The environmental dimension is the most essential, followed by the economic and social dimensions, and protected area proximity is the most significant factor. A sensitivity analysis is also performed to validate the framework (Ecer, 2021).

Using Ward's method, CRITIC, and TOPSIS, Omurbek et al. intended to cluster Turkish localities based on their local agricultural production and rank them in terms of performance. Results indicate that only a handful of cities produce more than tens of them collectively, indicating that local governments should reconsider their agricultural programs (Ömürbek et al., 2020).

3.6 CONCLUSIONS

Agriculture sector is an important contributor to sustainable development of the country as it has significant economic, social, and environmental impacts. Sustainability assessment of agriculture should be based on indicators that quantify the economic, social, and environmental impacts of agricultural sector. To assess the social, economic, and environmental impacts of the agriculture systems, it is essential also to analyze the relationship between these issues and sustainable agriculture development and apply MCDM tools for trade-off between social, economic, and environmental impacts of agricultural activities.

The agricultural sector has negative environmental impacts such as air and water pollution, excessive pesticide use, and hydrological changes, so taking into account vital role of agriculture in food supply and population growth, implementation of sustainable development principles in agriculture sector is necessary. Agriculture is essential for attaining global food security and ensuring other important economic, social, and environmental objectives of sustainable development.

The concept of sustainable development is fundamental to agriculture contributing to the emergence of the circular economy principles in agriculture. Cyclic economy models in agriculture are based on penetration of renewables and emphasize the elimination of waste, the recycling of materials and products, production of energy from agriculture wastes, energy saving and GHG emission reduction, the protection of the natural environment, and the promotion of eco-friendly production and consumption practices.

Harmonized policies and measures to reduce carbon dioxide and toxic substance emissions, mitigate climate change, promote technological development in the agricultural sector, and encourage renewable energy solutions in agriculture are required for achieving sustainable agriculture development.

REFERENCES

Abdur Rashid Sarker, M., Alam, K., & Gow, J. (2013). Assessing the determinants of rice farmers' adaptation strategies to climate change in Bangladesh. *International Journal of Climate Change Strategies and Management*, 5(4), 382–403.

Adnan, N., Nordin, S. M., Rahman, I., Vasant, P. M., & Noor, A. (2017). A comprehensive review on theoretical framework-based electric vehicle consumer adoption research. *International Journal of Energy Research*, 41(3), 317–335.

Alaoui, A.; Barão, L.; Ferreira, C.S.S.; Hessel, R. (2022). An overview of sustainability assessment frameworks in agriculture. *Land*, 11, 537. https://doi.org/10.3390/land11040537

Agovino, M., Casaccia, M., Ciommi, M., Ferrara, M., & Marchesano, K. (2019). Agriculture, climate change and sustainability: The case of EU-28. *Ecological Indicators*, 105, 525–543.

Ahmadzai, H., Tutundjian, S., & Elouafi, I. (2021). Policies for sustainable agriculture and livelihood in marginal lands: A review. *Sustainability*, *13*(16), 8692.

Ajani, E., Mgbenka, R., & Okeke, M. (2013). Use of indigenous knowledge as a strategy for climate change adaptation among farmers in sub-Saharan Africa: Implications for policy.

Aleksandrova, M., Gain, A. K., & Giupponi, C. (2016). Assessing agricultural systems vulnerability to climate change to inform adaptation planning: An application in Khorezm, Uzbekistan. *Mitigation and Adaptation Strategies for Global Change*, *21*, 1263–1287.

Allen, T., Prosperi, P., Cogill, B., & Flichman, G. (2014). Agricultural biodiversity, social-ecological systems and sustainable diets. *Proceedings of the Nutrition Society*, *73*(4), 498–508.

Alrøe, H. F., & Noe, E. (2016). Sustainability assessment and complementarity. *Ecology and Society*, *21*(1), 30.

Altieri, M. (2011). Modern Agriculture: Ecological impacts and the possibilities for truly sustainable farming. *Agroecology in Action*. https://parrottlab.uga.edu/Tropag/CR2010/Pre-trip%20readings/Modern%20Ag%20&%20Sustainablity--Altieri.pdf

Altieri, M. A., & Nicholls, C. I. (2017). The adaptation and mitigation potential of traditional agriculture in a changing climate. *Climatic Change*, *140*, 33–45.

Altieri, M. A., Nicholls, C. I., Henao, A., & Lana, M. A. (2015). Agroecology and the design of climate change-resilient farming systems. *Agronomy for Sustainable Development*, *35*(3), 869–890.

Altieri, M. A., Nicholls, C. I., & Montalba, R. (2017). Technological approaches to sustainable agriculture at a crossroads: An agroecological perspective. *Sustainability*, *9*(3), 349.

Asante, W. A., Acheampong, E., Kyereh, E., & Kyereh, B. (2017). Farmers' perspectives on climate change manifestations in smallholder cocoa farms and shifts in cropping systems in the forest-savannah transitional zone of Ghana. *Land Use Policy*, *66*, 374–381.

Bathaei, A., & Štreimikienė, D. (2023). A systematic review of agricultural sustainability indicators. *Agriculture*, *13*, 241. https://doi.org/10.3390/agriculture13020241

Bardy, R., Rubens, A., & Massaro, M. (2015). The systemic dimension of sustainable development in developing countries. *Journal of Organisational Transformation & Social Change*, *12*(1), 22–41.

Barnes, A. P., Soto, I., Eory, V., Beck, B., Balafoutis, A. T., Sanchez, B., Vangeyte, J., Fountas, S., van der Wal, T., & Gómez-Barbero, M. (2019). Influencing incentives for precision agricultural technologies within European arable farming systems. *Environmental Science & Policy*, *93*, 66–74. https://doi.org/10.1016/j.envsci.2018.12.014

Bastan, M., Delshad Sisi, S., Nikoonezhad, Z., & Ahmadvand, A. (2016). *Sustainable development analysis of agriculture using system dynamics approach*. Paper presented at the The 34th International Conference of the System Dynamics Society, Delft, Netherlands.

Bathaei, A., & Štreimikienė, D. (2023). A systematic review of agricultural sustainability indicators. *Agriculture*, *13*(2), 241.

Beck & Wilms. (2004). Sustainable development is a powerful global contradiction to the contemporary western culture and lifestyle. Conversations with Ulrich Beck; Cambridge: Polity Press.

Bedoussac, L., Journet, E.-P., Hauggaard-Nielsen, H., Naudin, C., Corre-Hellou, G., Jensen, E. S., Prieur, L., & Justes, E. (2015). Ecological principles underlying the increase of productivity achieved by cereal-grain legume intercrops in organic farming. A review. *Agronomy for Sustainable Development*, *35*, 911–935.

Bell, S., & Morse, S. (2012). *Sustainability indicators: Measuring the immeasurable?* London: Routledge.

Ben Ayed, R., & Hanana, M. (2021). Artificial intelligence to improve the food and agriculture sector. *Journal of Food Quality*, *2021*, 1–7.

Biardeau, L., Crebbin-Coates, R., Keerati, R., Litke, S., & Rodríguez, H. (2016). *Soil health and carbon sequestration in US croplands: A policy analysis*. United States Department of Agriculture and the Berkeley Food Institute. Available at: https://food.berkeley.edu/wp-content/uploads/2016/05/GSPPCarbon_03052016_FINAL.pdf

Blaydes, L., & Kayser, M. A. (2011). Counting calories: Democracy and distribution in the developing world. *International Studies Quarterly*, 55(4), 887–908.

Blickensdörfer, L., Schwieder, M., Pflugmacher, D., Nendel, C., Erasmi, S., & Hostert, P. (2022). Mapping of crop types and crop sequences with combined time series of Sentinel-1, Sentinel-2 and Landsat 8 data for Germany. *Remote Sensing of Environment*, 269, 112831.

Bockstaller, C., Feschet, P., & Angevin, F. (2015). Issues in evaluating sustainability of farming systems with indicators. *OCL Oléagineux Corps Gras Lipides*, 1(22), 1–12.

Borowy, I. (2013). *Defining sustainable development for our common future: A history of the World Commission on Environment and Development (Brundtland Commission)*. Routledge.

Boyer, J. (2020). Toward an evolutionary and sustainability perspective of the innovation ecosystem: Revisiting the panarchy model. *Sustainability*, 12(8), 3232.

Bozdağ, A., Yavuz, F., & Günay, A. S. (2016). AHP and GIS based land suitability analysis for Cihanbeyli (Turkey) county. *Environmental Earth Sciences*, 75, 1–15.

Bravo-Monroy, L., Potts, S. G., & Tzanopoulos, J. (2016). Drivers influencing farmer decisions for adopting organic or conventional coffee management practices. *Food Policy*, 58, 49–61.

Brussaard, L., Caron, P., Campbell, B., Lipper, L., Mainka, S., Rabbinge, R., Babin, D., & Pulleman, M. (2010). Reconciling biodiversity conservation and food security: Scientific challenges for a new agriculture. *Current Opinion in Environmental Sustainability*, 2(1–2), 34–42.

Byerlee, D., Stevenson, J., & Villoria, N. (2014). Does intensification slow crop land expansion or encourage deforestation? *Global Food Security*, 3(2), 92–98.

Carayannis, E. G., & Rakhmatullin, R. (2014). The quadruple/quintuple innovation helixes and smart specialisation strategies for sustainable and inclusive growth in Europe and beyond. *Journal of the Knowledge Economy*, 5, 212–239.

Cario, G., Casavola, A., Gjanci, P., Lupia, M., Petrioli, C., & Spaccini, D. (2017). *Long lasting underwater wireless sensors network for water quality monitoring in fish farms*. Paper presented at the Oceans 2017, Aberdeen.

Chan, A. W., Hon, K., Leung, T., Ho, M. H., Sou Da Rosa Duque, J., & Lee, T. (2018). The effects of global warming on allergic diseases. *Hong Kong Medical Journal*. 24(3), 277–284. https://doi.org/10.12809/hkmj177046.

Chen, C.-C., & Lin, Y.-C. (2018). What drives live-stream usage intention? The perspectives of flow, entertainment, social interaction, and endorsement. *Telematics and Informatics*, 35(1), 293–303.

Cheney, G., Santa Cruz, I., Peredo, A. M., & Nazareno, E. (2014). Worker cooperatives as an organizational alternative: Challenges, achievements and promise in business governance and ownership. *Organization*, 21(5), 591–603.

Clerino, P., Fargue-Lelièvre, A. & Meynard, J. M. (2023). Stakeholder's practices for the sustainability assessment of professional urban agriculture reveal numerous original criteria and indicators. *Agronomy for Sustainable Development*. 43, 3. https://doi.org/10.1007/s13593-022-00849-6

Coquil, X., Cerf, M., Auricoste, C., Joannon, A., Barcellini, F., Cayre, P., Chizallet, M., Dedieu, B., Hostiou, N., Hellec, F., & Lusson, J. M. (2018). Questioning the work of farmers, advisors, teachers and researchers in agro-ecological transition. A review. *Agronomy for Sustainable Development*, 38, 1–12.

Costanza, R., Cumberland, J. H., Daly, H., Goodland, R., Norgaard, R. B., Kubiszewski, I., & Franco, C. (2014). *An introduction to ecological economics*. London: CRC Press.

Coteur, I., Wustenberghs, H., Debruyne, L., Lauwers, L., & Marchand, F. (2020). How do current sustainability assessment tools support farmers' strategic decision making? *Ecological Indicators*, *114*, 106298.

D'Odorico, P., Davis, K. F., Rosa, L., Carr, J. A., Chiarelli, D., Dell'Angelo, J., Gephart, J., MacDonald, G. K., Seekell, D. A., Suweis, S., & Rulli, M. C. (2018). The global food-energy-water nexus. *Reviews of Geophysics*, *56*(3), 456–531.

Daghagh Yazd, S., Wheeler, S. A., & Zuo, A. (2019). Key risk factors affecting farmers' mental health: A systematic review. *International Journal of Environmental Research and Public Health*, *16*(23), 4849.

Dantsis, T., Douma, C., Giourga, C., Loumou, A., & Polychronaki, E. A. (2010). A methodological approach to assess and compare the sustainability level of agricultural plant production systems. *Ecological Indicators*, *10*(2), 256–263.

Dabkiene, V., Balezentis, T., & Streimikiene, D. (2021). Development of agri-environmental footprint indicator using the FADN data: Tracking development of sustainable agriculturaldevelopment in Eastern Europe. *Sustainable Production and Consumption*, *27*, 2121–2133.

De Bon, H., Parrot, L., & Moustier, P. (2010). Sustainable urban agriculture in developing countries. A review. *Agronomy for Sustainable Development*, *30*, 21–32.

De Clercq, D., & Brieger, S. A. (2021). When discrimination is worse, autonomy is key: How women entrepreneurs leverage job autonomy resources to find work-life balance. *Journal of Business Ethics*, 1–18.

De Olde, E. M., Moller, H., Marchand, F., McDowell, R., MacLeod, C., Sautier, M., Halloy, S., Barber, A., Benge, J., Bockstaller, C., & Bokkers, E. A. (2017). When experts disagree: The need to rethink indicator selection for assessing sustainability of agriculture. *Environmental, Development and. Sustainability.* *19*, 1327–1342.

Deichmann, U., Goyal, A., & Mishra, D. (2016). Will digital technologies transform agriculture in developing countries? *Agricultural Economics*, *47*(S1), 21–33.

Dempsey, N., Bramley, G., Power, S., & Brown, C. (2011). The social dimension of sustainable development: Defining urban social sustainability. *Sustainable Development*, *19*(5), 289–300.

Desiderio, E., García-Herrero, L., Hall, D., Segrè, A., & Vittuari, M. (2022). Social sustainability tools and indicators for the food supply chain: A systematic literature review. *Sustainable Production and Consumption*, *30*, 527–540.

Devi, R. M. (2023). Toward smart agriculture for climate change adaptation. In Chaitanya B. Pande, Kanak N. Moharir, Sudhir Kumar Singh, Quoc Bao Pham, & Ahmed Elbeltagi (Eds.), *Climate change impacts on natural resources, ecosystems and agricultural systems* (pp. 469–482). Cham: Springer.

Dhakal, B., & Kattel, R. R. (2019). Effects of global changes on ecosystems services of multiple natural resources in mountain agricultural landscapes. *Science of the Total Environment*, *676*, 665–682. https://doi.org/10.1016/j.scitotenv.2019.04.276

Di Bene, C., Dolores Gomez-Lopez, M., Francaviglia, R., Farina, R., Blasi, E., Martínez-Granados, D., & Calatrava, J. (2022). Barriers and opportunities for sustainable farming practices and crop diversification strategies in Mediterranean cereal-based systems. *Frontiers in Environmental Science*, *10*, 861225.

Di Maio, F., & Rem, P. C. (2015). A robust indicator for promoting circular economy through recycling. *Journal of Environmental Protection*, *6*(10), 1095.

Díaz Bonilla, E., Saini, E., Henry, G., Creamer, B., & Trigo, E. (2014). *Global strategic trends and agricultural research and development in Latin America and the Caribbean: A framework for analysis*. Cali, DC: CIAT Publication.

Díaz, S., Demissew, S., Carabias, J., Joly, C., Lonsdale, M., Ash, N., Larigauderie, A., Adhikari, J. R., Arico, S., Báldi, A., & Bartuska, A. (2015). The IPBES conceptual framework-connecting nature and people. *Current Opinion in Environmental Sustainability*, *14*, 1–16.

Dillon, E. J., Hennessy, T., & Hynes, S. (2010). Assessing the sustainability of Irish agriculture. *International Journal of Agricultural Sustainability*, 8(3), 131–147.
Dora, M., Biswas, S., Choudhary, S., Nayak, R., & Irani, Z. (2021). A system-wide interdisciplinary conceptual framework for food loss and waste mitigation strategies in the supply chain. *Industrial Marketing Management*, 93, 492–508.
Duesberg, S., O'Connor, D., & Dhubháin, Á. N. (2013). To plant or not to plant-Irish farmers' goals and values with regard to afforestation. *Land Use Policy*, 32, 155–164.
Duran, D. C., Gogan, L. M., Artene, A., & Duran, V. (2015). The components of sustainable development-a possible approach. *Procedia Economics and Finance*, 26, 806–811.
Ecer, F. (2021). Sustainability assessment of existing onshore wind plants in the context of triple bottom line: A best-worst method (BWM) based MCDM framework. *Environmental Science and Pollution Research*, 28(16), 19677–19693.
Einarsson, R. (2020). *Agricultural nutrient budgets in Europe: Data, methods, and indicators*. Gothenburg: Chalmers Tekniska Hogskola.
Eizenberg, E., & Jabareen, Y. (2017). Social sustainability: A new conceptual framework. *Sustainability*, 9(1), 68.
El Bilali, H., & Allahyari, M. S. (2018). Transition towards sustainability in agriculture and food systems: Role of information and communication technologies. *Information Processing in Agriculture*, 5(4), 456–464.
El Bilali, H., Bottalico, F., Ottomano Palmisano, G., & Capone, R. (2020). *Information and communication technologies for smart and sustainable agriculture*. Paper presented at the 30th Scientific-Experts Conference of Agriculture and Food Industry: Answers for Forthcoming Challenges in Modern Agriculture, Sarajavo, Serbia.
Elawama, A. (2016). *The influence of natural and human factors on the sustainability of agriculture in azzawia Libya*. Penang: Universiti Sains Malaysia Penang.
El-Ramady, H. R., Alshaal, T., Amer, M., Domokos-Szabolcsy, É., Elhawat, N., Prokisch, J., & Fári, M. (2014). Soil quality and plant nutrition. *Sustainable Agriculture Reviews*, 14, 345–447.
ESCAP & Scientific (2015). Integrating the three dimensions of sustainable development: A framework and tools. Available at: https://www.unescap.org/sites/default/files/Integrating%20the%20three%20dimensions%20of%20sustainable%20development%20A%20framework.pdf
European Commission (2023). EU datasheets [Data set]. Eurostat. https://ec.europa.eu/eurostat/databrowser/explore/all/agric?lang=en&subtheme=agric&display=list&sort=category
EUROSTAT (2023). Agri-Environmnetal Indicators. [Data set]. https://ec.europa.eu/eurostat/web/agriculture/agri-environmental-indicators
Ferdushi, K. F., Ismail, M. T., & Kamil, A. A. (2019). Perceptions, knowledge and adaptation about climate change: A study on farmers of Haor areas after a flash flood in Bangladesh. *Climate*, 7(7), 85.
French, E., Kaplan, I., Iyer-Pascuzzi, A., Nakatsu, C. H., & Enders, L. (2021). Emerging strategies for precision microbiome management in diverse agroecosystems. *Nature Plants*, 7(3), 256–267.
Fu Jia, L., Dong, S. C., Li, F. (2012). A system dynamics model for analyzing the eco-agriculture system with policy recommendations. *Ecological Modelling*, 227, 34–45. https://doi.org/10.1016/j.ecolmodel.2011.12.005.
Gaba, S., Lescourret, F., Boudsocq, S., Enjalbert, J., Hinsinger, P., Journet, E.-P., Malézieux, E., & Pelzer, E. (2015). Multiple cropping systems as drivers for providing multiple ecosystem services: From concepts to design. *Agronomy for Sustainable Development*, 35, 607–623.

Gan, Y., Liang, C., Chai, Q., Lemke, R. L., Campbell, C. A., & Zentner, R. P. (2014). Improving farming practices reduces the carbon footprint of spring wheat production. *Nature Communications, 5*(1), 5012.

Garibaldi, L. A., Gemmill-Herren, B., D'Annolfo, R., Graeub, B. E., Cunningham, S. A., & Breeze, T. D. (2017). Farming approaches for greater biodiversity, livelihoods, and food security. *Trends in Ecology & Evolution, 32*(1), 68–80.

Gobarah, M. E., Tawfik, M., Thalooth, A., & Housini, E. A. E. (2015). Water conservation practices in agriculture to cope with water scarcity. *International Journal of Water Resources and Arid Environments, 4*(1), 20–29.

Gómez-Limón, J. A., & Sanchez-Fernandez, G. (2010). Empirical evaluation of agricultural sustainability using composite indicators. *Ecological Economics, 69*(5), 1062–1075.

Gomiero, T. (2018). Agriculture and degrowth: State of the art and assessment of organic and biotech-based agriculture from a degrowth perspective. *Journal of Cleaner Production, 197,* 1823–1839.

Goss, M., Swain, D., Abatzoglou, J., Sarhadi, A., Kolden, C., Williams, A., & Diffenbaugh, N. (2020). Climate change is increasing the risk of extreme autumn wildfire conditions across California. *Environmental Research Letters, 15,* 094016. https://doi.org/10.1088/1748-9326/ab83a7

Govindan, K. (2018). Sustainable consumption and production in the food supply chain: A conceptual framework. *International Journal of Production Economics, 195,* 419–431.

Greiner, R. (2015). Motivations and attitudes influence farmers' willingness to participate in biodiversity conservation contracts. *Agricultural Systems, 137,* 154–165.

Gruskin, S., Ferguson, L., Alfven, T., Rugg, D., & Peersman, G. (2013). Identifying structural barriers to an effective HIV response: Using the National Composite Policy Index data to evaluate the human rights, legal and policy environment. *Journal of the International AIDS Society, 16*(1), 18000.

Hansson, H., Ferguson, R., Olofsson, C., & Rantamäki-Lahtinen, L. (2013). Farmers' motives for diversifying their farm business – The influence of family. *Journal of Rural Studies, 32,* 240–250.

Haque, M. H., Sarker, S., Islam, M. S., Islam, M. A., Karim, M. R., Kayesh, M. E. H., Shiddiky, M. J., & Anwer, M. S. (2020). Sustainable antibiotic-free broiler meat production: Current trends, challenges, and possibilities in a developing country perspective. *Biology, 9*(11), 411.

Harris, K. (2014). Bread and freedom: Linking democracy and food security in Sub-Saharan Africa. *African Studies Quarterly, 15*(1), 13–35.

Harwood, R. R. (1990). A history of sustainable agriculture. In: C. A. Edwards et al. (Eds.), *Sustainable agricultural systems* (pp. 3–19). Ankery, IA: Soil and Water Conservation Society.

Hayden, J., Rocker, S., Phillips, H., Heins, B., Smith, A., & Delate, K. (2018). The importance of social support and communities of practice: Farmer perceptions of the challenges and opportunities of integrated crop-livestock systems on organically managed farms in the northern US. *Sustainability, 10*(12), 4606.

Headey, D., & Ecker, O. (2013). Rethinking the measurement of food security: From first principles to best practice. *Food Security, 5,* 327–343.

Heeb, L., Jenner, E. J., & Cock, M. J. W. (2019). Climate-smart pest management: Building resilience of farms and landscapes to changing pest threats. *Journal of Pest Science, 92.* https://doi.org/10.1007/s10340-019-01083-y

Hicks, C. C., Levine, A., Agrawal, A., Basurto, X., Breslow, S. J., Carothers, C., Charnley, S., Coulthard, S., Dolsak, N., Donatuto, J., & Garcia-Quijano, C. (2016). Engage key social concepts for sustainability. *Science, 352*(6281), 38–40.

Hildén, M., Jokinen, P., & Aakkula, J. (2012). The sustainability of agriculture in a northern industrialized country-From controlling nature to rural development. *Sustainability, 4*(12), 3387–3403.

Himeur, Y., Rimal, B., Tiwary, A., & Amira, A. (2022). Using artificial intelligence and data fusion for environmental monitoring: A review and future perspectives. *Information Fusion, 86,* 44–75.

Ikerd, J. E. (1993). The need for a system approach to sustainable agriculture. *Agriculture, Ecosystems & Environment, 46*(1–4), 147–160.

Ikram, M., Ferasso, M., Sroufe, R., & Zhang, Q. (2021). Assessing green technology indicators for cleaner production and sustainable investments in a developing country context. *Journal of Cleaner Production, 322,* 129090.

Ingram, J. (2018). Agricultural transition: Niche and regime knowledge systems' boundary dynamics. *Environmental Innovation and Societal Transitions, 26,* 117–135.

Iocola, I., Angevin, F., Bockstaller, C., Catarino, R., Curran, M., Messéan, A., Schader, C., Stilmant, D., Van Stappen, F., Vanhove, P., & Ahnemann, H. (2020). An actor-oriented multi-criteria assessment framework to support a transition towards sustainable agricultural systems based on crop diversification. *Sustainability, 12,* 5434.

IUCN, UNDP & WWF (1991). *International Union for conservation of nature and natural resources.* Gland: United Nations Environmental Programme & World Wildlife Fund for Nature.

Janker, J., Mann, S., & Rist, S. (2019). Social sustainability in agriculture – A system-based framework. *Journal of Rural Studies, 65,* 32–42.

Johns, T., Powell, B., Maundu, P., & Eyzaguirre, P. B. (2013). Agricultural biodiversity as a link between traditional food systems and contemporary development, social integrity and ecological health. *Journal of the Science of Food and Agriculture, 93*(14), 3433–3442.

Jost, C., Kyazze, F., Naab, J., Neelormi, S., Kinyangi, J., Zougmore, R., Aggarwal, P., Bhatta, G., Chaudhury, M., Tapio-Bistrom, M. L., & Nelson, S. (2015). Understanding gender dimensions of agriculture and climate change in smallholder farming communities. *Climate and Development, 8,* 133–144. https://doi.org/10.1080/17565529.2015.1050978

Juhasz, M. (2014). *Agri-environmental management in Southern Ontario: Enhanced program participation through better understanding of dairy farmers' social dynamics.* Doctoral dissertation, University of Guelph, Guelph, ON.

Kalinowska, B., Bórawski, P., Bełdycka-Bórawska, A., Klepacki, B., Perkowska, A., & Rokicki, T. (2022). Sustainable development of agriculture in member states of the European Union. *Sustainability, 14*(7), 4184.

Keenan, R. J. (2015). Climate change impacts and adaptation in forest management: A review. *Annals of Forest Science, 72,* 145–167.

Kernecker, M., Seufert, V., & Chapman, M. (2021). Farmer-centered ecological intensification: Using innovation characteristics to identify barriers and opportunities for a transition of agroecosystems towards sustainability. *Agricultural Systems, 191,* 103142.

Khan, M. N., & Mohammad, F. (2014). Eutrophication: Challenges and solutions. *Eutrophication: Causes, Consequences and Control. 2,* 1–15.

Khanna, M., Atallah, S. S., Kar, S., Sharma, B., Wu, L., Yu, C., Chowdhary, G., Soman, C., & Guan, K. (2022). Digital transformation for a sustainable agriculture in the United States: Opportunities and challenges. *Agricultural Economics, 53*(6), 924–937.

Khatri-Chhetri, A., Pant, A., Aggarwal, P. K., Vasireddy, V. V., & Yadav, A. (2019). Stakeholders prioritization of climate-smart agriculture interventions: Evaluation of a framework. *Agricultural Systems, 174,* 23–31.

Kopnina, H. (2014). Revisiting education for sustainable development (ESD): Examining anthropocentric bias through the transition of environmental education to ESD. *Sustainable Development, 22*(2), 73–83.

Kreitzman, M., Chapman, M., Keeley, K. O., & Chan, K. M. (2022). Local knowledge and relational values of Midwestern woody perennial polyculture farmers can inform tree-crop policies. *People and Nature*, *4*(1), 180–200.

Kuhlicke, C., Steinführer, A., Begg, C., Bianchizza, C., Bründl, M., Buchecker, M., De Marchi, B., Tarditti, M. D. M., Höppner, C., Komac, B., & Lemkow, L. (2011). Perspectives on social capacity building for natural hazards: Outlining an emerging field of research and practice in Europe. *Environmental Science & Policy*, *14*(7), 804–814.

Läpple, D. (2013). Comparing attitudes and characteristics of organic, former organic and conventional farmers: Evidence from Ireland. *Renewable Agriculture and Food Systems*, *28*(4), 329–337.

Latruffe, L., Diazabakana, A., Bockstaller, C., Desjeux, Y., Finn, J., Kelly, E., Ryan, M., & Uthes, S. (2016). Measurement of sustainability in agriculture: A review of indicators. *Studies in Agricultural Economics*, *118*(3), 123–130.

Lawson, T., & Blatt, M. R. (2014). Stomatal size, speed, and responsiveness impact on photosynthesis and water use efficiency. *Plant Physiology*, *164*(4), 1556–1570.

Lebacq, T., Baret, P. V., & Stilmant, D. (2013). Sustainability indicators for livestock farming. A review. *Agronomy for Sustainable Development*, *33*, 311–327.

Lee, D. R. (2005). Agricultural sustainability and technology adoption: Issues and policies for developing countries. *American Journal of Agricultural Economics*, *87*(5), 1325–1334.

Li, F. J., Dong, S. C., & Li, F. (2012). A system dynamics model for analyzing the eco-agriculture system with policy recommendations. *Ecological Modelling*, *227*, 34–45.

Lipper, L., McCarthy, N., Zilberman, D., Asfaw, S., & Branca, G. (2017). *Climate smart agriculture: Building resilience to climate change*. Berlin: Springer Nature.

Lipper, L., Thornton, P., Campbell, B. M., Baedeker, T., Braimoh, A., Bwalya, M., Caron, P., Cattaneo, A., Garrity, D., Henry, K., & Hottle, R. (2014). Climate-smart agriculture for food security. *Nature Climate Change*, *4*, 1068–1072. https://doi.org/10.1038/nclimate2437

Littig, B., & Griessler, E. (2005). Social sustainability: A catchword between political pragmatism and social theory. *International Journal of Sustainable Development*, *8*(1–2), 65–79.

Liu, J. (2019). China's renewable energy law and policy: A critical review. *Renewable and Sustainable Energy Reviews*, *99*, 212–219.

Longoni, A., & Cagliano, R. (2015). Environmental and social sustainability priorities: Their integration in operations strategies. *International Journal of Operations & Production Management*, *35*(2), 216–245.

Lowenberg-DeBoer, J., & Erickson, B. (2019). Setting the record straight on precision agriculture adoption. *Agronomy Journal*, *111*(4), 1552–1569.

Lyulyov, O., Pimonenko, T., Stoyanets, N., & Letunovska, N. (2019). Sustainable development of agricultural sector: Democratic profile impact among developing countries. *Research in World Economy*, *10*(4), 97–105.

Maddikunta, P. K. R., Hakak, S., Alazab, M., Bhattacharya, S., Gadekallu, T. R., Khan, W. Z., & Pham, Q.-V. (2021). Unmanned aerial vehicles in smart agriculture: Applications, requirements, and challenges. *IEEE Sensors Journal*, *21*(16), 17608–17619.

Magee, C. A., Miller, L. M., & Heaven, P. C. (2013). Personality trait change and life satisfaction in adults: The roles of age and hedonic balance. *Personality and Individual Differences*, *55*(6), 694–698.

Magis, K. (2010). Community resilience: An indicator of social sustainability. *Society and Natural Resources*, *23*(5), 401–416.

Magrach, A., & Sanz, M. J. (2020). Environmental and social consequences of the increase in the demand for 'superfoods' world-wide. *People and Nature*, *2*(2), 267–278.

Makate, C., Wang, R., Makate, M., & Mango, N. (2016). Crop diversification and livelihoods of smallholder farmers in Zimbabwe: Adaptive management for environmental change. *SpringerPlus*, *5*, 1–18.

Malhi, G. S., Kaur, M., & Kaushik, P. (2021). Impact of climate change on agriculture and its mitigation strategies: A review. *Sustainability, 13*(3), 1318.

Mani, V., Agarwal, R., Gunasekaran, A., Papadopoulos, T., Dubey, R., & Childe, S. J. (2016). Social sustainability in the supply chain: Construct development and measurement validation. *Ecological Indicators, 71*, 270–279.

Mantino, F., & Vanni, F. (2018). The role of localized agri-food systems in the provision of environmental and social benefits in peripheral areas: Evidence from two case studies in Italy. *Agriculture, 8*(8), 120.

Marin, C., Dorobantu, R., Codreanu, D., & Mihaela, R. (2012). The fruit of collaboration between local government and private partners in the sustainable development community case study: County Valcea. *Economy Transdisciplinarity Cognition, 15*(2), 93.

Meadows, D. H. (1998). *Indicators and information systems for sustainable development*. Hartland, VT: Sustainability Institute.

Meier, W. N., Hovelsrud, G. K., Van Oort, B. E., Key, J. R., Kovacs, K. M., Michel, C., Haas, C., Granskog, M. A., Gerland, S., Perovich, D. K., & Makshtas, A. (2014). Arctic sea ice in transformation: A review of recent observed changes and impacts on biology and human activity. *Reviews of Geophysics, 52*(3), 185–217.

Meuwissen, M., Feindt, P., Spiegel, A., Termeer, C. J. A. M., Mathijs, E., De Mey, Y., Finger, R., Balmann, A., Wauters, E., Urquhart, J., & Vigani, M. (2019). A framework to assess the resilience of farming systems. *Agricultural Systems, 176*, 102656. https://doi.org/10.1016/j.agsy.2019.102656

Migliorini, P., Galioto, F., Chiorri, M., & Vazzana, C. (2018). An integrated sustainability score based on agro-ecological and socioeconomic indicators. A case study of stockless organic farming in Italy. *Agroecology and Sustainable Food Systems, 42*(8), 859–884.

Mir, M. A., Ghazvinei, P. T., Sulaiman, N., Basri, N., Saheri, S., Mahmood, N., Jahan, A., Begum, R. A., & Aghamohammadi, N. (2016). Application of TOPSIS and VIKOR improved versions in a multi criteria decision analysis to develop an optimized municipal solid waste management model. *Journal of Environmental Management, 166*, 109–115.

Missimer, M., Robèrt, K.-H., & Broman, G. (2017). A strategic approach to social sustainability – Part 1: Exploring the social system. *Journal of Cleaner Production, 140*, 32–41.

Moraine, M., Duru, M., Nicholas, P., Leterme, P., & Therond, O. (2014). Farming system design for innovative crop-livestock integration in Europe. *Animal, 8*(8), 1204–1217.

More, N., Verma, A., Bharagava, R. N., Kharat, A. S., Gautam, R., & Navaratna, D. (2022). Sustainable development in agriculture by revitalization of PGPR. In: R. N. Bharagava, S. Mishra, G. D. Saratale, R. G. Saratale, & L. F. R. Ferreira (Eds.), *Bioremediation: Green approaches for a clean and sustainable environment* (pp. 127–142). Boca Raton, FL: CRC Press.

Morton, S., Pencheon, D., & Squires, N. (2017). Sustainable Development Goals (SDGs), and their implementation: A national global framework for health, development and equity needs a systems approach at every level. *British Medical Bulletin, 124*(1), 81–90.

Mowlds, S. (2020). The EU's Farm to Fork strategy: Missing links for transformation. *Acta Innovations, 36*, 17–30.

Mugambiwa, S. S., & Tirivangasi, H. M. (2017). Climate change: A threat towards achieving 'Sustainable Development Goal number two' (end hunger, achieve food security and improved nutrition and promote sustainable agriculture) in South Africa. *Jàmbá: Journal of Disaster Risk Studies, 9*(1), 1–6.

Muller, A., Schader, C., El-Hage Scialabba, N., Brüggemann, J., Isensee, A., Erb, K.-H., Smith, P., Klocke, P., Leiber, F., Stolze, M., & Niggli, U. (2017). Strategies for feeding the world more sustainably with organic agriculture. *Nature Communications, 8*(1), 1–13.

Munda, G. (2016). Multiple criteria decision analysis and sustainable development. In: S. Greco, M. Ehrgott, & J. Figueira (Eds.), *Multiple criteria decision analysis: State of the art surveys* (pp. 1235–1267). New York: Springer.

Musvoto, C., Nortje, K., De Wet, B., Mahumani, B. K., & Nahman, A. (2015). Imperatives for an agricultural green economy in South Africa. *South African Journal of Science, 111*(1–2), 01–08.

Nachum, L. (1999). The productivity of intangible factors of production: Some measurement issues applied to Swedish management consulting firms. *Journal of Service Research, 2*(2), 123–137.

Nadaraja, D., Lu, C., & Islam, M. M. (2021). The sustainability assessment of plantation agriculture – A systematic review of sustainability indicators. *Sustainable Production and Consumption, 26*, 892–910.

Napitupulu, J., & Hasibuan, D. (2017). Study approach ELimination Et Choix Traduisant la REalite (ELECTRE) for dynamic multi-criteria decision. *International Journal of Scientific Research in Science and Technology, 3*(3), 460–465.

Nnadi, F., Chikaire, J., Echetama, J., Ihenacho, R., Umunnakwe, P., & Utazi, C. (2013). Agricultural insurance: A strategic tool for climate change adaptation in the agricultural sector. *Net Journal of Agricultural Science, 1*(1), 1–9.

Notarnicola, B., Sala, S., Anton, A., McLaren, S. J., Saouter, E., & Sonesson, U. (2017). The role of life cycle assessment in supporting sustainable agri-food systems: A review of the challenges. *Journal of Cleaner Production, 140*, 399–409.

Nunes, A. T., Paivade Lucena, R. F., Ferreira dos Santos, M. V., & Albuquerque, U. P. (2015). Local knowledge about fodder plants in the semi-arid region of Northeastern Brazil. *Journal of Ethnobiology and Ethnomedicine, 11*(1), 1–12.

Ömürbek, N., Okan, D., & Hande, E. (2020). EM algoritmasına göre kümelenen havalimanlarının borda sayım yöntemi ile değerlendirilmesi. *Atatürk Üniversitesi İktisadi ve İdari Bilimler Dergisi, 34*(2), 491–514.

Pani, S. K., Jena, D., & Parida, N. R. (2020). Agricultural sustainability and sustainable agribusiness model: A review on economic and environmental perspective. *International Journal of Modern Agriculture, 9*(4), 875–883.

Patel, S. K., Sharma, A., & Singh, G. S. (2020). Traditional agricultural practices in India: An approach for environmental sustainability and food security. *Energy, Ecology and Environment, 5*, 253–271.

Pearce, D., Atkinson, G., & Hamilton, K. (1998). The measurement of sustainable development. In J. C. J. M. van den Bergh & M. W. Hofkes (Eds.), *Theory and implementation of economic models for sustainable development* (pp. 175–193). Dordrecht: Springer Netherlands.

Petti, L., Münzenrieder, N., Vogt, C., Faber, H., Büthe, L., Cantarella, G., Bottacchi, F., Anthopoulos, T. D., & Tröster, G. (2016). Metal oxide semiconductor thin-film transistors for flexible electronics. *Applied Physics Reviews, 3*(2), 021303.

Phatak, S. C. (1992). An integrated sustainable vegetable production system. *HortScience, 27*(7), 738–741.

Pilgeram, R. (2011). "The only thing that isn't sustainable… is the farmer": Social sustainability and the politics of class among Pacific Northwest farmers engaged in sustainable farming. *Rural Sociology, 76*(3), 375–393.

Polasky, S., Kling, C. L., Levin, S. A., Carpenter, S. R., Daily, G. C., Ehrlich, P. R., Heal, G. M., & Lubchenco, J. (2019). Role of economics in analyzing the environment and sustainable development. *Proceedings of the National Academy of Sciences, 116*(12), 5233–5238.

Popkin, B. M., & Reardon, T. (2018). Obesity and the food system transformation in Latin America. *Obesity Reviews, 19*(8), 1028–1064.

Pramanik, M. K. (2016). Site suitability analysis for agricultural land use of Darjeeling district using AHP and GIS techniques. *Modeling Earth Systems and Environment, 2*, 1–22.

Prasad, R., Bhattacharyya, A., & Nguyen, Q. D. (2017). Nanotechnology in sustainable agriculture: Recent developments, challenges, and perspectives. *Frontiers in Microbiology, 8*, 1014.

Qaim, M. (2020). Role of new plant breeding technologies for food security and sustainable agricultural development. *Applied Economic Perspectives and Policy, 42*, 129–150. https://doi.org/10.1002/aepp.13044

Qin, G., Niu, Z., Yu, J., Li, Z., Ma, J., & Xiang, P. (2021). Soil heavy metal pollution and food safety in China: Effects, sources and removing technology. *Chemosphere, 267*, 129205.

Quisumbing, A. R., Rubin, D., Manfre, C., Waithanji, E., Van den Bold, M., Olney, D., Johnson, N., & Meinzen-Dick, R. (2015). Gender, assets, and market-oriented agriculture: Learning from high-value crop and livestock projects in Africa and Asia. *Agriculture and Human Values, 32*, 705–725.

Radeny, M., Desalegn, A., Mubiru, D., Kyazze, F., Mahoo, H., Recha, J., Kimeli, P., & Solomon, D. (2019). Indigenous knowledge for seasonal weather and climate forecasting across East Africa. *Climatic Change, 156*, 509–526.

Raimi, M. O., Vivien, O. T., & Oluwatoyin, O. A. (2021). Creating the healthiest nation: Climate change and environmental health impacts in Nigeria: A narrative review. *Scholink Sustainability in Environment, 6*, 61–122.

Rakhra, M., Soniya, P., Tanwar, D., Singh, P., Bordoloi, D., Agarwal, P., Takkar, S., Jairath, K., & Verma, N. (2021). WITHDRAWN: Crop price prediction using random forest and decision tree regression – A review. Elsevier.

Ramcilovic-Suominen, S., & Pülzl, H. (2018). Sustainable development – A 'selling point' of the emerging EU bioeconomy policy framework? *Journal of Cleaner Production, 172*, 4170–4180.

Rashid, C. A. (2021). The efficiency of financial ratios analysis to evaluate company's profitability. *Journal of Global Economics and Business, 2*(4), 119–132.

Raven, P. H., & Wagner, D. L. (2021). Agricultural intensification and climate change are rapidly decreasing insect biodiversity. *Proceedings of the National Academy of Sciences, 118*(2), e2002548117.

Redlich, S. (2020). *Opportunities and obstacles of ecological intensification: Biological pest control in arable cropping systems*. Doctoral dissertation, Universität Würzburg, Würzburg.

Reed, M., & Stringer, L. C. (2015). *Climate change and desertification: Anticipating, assessing & adapting to future change in drylands*. Montpellier: Agropolis International.

Reed, M. S., Ferré, M., Martin-Ortega, J., Blanche, R., Lawford-Rolfe, R., Dallimer, M., & Holden, J. (2021). Evaluating impact from research: A methodological framework. *Research Policy, 50*(4), 104147.

Regan, C. M., Bryan, B. A., Connor, J. D., Meyer, W. S., Ostendorf, B., Zhu, Z., & Bao, C. (2015). Real options analysis for land use management: Methods, application, and implications for policy. *Journal of Environmental Management, 161*, 144–152.

Reganold, J. P., Papendick, R. I., & Parr, J. F. (1990). Sustainable agriculture. *Scientific American, 262*(6), 112–121.

Reganold, J. P., & Wachter, J. M. (2016). Organic agriculture in the twenty-first century. *Nature Plants, 2*(2), 1–8.

Reig-Martínez, E., Gómez-Limón, J. A., & Picazo-Tadeo, A. J. (2011). Ranking farms with a composite indicator of sustainability. *Agricultural Economics, 42*(5), 561–575.

Riccaboni, A., Neri, E., Trovarelli, F., & Pulselli, R. M. (2021). Sustainability-oriented research and innovation in 'Farm to Fork' value chains. *Current Opinion in Food Science, 42*, 102–112.

Richard, A., Casagrande, M., Jeuffroy, M.-H., & David, C. (2020). A farmer-oriented method for co-designing groundwater-friendly farm management. *Agronomy for Sustainable Development, 40*, 1–11.

Rigby, D., Woodhouse, P., Young, T., & Burton, M. (2001). Constructing a farm level indicator of sustainable agricultural practice. *Ecological Economics, 39*(3), 463–478.

Rosset, P. M., & Altieri, M. A. (1997). Agroecology versus input substitution: A fundamental contradiction of sustainable agriculture. *Society & Natural Resources, 10*(3), 283–295.

Rossignoli, D., & Balestri, S. (2018). Food security and democracy: Do inclusive institutions matter? *Canadian Journal of Development Studies/Revue canadienne d'*études du développement, *39*(2), 215–233. https://doi.org/10.1080/02255189.2017.1382335

Rozman, Č., Pažek, K., Prišenk, J., Škraba, A., & Kljajić, M. (2012). System dynamics model for policy scenarios of organic farming development. *Organizacija, 45*(5), 212–218.

Ruggerio, C. A. (2021). Sustainability and sustainable development: A review of principles and definitions. *Science of the Total Environment, 786*, 147481.

Ruzzante, S., Labarta, R., & Bilton, A. (2021). Adoption of agricultural technology in the developing world: A meta-analysis of the empirical literature. *World Development, 146*, 105599.

Sachs, J. D., Schmidt-Traub, G., Mazzucato, M., Messner, D., Nakicenovic, N., & Rockström, J. (2019). Six transformations to achieve the sustainable development goals. *Nature Sustainability, 2*(9), 805–814.

Sadok, W., Angevin, F., Bergez, J.-E., Bockstaller, C., Colomb, B., Guichard, L., Reau, R., & Doré, T. (2008). Ex ante assessment of the sustainability of alternative cropping systems: Implications for using multi-criteria decision-aid methods. A review. *Agronomy for Sustainable Development, 28*, 163–174.

Saiz-Rubio, V., & Rovira-Más, F. (2020). From smart farming towards agriculture 5.0: A review on crop data management. *Agronomy, 10*(2), 207.

Sanderson, M. A., Archer, D., Hendrickson, J., Kronberg, S., Liebig, M., Nichols, K., Schmer, M., Tanaka, D., & Aguilar, J. (2013). Diversification and ecosystem services for conservation agriculture: Outcomes from pastures and integrated crop-livestock systems. *Renewable Agriculture and Food Systems, 28*(2), 129–144.

Santeramo, F. G. (2015). On the composite indicators for food security: Decisions matter! *Food Reviews International, 31*(1), 63–73.

Schröder, P., Lemille, A., & Desmond, P. (2020). Making the circular economy work for human development. *Resources, Conservation and Recycling, 156*, 104686.

Scott, N. R., Chen, H., & Cui, H. (2018). Nanotechnology applications and implications of agrochemicals toward sustainable agriculture and food systems. *Nanomaterials, 66*, 6451–6456.

Seman, N. A. A., Govindan, K., Mardani, A., Zakuan, N., Saman, M. Z. M., Hooker, R. E., & Ozkul, S. (2019). The mediating effect of green innovation on the relationship between green supply chain management and environmental performance. *Journal of Cleaner Production, 229*, 115–127.

Senderowicz, L. (2020). Contraceptive autonomy: Conceptions and measurement of a novel family planning indicator. *Studies in Family Planning, 51*(2), 161–176.

Serebryakova, N.A., Dorokhova, N.V., & Isaenko, M.I. (2017). Formation of the system of clustering as a means of perspective development of innovational infrastructure of region. In: E.G. Popkova, V.E. Sukhova, A.F. Rogachev, Y.G. Tyurina, O.A. Boris, & V.N. Parakhina (Eds.), *Integration and Clustering for Sustainable Economic Growth. Contributions to Economics*. Cham: Springer. https://doi.org/10.1007/978-3-319-45462-7_15

Seufert, V., & Ramankutty, N. (2017). Many shades of gray-The context-dependent performance of organic agriculture. *Science Advances, 3*(3), e1602638.

Seufert, V., Ramankutty, N., & Mayerhofer, T. (2017). What is this thing called organic?-How organic farming is codified in regulations. *Food Policy*, *68*, 10–20.

Shah, F., & Wu, W. (2019). Soil and crop management strategies to ensure higher crop productivity within sustainable environments. *Sustainability*, *11*(5), 1485.

Sharifi, A. (2020). A typology of smart city assessment tools and indicator sets. *Sustainable Cities and Society*, *53*, 101936.

Sharifi, A., & Murayama, A. (2013). A critical review of seven selected neighborhood sustainability assessment tools. *Environmental Impact Assessment Review*, *38*, 73–87.

Sheng, Y., Tian, X., Qiao, W., & Peng, C. (2020). Measuring agricultural total factor productivity in China: Pattern and drivers over the period of 1978-2016. *Australian Journal of Agricultural and Resource Economics*, *64*(1), 82–103.

Silveira, A., Ferrão, J., Muñoz-Rojas, J., Pinto-Correia, T., Guimarães, M., & Schmidt, L. (2018). The sustainability of agricultural intensification in the early 21st century: Insights from the olive oil production in Alentejo (Southern Portugal). In A. Delicado, N. Domingos, & L. de Sousa, (Eds.), *Changing Societies: Legacies and Challenges. Vol. 3. The Diverse Worlds of Sustainability* (pp. 247–275). Lisbon: Imprensa de Ciências Sociais.

Singh, R., & Singh, G. S. (2017). Traditional agriculture: A climate-smart approach for sustainable food production. *Energy, Ecology and Environment*, *2*(5), 296–316. https://doi.org/10.1007/s40974-017-0074-7

Singh, R. P., Handa, R., & Manchanda, G. (2021). Nanoparticles in sustainable agriculture: An emerging opportunity. *Journal of Controlled Release*, *329*, 1234–1248.

Slámová, M., & Belčáková, I. (2019). The role of small farm activities for the sustainable management of agricultural landscapes: Case studies from Europe. *Sustainability*, *11*(21), 5966.

Stagnari, F., Maggio, A., Galieni, A., & Pisante, M. (2017). Multiple benefits of legumes for agriculture sustainability: An overview. *Chemical and Biological Technologies in Agriculture*, *4*(1), 1–13.

Steenwerth, K. L., Hodson, A. K., Bloom, A. J., Carter, M. R., Cattaneo, A., Chartres, C. J., Hatfield, J. L., Henry, K., Hopmans, J. W., Horwath, W. R., & Jenkins, B. M. (2014). Climate-smart agriculture global research agenda: Scientific basis for action. *Agriculture & Food Security*, *3*(1), 1–39.

Sterling, S. (2010). *Sustainability education: Perspectives and practice across higher education*. Philadelphia, PA: Taylor & Francis.

Stringer, L. C., Fraser, E. D., Harris, D., Lyon, C., Pereira, L., Ward, C. F., & Simelton, E. (2020). Adaptation and development pathways for different types of farmers. *Environmental Science & Policy*, *104*, 174–189.

Streimikiene, D., & Mikalauskiene, A. (2023). Assessment of agricultural sustainability: Case study of Baltic States, *Contemporary Economics*, *17*(2), 128–141.

Sulewski, P., Kłoczko-Gajewska, A., & Sroka, W. (2018). Relations between agri-environmental, economic and social dimensions of farms' sustainability. *Sustainability*, *10*(12), 4629.

Suresh, H., Gomez, S. R., Nam, K. K., & Satyanarayan, A. (2021). *Beyond expertise and roles: A framework to characterize the stakeholders of interpretable machine learning and their needs*. Paper presented at the Proceedings of the 2021 CHI Virtual Conference on Human Factors in Computing Systems. Available at https://dl.acm.org/doi/10.1145/3411764.3445088

Swinnen, J., & Vandevelde, S. (2018). The political economy of food security and sustainability. *Encyclopedia of Food Security and Sustainability*, *1*, 9–16

Tal, A. (2018). Making conventional agriculture environmentally friendly: Moving beyond the glorification of organic agriculture and the demonization of conventional agriculture. *Sustainability*, *10*(4), 1078.

Teng, C.-C., & Wang, Y.-M. (2015). Decisional factors driving organic food consumption: Generation of consumer purchase intentions. *British Food Journal*, *117*(3), 1066–1081.

Terrier, L., & Marfaing, B. (2015). Using social norms and commitment to promote pro-environmental behavior among hotel guests. *Journal of Environmental Psychology*, *44*, 10–15.

Tesfaye, W., & Seifu, L. (2016). Climate change perception and choice of adaptation strategies: Empirical evidence from smallholder farmers in east Ethiopia. *International Journal of Climate Change Strategies and Management*, *8*, 253–270.

Testa, F., Annunziata, E., Iraldo, F., & Frey, M. (2016). Drawbacks and opportunities of green public procurement: An effective tool for sustainable production. *Journal of Cleaner Production*, *112*, 1893–1900.

Thomson, A. M., Ramsey, S., Barnes, E., Basso, B., Eve, M., Gennet, S., Grassini, P., Kliethermes, B., Matlock, M., McClellen, E., & Spevak, E. (2017). Science in the supply chain: Collaboration opportunities for advancing sustainable agriculture in the United States. *Agricultural & Environmental Letters*, *2*(1), 170015.

Tikkinen-Piri, C., Rohunen, A., & Markkula, J. (2018). EU General Data Protection Regulation: Changes and implications for personal data collecting companies. *Computer Law & Security Review*, *34*(1), 134–153.

Tomislav, K. (2018). The concept of sustainable development: From its beginning to the contemporary issues. *Zagreb International Review of Economics & Business*, *21*(1), 67–94.

Tomlinson, I. (2013). Doubling food production to feed the 9 billion: A critical perspective on a key discourse of food security in the UK. *Journal of Rural Studies*, *29*, 81–90.

Too, E. G., & Weaver, P. (2014). The management of project management: A conceptual framework for project governance. *International Journal of Project Management*, *32*(8), 1382–1394.

Torres-Delgado, A., & Saarinen, J. (2014). Using indicators to assess sustainable tourism development: A review. *Tourism Geographies*, *16*(1), 31–47.

Tubiello, F., Salvatore, M., Rossi, S., Ferrara, A., Fitton, N., & Smith, P. (2013). The FAOSTAT database of greenhouse gas emissions from agriculture. *Environmental Research Letters*, *8*, 015009. https://doi.org/10.1088/1748-9326/8/1/015009

Valdes-Vasquez, R., & Klotz, L. E. (2013). Social sustainability considerations during planning and design: Framework of processes for construction projects. *Journal of Construction Engineering and Management*, *139*(1), 80–89.

Valentine, J., Clifton-Brown, J., Hastings, A., Robson, P., Allison, G., & Smith, P. (2012). Food vs. fuel: The use of land for lignocellulosic 'next generation' energy crops that minimize competition with primary food production. *Gcb Bioenergy*, *4*(1), 1–19.

Vallance, S., Perkins, H. C., & Dixon, J. E. (2011). What is social sustainability? A clarification of concepts. *Geoforum*, *42*(3), 342–348.

Van Calker, K., Berentsen, P., De Boer, I., Giesen, G., & Huirne, R. (2007). Modelling worker physical health and societal sustainability at farm level: An application to conventional and organic dairy farming. *Agricultural Systems*, *94*(2), 205–219.

Van Cauwenbergh, N., Biala, K., Bielders, C., Brouckaert, V., Franchois, L., Cidad, V. G., Hermy, M., Mathijs, E., Muys, B., Reijnders, J., & Sauvenier, X. (2007). SAFE – A hierarchical framework for assessing the sustainability of agricultural systems. *Agriculture, Ecosystems & Environment*, *120*(2–4), 229–242.

Vander-Merwe, I., & Van-der-Merwe, J. (1999). Sustainable development at the local level: An introduction to local agenda 21. Pretoria: Department of environmental affairs and tourism. *Duran, CD, Gogan, LM, Artene, A. & Duran*, 2015, 806–811.

Vare, P., & Scott, W. (2007). Learning for a change: Exploring the relationship between education and sustainable development. *Journal of Education for Sustainable Development*, *1*(2), 191–198.

Velten, S., Leventon, J., Jager, N., & Newig, J. (2015). What is sustainable agriculture? A systematic review. *Sustainability*, *7*(6), 7833–7865.

Vifell, Å., & Soneryd, L. (2012). Organizing matters: How 'the social dimension' gets lost in sustainability projects. *Sustainable Development*, *20*, 18–27. https://doi.org/10.1002/sd.461

Vinuesa, R., Azizpour, H., Leite, I., Balaam, M., Dignum, V., Domisch, S., Felländer, A., Langhans, S. D., Tegmark, M., & Fuso Nerini, F. (2020). The role of artificial intelligence in achieving the Sustainable Development Goals. *Nature Communications*, *11*(1), 1–10.

Walter, A., Finger, R., Huber, R., & Buchmann, N. (2017). Smart farming is key to developing sustainable agriculture. *Proceedings of the National Academy of Sciences*, *114*(24), 6148–6150.

Wang, N., Zhu, H., Guo, Y., & Peng, C. (2018). The heterogeneous effect of democracy, political globalization, and urbanization on PM2. 5 concentrations in G20 countries: Evidence from panel quantile regression. *Journal of Cleaner Production*, *194*, 54–68.

Ward, P. S., Bell, A. R., Droppelmann, K., & Benton, T. G. (2018). Early adoption of conservation agriculture practices: Understanding partial compliance in programs with multiple adoption decisions. *Land Use Policy*, *70*, 27–37.

Wasieleski, D., Waddock, S., Fort, T., & Guimarães-Costa, N. (2021). Natural sciences, management theory, and system transformation for sustainability. *Business & Society*, *60*(1), 7–25.

Weingaertner, C., & Moberg, Å. (2014). Exploring social sustainability: Learning from perspectives on urban development and companies and products. *Sustainable Development*, *22*(2), 122–133.

Wezel, A., Brives, H., Casagrande, M., Clement, C., Dufour, A., & Vandenbroucke, P. (2016). Agroecology territories: Places for sustainable agricultural and food systems and biodiversity conservation. *Agroecology and Sustainable Food Systems*, *40*(2), 132–144.

Wingeyer, A. B., Amado, T. J., Pérez-Bidegain, M., Studdert, G. A., Perdomo Varela, C. H., Garcia, F. O., & Karlen, D. L. (2015). Soil quality impacts of current South American agricultural practices. *Sustainability*, *7*(2), 2213–2242.

Withers, P. J., Neal, C., Jarvie, H. P., & Doody, D. G. (2014). Agriculture and eutrophication: Where do we go from here? *Sustainability*, *6*(9), 5853–5875.

Wolsko, C., Ariceaga, H., & Seiden, J. (2016). Red, white, and blue enough to be green: Effects of moral framing on climate change attitudes and conservation behaviors. *Journal of Experimental Social Psychology*, *65*, 7–19.

Yu, S., & Mu, Y. (2022). Sustainable agricultural development assessment: A comprehensive review and bibliometric analysis. *Sustainability*, *14*, 11824. https://doi.org/10.3390/su141911824

Youngberg, G., & DeMuth, S. P. (2013). Organic agriculture in the United States: A 30-year retrospective. *Renewable Agriculture and Food Systems*, *28*(4), 294–328.

Zamani, R., Ali, A. M. A., & Roozbahani, A. (2020). Evaluation of adaptation scenarios for climate change impacts on agricultural water allocation using fuzzy MCDM methods. *Water Resources Management*, *34*, 1093–1110.

Zhang, Y., Min, Q., Li, H., He, L., Zhang, C., & Yang, L. (2017). A conservation approach of globally important agricultural heritage systems (GIAHS): Improving traditional agricultural patterns and promoting scale-production. *Sustainability*, *9*(2), 295.

4 Use of Renewables in Agriculture

Dalia Streimikiene

4.1 RENEWABLE ENERGY PRODUCTION IN AGRICULTURE

Agricultural production relies upon the natural process of photosynthesis, which converts sunlight into plant proteins with the help of external energy inputs. Traditionally, manual labor and animal power have been used to provide such energy inputs, and the inefficient combustion of biomass has been used for cooking and hot water preparation. Additional energy contributions are required for storing, processing, manufacturing, transporting, and distribution of food products via entire agri-food supply chains. Fossil fuels were applied as these traditional energy inputs as the agriculture sector has become increasingly industrialized and agricultural and food production turned into more energy-intensive processes. Consequently, up-to-date energy services, essential for the entire agri-food chain and its related businesses, have become primarily dependent on fossil fuel suppliers. Such services include heating, ventilation, transportation, water processing and irrigation, animal comfort, mechanical power, and lighting. Power is a vital energy carrier in various farm activities and food processing facilities. Electricity is necessary to produce fertilizers, pesticides, machinery, and construction materials used in agriculture. Over two-thirds of the world's total power generation depends on fossil fuels such as heavy fuel oil, natural gas, and coal. The remaining one-third of the entire world's power generation is fairly evenly split between nuclear and renewable energy systems, primarily hydropower plants.

Due to inadequate electricity grid and road infrastructure in rural areas, many small, remote rural communities still lack access to modern, reliable energy services. In rural areas, problems are linked to frequent power supply disruptions and fluctuating power quality. In remote locations, diesel generators were frequently used to generate power, and more recently, renewable energy technologies such as micro-hydro, geothermal, wind, and solar power systems have been developed in rural communities. Farms and local enterprises can use power generated by renewable energy sources (RES) for food production, storage, handling, and processing (IEA, 2011).

Due to high transportation costs, liquefied fuels purchased in remote rural areas are relatively expensive. Consequently, there can be greater incentives to develop local renewable energy resources for small and medium-sized businesses that process food locally. However, for the use of renewables, skilled labor is necessary for the effective and secure operation of renewable energy plants, as well as for repairs and maintenance. Consequently, capacity building is often essential for the lasting success of renewable energy deployment in rural areas (Gorjian et al., 2022).

In conjunction with energy conservation techniques, farmers can generate energy from renewables to become even more self-sufficient by reducing their reliance on external energy suppliers. Renewable energy not only helps farmers save money but also mitigates climate change (Banos et al., 2011). Biomass, geothermal, hydro, solar, and wind energy can produce power for heating, lighting, and mechanical energy use on a farm. Numerous farmers started cultivating maize for ethanol production. Therefore, many farmers and landowners can increase their incomes by converting wind, solar, or biomass into power. Moreover, new renewable energy alternatives are emerging due to the reduction of advanced renewable energy technologies costs (Frey & Linke, 2002).

Therefore, agriculture and renewable energy can provide a successful combination, especially for remote rural areas, by ensuring the security of the energy supply, reducing energy supply dependency, climate change mitigation, and saving expenses of farmers used to pay for external energy inputs. In addition, renewable energy can provide producers with a permanent source of revenue and substitute other income resources (Rosa & Gabrielli, 2023a).

Bioenergy is most obviously linked to rural areas as the basic resources used to produce bioenergy are extracted almost exclusively from rural areas. Cultivating and obtaining agricultural and forest biomass provides more and more prospects for diversification of farm revenue, creation of new jobs, and new business opportunities in remote areas. RESs, like solar, wind, and hydro, can produce energy for free with minimal initial investment. Before constructing a renewable energy system on the property, ensure it complies with local zoning regulations. Also, renewable energy technologies on the farm can be connected to the power grid and generate additional income for the farmer for electricity generated by renewables at farms (Rosa & Gabrielli, 2023a).

Biomass refers to the use of plant and animal resources to generate energy. Biomass, such as carbohydrates and lipids derived from plants, can produce biofuel and biodiesel for cars, and biomass combustion used to generate heat or power is known as bio-power. Biomass is obtained from vegetation and organic detritus, including cereals, trees, crop residues, and manure. Similar to sustenance crops, energy crops can be cultivated in vast quantities (IEA Bioenergy, 2023). The most popular energy crops are: corn and switchgrass, but fast-growing trees like poplar and willow will likely surpass them soon. These crops need less preservation and less effort than typical crops like maize, making them more cost-effective and sustainable (Cardoso et al., 2021).

Crops and biomass residues can be transformed into energy on the farm or offered to enterprises processing fuel for cars and tractors, as well as producing and supplying heat and electricity for households and industries. Biofuels are processed from farm-grown cereals that can be used to power vehicles. Utilizing valued agricultural land to grow energy crops instead of nutrition crops is a source of biofuels-related controversy (Jeswani et al., 2020). To counteract this argument, some land owners and farmers cultivate crops like sunflower, canola that can be used for biodiesel production and after pulverizing the seed for oil extraction, they obtain meal by-products having a higher price on the market than the processed oil (Kumar et al., 2020). Even though oil presses are high-priced, several farmers have discovered that cooperatively purchasing presses with other farmers is effective. Long-term storage of biodiesel can result in the formation of sediments. Additionally, biodiesel performs poorly in frigid conditions. Farmers who power their machinery with biofuel can recycle used vegetable oil obtained as food waste from local eateries (Jeswani et al., 2020).

Biopower is the generation of electricity from biomass-derived steam or biodigester-produced methane. Biodigesters capture and combust the by-products of the microbial degeneration of biomass as manure. Though biodigester technology is relatively straightforward and has been utilized for a long time, new advanced systems have been developed recently. Biomass commodities like switchgrass, maize, and fast-growing trees can be incinerated for heating greenhouses or transformed to power through steam. There are agricultural by-products like maize that generate energy without reducing soil quality or accelerating erosion. The pelletizers are quite expensive on the markets, and the cooperative purchase of a pelletizer can help farmers who desire to use pellets as a heat source. Power generation from biomass for public use typically necessitates a large-scale system owned by utilities, and farmers can sell them biomass products. Transportation costs usually impede large-scale production of power from biomass, but small-scale power generation on farms is possible (Kumar et al., 2002).

While it is considered that geothermal pools and naturally heated water are the sources of geothermal energy, a much more common source is the constant ground temperature. Agricultural and residential structures can apply geothermal heat exchangers to interchange air and earth temperatures throughout the year, keeping facilities calm in the summer and heating in the winter. Even though geothermal systems and heat pumps are more costly than conventional fossil fuel-based furnaces, the payback period is quite short due to the free geothermal energy. Geothermal systems require extensive excavation works and thus fit better for new constructions.

In order to install a micro hydro power plant on a farm, a sustainable water source with a continuous water flow is essential. If these natural conditions are available and water debit requirements are satisfied, the farmer can prefer various types of water turbines or waterwheels. Although waterwheels have lower efficiency than all types of water turbines and are best suited for propelling mills, they are straightforward to operate. They can accommodate a wide variety of water flows. At the same time, water turbines necessitate an exact and continuous water flow rate and an increased head or elevation difference from the water source to the blades of the turbines. Micro hydropower plants have long payback periods; consequently, it is necessary to plan accordingly. The larger the hydropower plant is, the cheaper each kilowatt-hour to produce (Chel et al., 2009).

Each day, an immense quantity of solar energy reaches the Earth. Solar energy can be utilized in cultivation in numerous modes to save revenues, increase self-sufficiency, decrease air pollution, and mitigate climate change. Solar panels can significantly reduce the electricity and heating costs of a farm. Solar collectors can be used to dry crops and heat livestock buildings, greenhouses, and farmers' residences. Solar water heaters can be used to supply hot water for dairy farms, enclosure washing, and residential use. Photovoltaics (PVs) can be used to generate electricity necessary for farm functions. Also, buildings and outbuildings can be reconstructed to capture natural light. There are many applications of solar energy in rural areas.

Solar energy can be used for heating passive buildings, solar thermal heating for hot water supply, and PVs to generate electricity. Although photovoltaic (PV) systems are more costly; technological advances make them less expensive. PV can be applied for lighting, small motors, circulating water, drying, and charging batteries. PV may be the only option in remote rural areas or sectors of a farm without access to electricity lines (Kwon et al., 2021).

Small producers interested in prolonging the growing season can save money by heating greenhouses with passive solar systems. Passive solar greenhouses have good insulation, apply natural ventilation systems, and have glass panels right-angled to collect the highest amount of solar heat depending on latitude or location. Solar thermal water heating technologies also operate similarly to passive solar systems by accumulating and storing heat. Such solar water heating systems can reduce annual heating expenditures by almost 90% on a dairy farm, using high energy for water heating (Eker, 2005).

Wind energy alone could generate new jobs and huge additional incomes for farmers and rural landowners. Additionally, renewable energy can help reduce pollution, global warming, and reliance on imported fuels. Wind energy is an attractive energy generation alternative for farmers and is a rising source of energy and revenue in rural areas. Farms have used wind energy for water supply, irrigation, and power generation for decades. Developers are establishing large wind turbines on farms to supply electricity to the grid. Wind project developers pay the farmers almost $5,000 per year for each installed wind turbine. Such wind turbines require about 0.2 ha of land, allowing farmers to cultivate crops and raise cattle. Some farmers have invested in wind turbines by themselves, while others are beginning to join wind power cooperatives and share investment costs (IRENA, 2021).

Wind turbines can deliver a significant share of a farm's average electricity requirements; however, they must be situated in high wind speed areas and require at least 0.5 ha of land to establish this technology. Farmers having land in high wind velocities areas could also lease their land for energy utilities to build wind turbines or solar panels while still using the land for their livestock (Reisinger et al., 2021). To operate a grid-connected system, the landowner must have an average annual wind speed of at least 10 miles per hour. A residential wind energy system costs between $13,000 and $40,000 to install, a long-term investment. Nevertheless, the benefits of wind power plants counterweigh the installation of wind technologies costs for farmers if electricity costs are high in the country or infrastructure needs to be well-developed in remote rural areas and electricity supply security is under concern (Sampson et al., 2020).

Therefore, the wind parks can be a valuable addition to a farm PV system as wind power is accessible when solar power is unavailable. The wind is usually the strongest in the Fall, Winter, and Spring, as well as at night, and solar panels generate the highest amounts of energy in the Summer and during the day.

The farms can benefit from all available renewable energy technologies ranging from biomass to solar, wind, hydro, and geothermal (Omer, 2008). All these technologies require land for construction and the best conditions for them can be found in rural areas. In addition, the use of biomass for energy generation is one of the best examples of circular economy applications in agriculture (Table 4.1).

TABLE 4.1
The Main RES Technologies in Agriculture

RES Source	RES Technology	Scale
Solar energy	Domestic PV water heaters	Small
Solar energy	PV for water heating	Medium and large scale
Solar energy	PV roof grid-connected systems generating electric energy	Medium and large scale
Solar	Agri-voltaic systems in greenhouses etc.	Medium and small scale
Solar energy	Solar-Powered Agricultural Machinery and Farm Robots	Small scale
Solar energy	PV grid-connected large PP	Large scale
Wind energy	Wind Turbines (grid-connected)	Medium-large scale
Wind energy	Small wind turbines	Small scale
Hydro energy	Hydro power plants in derivation schemes	Medium-small scale
Hydro energy	Hydro power plants in existing water distribution networks	Medium-small scale
Biomass	High-efficiency wood boilers	Small scale
Biomass	CHP plants burning agricultural wastes or energy crops	Medium scale
Biomass	Animal manure CHP plants burning biogas	Small scale
Biomass	Combined heat and power (CHP) High-efficiency lighting	Large scale
Geothermal	Heat pumps	Small scale

Source: Created by authors.

Renewable energy is expected to deliver in three main areas important for agricultural sector development: energy security, climate change mitigation, and job creation. However, there are important trade-offs among these three areas. For example, large biomass power plants can provide big employment opportunities in rural areas but may negatively impact GHG emissions due to land-use change and transportation of feedstock over long distances. Also, RES is, in most cases, a capital-intensive activity, and energy represents a small share of employment in local communities. Small-scale RES installations are mainly the source of labor and equipment from international suppliers, so the impact on job creation in local communities can be insignificant.

The main drivers and barriers to renewable energy deployment in the agriculture sector and rural communities are summarized in Table 4.2.

As one can notice from Table 4.1, some factors are helping or hindering the penetration of renewable energy in achieving its three main goals. Focusing on supportive aspects will be a step forward in putting RES to work better for rural communities.

In the next section, the benefits and opportunities of RES usage in the agricultural sector are discussed.

4.2 RENEWABLE ENERGY USAGE IN AGRICULTURE

By replacing fossil fuels with solid biomass, biogas, or biomethane, it is possible to achieve significant GHG emission reduction in agriculture (Barros et al., 2020). This replacement of fossil fuel in agriculture by biomass-based fuels depends on

TABLE 4.2
The Main Drivers and Barriers of Renewable Energy Deployment in the Agriculture Sector

Drivers	Barriers
High-quality new RES technologies	Low quality of available RES
Expensive fossil fuels	Low-cost conventional energy
Provision of small subsidies for RES by governments	Provision of subsidies for fossil fuel resources
Possibilities to link renewable energy sources to existing agricultural; activities	RES is a standalone sector within the regional economy
Existing energy transport/transmission infrastructure	RES produces stranded energy that cannot be exported
Strong local community support to RES	Significant local opposition like Not in My Back Yard) NYMBY syndrome
Possibilities to integrate RES within a broader energy framework that facilitates dispatch	Inadequate backstop energy for intermittent power generation sources
Mature RES technologies	Novel or infant RES technology
RES relies on local energy inputs that have limited current uses	High opportunity cost of RES projects
RES policy support for producing cheap energy from RES	Excessive focus on job creation absorbs large quantity of public resources that could be better spent connecting RES to the rural economy
RES provides many opportunities for application of circular economy practices in agriculture	Lack of know-how and competencies to install and use RES technologies
Increase security of energy supply due to distributed energy generation based on RES	High bureaucratic burdens for installation of RES technologies

Source: Created by authors.

the available agricultural and bioenergy production systems. Nevertheless, biomass-based energy poses some important socioeconomic and ecological hazards to rural zones. For instance, changes in land use, intensification of forestation, and cultivation of energy crops can decrease biodiversity and cause soil degradation, water stress, and pollution. In addition, the combustion of wood can also provide for an increase in harmful air pollutants, and there are serious concerns and debates about whether wood biomass is truly carbon neutral as based on UNFCCC rules and policies (Pellegrini & Fernández, 2018; Gorjian et al., 2022).

Most of RESs, such as solar, wind, and hydro, can be used for distributed electricity generation, which is very important in urban areas. Solar energy is important not only for growing crops in agriculture but also for power generation and replacing fossil fuels in agriculture (Bordoloi, 2021). Agri-voltaic systems were developed, including PV systems located on agricultural land, and allocate sunlight to vegetation optimally, maximizing food and energy production to the possible extent (Leger et al., 2021; Riaz et al., 2022). The study by Gonocruz et al. (2021) assessed the shading proportions of PV systems deployed over rice crops in Japan, including other reasons impacting rice yields, like fertilization, weather conditions, and solar radiation. The obtained results demonstrated that the highest allowable shading rate boundary for agri-voltaic systems ranges from 27% to 40%, allowing to achieve at least 80% of the rice yield and generate nearly 30% of the power demand which is necessary for rice cultivation then applying agri-voltaic systems in rice crops in Japan. Agri-voltaic systems have proved to decrease water consumption and boost overall productivity, including water use efficiency of various crops (Barron-Gafford et al 2019). AL-agele et al. (2021) investigated the multiple reasons for crop productivity growth in crop fields due to the installation of agri-voltaic systems. It was found that water productivity was higher in inter-row treatments, but total crop yield was diminishing with the increase of shade. The findings of this study suggest that PV systems can increase water productivity even for crops with low shade tolerance.

Trommsdoriff et al. (2021) analyzed four arable and vegetable crops cultivated by using the agri-voltaic system. The performed measurements of land productivity demonstrated that the agri-voltaics increased land productivity by 90%. Taking into account climate change effects, and the growing land scarcity problem, the study indicates that agri-voltaics has a great deal of potential as an environmental and effective technology for addressing the main challenges of the agriculture sector in the current century.

Amjith and Bavanish (2022) analyzed the possibility of wind and biomass-based hybrid agro systems and assessed the most important characteristics impacting the productivity and efficiency of these configurations. The electricity generation from hybrid systems powered by biofuels derived from biomass (bioethanol, biodiesel, biomethane, biohydrogen, and biomethane) and wind power plants was evaluated. The results demonstrated that hybrid energy systems based on biomass and wind turbines are economical and ecologically favorable options for rural areas and allow to ensure electrification necessary for agricultural activities without having access to the grid in remote areas.

The use of renewables for greenhouse heating is a desirable option for farms. Various hybrid renewable energy-based systems were developed for greenhouse cultivation. Greenhouse cultivation is one form of modern agriculture in which plant development is manipulated to produce higher yields of superior quality and quantity. In a study, Kumar et al. (2022) investigated PV systems used for greenhouses. The spirited performance of PV modules provides the necessary power supply and ensures adequate crop cultivation in greenhouses. Using PV modules in greenhouses has increased the product's quantity and quality. The study by Gorjian, Ebadi et al. (2021) outlined the advancements in PV technologies applied to greenhouses. It examined the possibilities of using sun detectors and solar storage modules to solve problems such as shading. Thermal energy storage (TES) options were reviewed as a crucial element to ensure the security of energy supply from solar systems. Applying TES technologies can increase by almost 30% the thermal efficiency of solar greenhouses. Therefore, wind and solar energy systems have numerous financial and ecological advantages in rural areas. Vourdoubas (2020) identified the territory of Crete in Greece as having a concentration of greenhouses and high average annual wind velocities. The wind parks were developed to meet the power needs of greenhouses in Crete, and it was determined that the use of wind energy in Crete, which has high annual power demand, is both cost-effective and ecologically viable.

In a study by Sanchez-Molina et al. (2014) innovative biomass-based heating, carbon storage, and enrichment systems were developed. The dual-mode system utilized low-cost biomass-based pellets (made from tomatoes and peppers and was designed to decrease the requirement for gasoline and associated expenses. The system involved a furnace with carbon dioxide recovery from discharge gases to enrich the greenhouse and supply heat via pelletized wood chips. The findings of this study showed a significant increase in energy efficiency. The biomass was produced from agricultural waste, and the carbon dioxide discharged by the furnace was recycled.

Renewable energy can replace fossil fuels for cooling and heating space in agricultural buildings. Yadav et al. (2022) developed a new greenhouse thermal model that combines a semi-transparent PVT system with a ground-air heat exchanger (GAHE) system. The impact of numerous parameters, such as airflow speed, heat capacity, and air temperature in the greenhouse., was studied for severe winter conditions. The results demonstrated that the GAHE system is only economically viable during sunny hours. The greenhouse temperatures were elevated by 4°, and 0.5 kg/s increased the airflow rate through the GAHE system was obtained.

Faridi et al. (2019) used soil temperature modeling to investigate the feasibility of applying soil-air heat exchangers (SAHE) in agricultural constructions. In this study, the changes in soil temperature at various depths were examined, and the results showed that it is possible to use soil temperature to heat or chill agricultural structures such as greenhouses, etc. Ihoume et al. (2022) studied the effectiveness of heating a greenhouse with solar copper coil heating technologies. In this heating system, water was circulated in a locked cycle on the roof of a greenhouse and used as a heat transfer medium for accumulating heat during the day and to use it during the night for heating the greenhouse. The conducted parametric study yielded data beneficial for optimizing the heating system and estimating the necessary water flow to pass through the copper coil heat exchanger due to atmospheric temperature fluctuations. This advanced heating system was discovered to have just two years payback period, a slight ecological impact, and significant GHG emission reduction compared to conventional heating systems, as the carbon footprint of solar copper coil heating technology is 176 g/day paralleled to 41,000 g/day carbon footprint of traditional heating systems.

Gourdo et al. (2019) created a technology that stores solar energy in the limestone substrate to preheat the greenhouse's air to maintain the optimal plant-growing environment. This system accumulates additional greenhouse heat during the day and dissipates it during the evening. Experiment results indicate that the ambient air temperature in the greenhouse with a gravel substrate is an average of 1.9°C cooler throughout the day and 3°C warmer during the nighttime compared to the conventional greenhouse. Furthermore, this novel system increases tomato harvest by approximately 22%, paralleled to traditional greenhouses.

Guo et al. (2022) proposed an environmentally favorable strategy to recycle used face masks discarded due to the COVID-19 outbreak to build photothermal evaporators and apply them for high-efficiency solar desalination and solar ocean aquaculture technologies. The face masks were made from polyvinyl alcohol, increased the solar efficiency of desalination evaporators to 91.5%, and provided a lasting desalination process. The obtained pure water is conducive to various plant growing, allowing for cultivating agricultural crops on the ocean's surface. The developed strategy to reprocess the face covering for sea-level cultivation and pure water processing is a favorable environmental sustainability resolution providing multiple benefits in the agriculture sector as well. The study by Ling et al. (2018) created a desalination system and tidal energy technologies to utilize the mechanical energy obtained straightforwardly with the reverse osmosis (RO) system via a hydraulic turbine. Comparing the system functioning and water costs of conventional and tidal energy RO technology, it was discovered that the tidal system RO could avoid more than 40% water extraction costs compared to the conventional RO system. Furthermore, as an efficient future water desalination method, the tidal system RO can be even more cost-efficient if the water extraction rate is higher. Elfasakhany (2020) presented and investigated eight renewable biofuel mixtures as energy sources for desalination facilities to solve such global problems as climate change, fossil fuel restrictions, and growing high energy demand.

The results of this study demonstrated that gasoline bio-ethanol and bio-methanol exhibited high heat recuperation and low atmospheric emissions. Regarding heat recovery and air emissions, gasoline-acetone was the optimal blend and, therefore, the most recommended for desalination facilities. Banerjee et al. (2021) discussed biofuel cells, their varieties, and their function in desalination. Banerjee et al. (2021) study defined that a combination of a microbial biofuel cell and electrodialysis can be employed for water desalination and renewable energy generation. In a conventional MCD, effluent substrate, and exo-electrogenic microorganisms are employed for water desalination.

Renewable energy can be efficiently applied for water pumping and irrigation systems, playing an important role in agricultural activities. Work by Khan et al. (2022) showed a feasibility study of deploying off-grid hybrid renewable energy-based water pumping systems for land irrigation in rural Sudan areas. The techno-economic assessments of such hybrid systems were performed in twelve dissimilar urban areas in Sudan. The few hybridization systems comprising the PV, wind, and storage technologies were assessed and paralleled, considering the water requirements of various crops. The obtained results demonstrated that the selection of the hybridized system is heavily influenced by wind speed, solar radiation, and other climate and geographical conditions, including system costs.

The El Hierro island in Spain is planning to become a 100% renewable energy island. In 2018, a wind-hydro power facility that could supply nearly 60% of the annual electricity requirement in has been installed. Melian-Martel et al. (2021) studied the hydrological cycle of El Hierro island, which comprises extraction of groundwater, desalination, and supply and distribution of water. Almost 35% of the island's annual power demand is necessary for water desalination and supply in the island. The objective of the study was to determine whether the entire hydrological cycle could be powered by wind energy alone. The study showed that the use of renewable energy can be increased on the island even more, and this is an excellent example of a 100% renewable energy island. Parmar et al. (2021) studied the choice of portable PV-powered pumps, like centrifugal and piston pumps, based on their efficiency and found that efficiency depends on the volume of water pumped at a given head.

Another way to use renewables in agriculture is the drying of agricultural products. The study by Sethi et al. (2021) investigated numerous solar drying technologies for sea products. Conducted and experimental examination showed that the fast, low-temperature dyeing procedures fit most for preserving the original color of dried marine products such as fish. As marine product dehumidification requires extra time, hybrid solar-powered dryers can outperform usual dryers. Considering health standards, preserved foods with a minimal sodium concentration (approximately 5%) are acceptable. The study confirmed that solar-powered drying systems for marine products have significant potential in certain regions of India, such as coastal regions. Asnaz and Dolcek (2021) examined several low-cost solar dryers, such as natural and forced displacement dryers, heated purge desiccant dryers, and convection dryers. The study applied an experiment on the mushrooms exposed to approximately 790 W/m^2 of average daily solar radiation. Cutting narrow segments decreased the necessary drying period, and the estimated standard thermal efficiency of the natural displacement dryer (NCD), forced displacement dryer (FCD), and heated purge desiccant dryers (HDP) were 77.55%, 67.66%, and 59.74%, respectively.

In order to reduce reliance on fossil fuels usually used for paddy drying, Yahya et al. (2022) devised, deployed, and evaluated an experimental biomass mixed-flow drying technology with a drying capacity of 400 kg/h. Biomass was the only source of energy necessary for heating and dehydrating air in this system. The study assessed the thermal energy efficiency, specific and thermal energy consumption, and evaporation speed of this biomass-based drying system and evaluated the possible enhancement potential. The experiment showed that the paddy's moisture content decreased from 20.90% to 13.30% on a damp basis.

Solar-Powered Agricultural Machinery and Farm Robots have been recently developed to support farm activities based on RESs. A comprehensive study by Gorjian et al. (2021) on PV-powered electric tractors for farming and agricultural robotics showed the diverse applications of solar-based robots in agriculture (Basso & Antle, 2020). It encouraged the integration of solar systems with modern electric agricultural machinery (Pearson et al., 2022). The majority of commercial types of equipment for farming were examined and discussed in this study. In addition, the necessity of transitioning from fossil-based farming machinery to PV-powered electric tractors was deliberated. The main barriers and challenges of commercially viable PV-powered robots in agriculture were discussed in this study as well. The two primary obstacles to the global deployment of PV-powered electric agricultural machinery were defined: high investment cost linked to PV systems and batteries for electricity storage and the effect of air temperature and humidity, shading, and dust gathering on the PV generation by solar panels.

Ghobadpour et al. (2019) identified pollution reduction, advanced control, and high capability as the most important issues to consider when employing PV-powered electric farm equipment. The study proved that PV-based electric machinery could promote the faster penetration of autonomously networked vehicles that improve work quality and motorist convenience. Furthermore, the electrification of agricultural machinery shows that generating renewable energy on-site provides improved energy efficiency and independence and is a forward-looking strategy.

Carroquino et al. (2019) analyzed a case then electricity generated by a PV farm was used to power a trickle irrigation system and other supplementary purposes, and the excess energy was recovered through the electrolysis of water. The excess electricity generated at PV farms produced hydrogen (H_2) on-site. The obtained H_2 was then utilized for hybrid fuel cell electric machinery used for transportation at the farm. Therefore, using RESs is a good example of power and hydrogen generation on-site to satisfy the farm's energy needs.

In an additional study, Gonzalez-de-Soto et al. (2019) modeled a hybrid energy system used for insect control and vegetation enhancement at farms by using PV-powered fuel cell robots and impacting fuel consumption and GHG emissions reduction. The developed hybrid energy system combined the internal combustion engine of a tractor with a novel electric motor powered by a hydrogen fuel cell. According to the Gonzalez-de-Soto et al. (2019) study, the hybrid energy system reduces GHG emissions by approximately 50%. Another survey by Quaglia et al. (2020) examined a new uncrewed ground vehicle for agriculture having 7° of freedom, a robotic limb, and vision sensors for precision. This robot could be placed on a solar cell-based pier platform for drones, allowing PV energy to be stored. This novel device developed for the agricultural sector can play a crucial role in more multifaceted automatic agricultural systems applied for the coordinated observation and preservation of crop fields.

4.3 RENEWABLE ENERGY SOURCES AND CIRCULAR ECONOMY IN AGRICULTURE

Considering the priorities of EU circular economy development, there is significant potential for developing biogas technology that addresses energy and climate change mitigation concerns and promotes a circular economy development in agriculture (Barros et al., 2020). However, the bioenergy industry confronts numerous obstacles, such as limited economic benefits due to the low efficiency of biogas production technologies, the short life of biogas tanks and problems with biogas digester operations, and the high failure rate of these technologies and difficulties of biogas residues usage (Horton et al., 2021).

Several researchers seeking to define the concept of circular agricultural production applied the following terms: Circular Bioeconomy (Zabaniotou, 2018; Mohan et al., 2016). Rural Eco-economy (Kristensen et al., 2016), Agro-industrial Ecology (Fernandez-Mena et al., 2016), Agroecology (Dumont et al., 2013), Industrial Symbiosis (Patricio et al., 2018; Alfaro & Miller, 2014; Zhu et al., 2007; Noya et al., 2017), Closed Loop and Close the loop (Boh & Clark, 2020) etc. These terminologies indicate disciplines that study circular production systems and can be viewed as components of a circular model, which includes inputs, outputs, exchanges with the surrounding, and the return of a portion of the material to the manufacturing process. The main objective is to create a production process in line with a closed circle by reducing waste and emissions into the environment.

The circular economy conception is relatively new, as evidenced by its appearance in the academic literature. Industrial ecology has been in the spotlight for some time. However, the circular economy includes ancient agricultural practices, particularly using animal manure as a crop fertilizer (Barros et al., 2020). The term industrial ecology refers to the concept of circularity, like an analogy with natural ecological systems functioning based on networking and relations. Chertow (2000), one of the most cited researchers in this field, claims that the best explanation of industrial symbiosis is the collaborative and cooperative opportunities provided by geographic closeness. According to the philosophy of industrial ecology, it is necessary to devise workable processes within the agricultural production cycle (Pagotto & Halog, 2016) to raise efficiency and productivity, diminish unwanted outcomes, and reduce non-renewable contributions. According to Trokanas et al. (2014), industrial ecology aims to improve resource efficiency and bring economic, ecological, and social benefits to the involved sectors of the economy, including agriculture.

An industrial symbiosis is an approach encouraging the expansion of a circular economy (Herczeg et al., 2018). Industrial symbiosis allows the definition of strengths and opportunities for stable continuous growth of a specific organization (Grant et al., 2010). Industrial symbiosis is characterized by its ability to contemplate multiple categories of industries, including farms and upstream and downstream counterparts (Fernandez-Mena et al., 2016).

The idea of the circular economy and its practical examples have expanded substantially. Agreeing with Murray et al. (2017), a circular economy is one of the most promising offers to promote economic development while at the same time preventing the depletion of raw materials and addressing energy scarcity issues, and developing new business expansion models. To achieve a zero-waste goal, certain steps must be taken in the agriculture sector, first of all, modifying the main agricultural practices. Nevertheless, 100% recycling and a circular economy are only possible with top-level management's initiatives. In their study, Bluemling and Wang (2018) demonstrated that establishing agricultural cooperatives could effectively facilitate recycling by exchanging materials between livestock farmers, crop farmers, and other cooperative members. In this scenario, the life cycle approach has been applied. The study has pursued various options supporting a circular economy applicable to the life cycle approach of agricultural processes, actions, and products. Such circular economy options can provide multiple ecological and economic benefits. Environmental impacts and waste management have the same growth pattern in the circular economy. Some case studies implemented circular economy-based management principles in the agriculture sector (Toop et al., 2017; Vega-Quezada et al., 2017; Hidalgo et al., 2019). The topic of circularity is closely linked to sustainability. For achieving sustainability, like in the case of a circular economy, reusing natural resources, reducing waste generation, and recycling existing materials is necessary. Consequently, the planet's future and achievement of sustainability in economic development has a solution known as a circular economy. The importance of energy derived from waste (primarily agricultural waste) in following the circular economy principles and closing the production cycle must be emphasized. The circular economy is a common issue in agriculture. However, using renewable energy opened more alternatives and ways to address circular economy principles in agriculture (Barros et al., 2020).

Agriculture has adopted a closed-loop approach to materials and wastes many years ago. The primary aspect relates to animal manure (bovine, swine, and poultry) used as crop fertilizer. However, as the circular economy has advanced and new cycles have been closed, new ways of energy generation from agricultural waste have recently emerged. Biogas is the most advantageous regarding the circular economy principles as new advanced technologies were developed to generate biogas from various organic wastes available in agri-food industries. Using biodigester systems allows producing diverse forms of energy (thermal, mechanical, and electricity). It provides extra value to the agricultural sector, developing a more robust, competitive, and sustainable agriculture sector (Barros et al., 2020).

It is possible to trace the changes and evolutions of the circular economy in the agricultural sector throughout history based on a scientific literature review. The content analysis was founded on a comprehensive circular economy in the agriculture framework. In this book section, findings from various studies were discussed by providing examples and contributions on the topic. Agriculture and circular economy are the two main topics analyzed in this book section. Therefore, these keywords were the most frequently employed by studies analyzed here. The concepts of "industrial ecology" and "ecology" became less used in contemporary research. They were more popular in the middle of 2010. Current studies focus on sustainable agriculture, climate-smart agriculture, and circular agriculture with more emphasis on the use of manure, biomass, biogas, bioenergy, and anaerobic digestion in the context of the circular economy approach (Balafoutis et al., 2020).

There is evidence that numerous studies have developed agricultural circular economy research. In recent years, the most significant advancements and findings in various contexts linked to the circular economy were developed for the use of bioenergy in agriculture. In their study, Grimm and Wosten (2018) developed and presented a simple model for a circular economy based on mushroom substrate applications. Moreover, Ingrao et al. (2018) assessed the life cycle of food waste recovery from the circular economy viewpoint. Zabaniotou (2018) provided a comprehensive assessment of circular economy models in agriculture established on sustainability concerns. It is necessary to stress that the roots of circular economy can be found in industrial ecology (Therond et al., 2017).

Various studies discussed and developed diverse analytical frameworks for municipal agricultural systems. The cascaded biomass utilization process was developed, suitable for the biorefinery design (Egea et al., 2018). Mohan et al. (2016) presented a holistic biorefinery flowchart for versatile applications in agriculture. Kougias and Angelidaki (2018) reviewed the scientific literature on biogas production and developed novel outlooks on European anaerobic digestion processes. Awasthi et al. (2019) examined the prospects and barriers of the latest biorefinery for recovering heat in China, which is relevant for a workable circular economy development. Nielsen (2017) described microbial biotechnology in effluent treatment using circular economy concepts. The main areas of the circular economy studies in agriculture were varied and included specific sectors like mushrooms (Patricio et al., 2018), agri-food and dairy products (Niutanen & Korhonen, 2003), and refining of sugar (Ometto et al., 2007).

Chen et al (2017) conducted an analysis of initial investment requirements and total costs necessary to install a biodigester system in a Chinese swine agribusiness.

Integrating biodigester systems in agriculture is linked to bioenergy generation and use. Using biomass, like animal fat, agro-industrial waste, manure, and culinary waste, to generate biogas is one of the main examples of a circular economy approach in agriculture. In economics, the biodigester can generate electricity used on the farm or sold to the power grid. In addition, the biogas facility generates biomethane that can be used to power vehicles and reduce fossil fuel consumption and GHG emissions. However, the support for biogas facilities that enhance energy generation has put pressure on the food crop markets, resulting in a debate over food versus energy (Balussou et al., 2018). This system is more prevalent in European countries, and Germany holds a position of great influence and power on a global scale in the biogas production field (Gao et al., 2019), tailed by the United Kingdom and Italy (Ramos-Suarez et al., 2019). Several reasons can be highlighted, including low-interest rates and, therefore cheaper initial investment, support by the government to agribusiness cooperatives, high awareness about waste recycling benefits in communities, and the high potential for wastes to provide economic and ecological benefits for rural communities and energy cooperatives.

New biogas engineering solutions and production technologies can also enhance the security of the energy supply in conjunction with the portfolio of other RESs like wind, solar, hydro, geothermal, etc. In contrast to solar energy, which requires the necessary energy storage options, biogas is a comparatively simple option for users. Biogass delivers ecological benefits (Lyytimaki et al., 2018), including power production opportunities, and expands the value chain of agriculture products (Kougias & Angelidaki, 2018).

The study by Vaneeckhaute et al. (2018) provided the sustainability assessment of biogas and biofertilizers by developing a life cycle assessment (LCA) approach for evaluating ecological impact. A questionnaire was conducted on the social prism to assess stakeholders' perceptions regarding biofertilizers. Another study by Mosquera-Losada et al. (2019) appraised the impacts of various bio-waste fertilizers as replacements for mineral fertilizers in agriculture and defined bio-waste fertilizers as a workable option for adopting the circular economy approach in the agricultural sector. Santos and Magrini (2018) developed an agro-industrial symbiosis network in Brazil by utilizing short-term, medium, and long-standing flowcharts within a biorefinery. The study by Ometto et al. (2007) demonstrated that switching from fossil fuels to bioethanol in various agricultural practices, including livestock, and food processing in sugarcane industries, was advantageous in ensuring economic gains, ecological benefits, and better social equity via utilizing the concept of industrial symbiosis.

Several studies also discuss the agri-food aspects in terms of circular principles. The study by Caruso et al. (2019) examined the feasibility of biogas generation from agri-food waste via anaerobic digestion. Research by Niutanen and Korhonen (2003) developed a model for local Finish agri-food industries based on the material and energy flows within the system. Work by Fernandez-Mena et al. (2015) developed nutrients, substances, and materials flowcharts based on the principles of industrial ecology for the French agro-industrial sector. Furthermore, Kristensen et al. (2016) emphasized two cases based on the integration of circular economy principles into the territorial agri-food system. Water utilization plays an important role on rural property, and the establishment of a closed water cycle provides many economic and environmental benefits for rural farms. Some authors (Molina-Moreno et al., 2017; Nielsen, 2017) applied the circular economy principles to the establishment of reusing wastewater facilities in the agricultural sector. Maap and Grundmann (2018) developed a wastewater reuse value chain. The Molina-Moreno et al. (2017) study showed agriculture's main closed-water cycle benefits, like decreased water consumption and increased biogas-generated electricity. Nielsen (2017) examined microorganisms' identity, physiology, and ecology principles to be applied for carbon footprint reduction and effluent recovery optimization due to the growth of bioenergy production and use in the agribusiness sector.

Definite techniques and approaches for soil enrichment, climate change mitigation, and up-to-date work on food security enhancement are required in the agriculture sector. The utilization of biochar in the agro-food sector has been widespread and provided numerous advantages. Some studies have assessed the significance and benefits of biochar usage in soil used for agricultural purposes. The technology for using biochar in agricultural soil allows to achieve of significant GHG emission reduction, but a continuous computer monitoring arrangement is required (Tan, 2019). As a large amount of food waste happens in the global agri-food supply chain, circular economy solutions can help to improve the situation considerably. According to Elkhalifa et al. (2019), it is possible to apply slow pyrolysis in the biochars production process, and the efficiency of this product varies subject to the waste type and parameters of the process and the biomass can be transformed into biofuels, Barbecue (BBQ) charcoal briquettes, and biochars, thereby providing the extra benefits.

There are environmental and socioeconomic benefits associated with the utilization of biochar. According to a study by Chiaramonti and Panoutsou (2019), rural producers can obtain sufficient income to maintain the agricultural land's productivity through biochar, and this approach aligns with the agricultural and energy policies of the European Union. Biochar is a material containing a great deal of carbon (Donner et al., 2020); however, it is moreover rich in oxygen, hydrogen, sulfur, and nitrogen, making it beneficial to enhance the growth of plants and ensuring better harvest of various vegetation (Ferjani et al., 2019). Moreover, this technology reduces the demand for standard chemical fertilizers and delivers a high organic matter and macronutrient index of the agricultural; soil (Kizito et al., 2019). In addition, biochar is able to sequester carbon dioxide over time and reduce greenhouse gas emissions, mitigate soil deprivation, and improve the security of food supply (Amoah-Antwi et al., 2020).

In addition, the studies conducted did not exhibit any negative patterns of biochar usage. There was no identification of research institution or university history or potential trends. Most studies are conducted in European countries, like Italy, Denmark, Finland, Spain, etc. Studies conducted outside Europe can be mainly found in China (Zhu et al., 2007; Bluemling & Wang, 2018). As agricultural contamination in China worsens, various organizations have adopted environmental protection measures in response to pressure from the Chinese government and society (Xu et al., 2018).

A European study (Toop et al., 2017) summarized the findings AgroCycle Project to provide the outlook of the circular economy applications in agriculture. It assessed the technical potential prospects of anaerobic digestion usage at micro-scale local farms. The project received financing from the Horizon 2020 Program for recycling and recycling agro-food waste. The project showed that incorporating the circular economy into the cooperative agro-food system of the European Union (Grandl et al., 2016) is one of the most promising practices necessary for achieving sustainable food production and supply in the EU. The same result was confirmed by the study of Egea et al. (2018) analyzing the agri-food production chain in the Spain region of Almeria. Egea et al. (2018) developed a bioeconomy-based analysis of the agri-food system's performance. To transition from a linear economy to a circular one in the agri-food industry, it is necessary to apply innovative business models, improve reverse logistics, expand the customer-supplier connection, and create a novel marketing strategy (Donner et al., 2020). Therefore, the term "bioeconomy" refers to various actions, processes, and services reliant on bio-resources (Filho, 2018). The circular economy is based on a zero-waste approach, as all natural resources are converted into products with added value and used sustainable way.

A circular economy requires a renewable resource base, sustainable production and consumption practices, and circular flows of all materials (Kumar et al., 2019). It is necessary to stress that Circular Economy and Bioeconomy aim to maximize the value of biological and other wastes. In order to establish a circular economy in the agriculture sector, all stakeholders throughout the agri-food value chain, from product development to waste recycling, must be aware of the credible consequences of using biomass (Wainaina et al., 2020).

The energy-water-food nexus approach indicates a different way. Kilki and Kilki (Chel, 2011) developed a complete application of educational innovations in the energy-water-food nexus, employing circular economy perspectives to the dairy industry. They propose that enhancements to the dairy farms are based on circular economy principles, including implementing of PV panels and wind turbines and utilizing biological waste in anaerobic digestion. So, applications of the circular economy can enhance agricultural production and make it more environmentally friendly and effective. From this viewpoint, the main trend is to diminish the potential negative ecological impacts like GHG emissions in the agriculture sector by applying circular economy principles (Salvador et al., 2019; Morseletto, 2020). According to Alfaro and Miller (2014) study, studies in the circular agriculture field can provide for the expansion of a sustainable network of producers in developing nations as well as allow to address growing environmental and food system pressures.

Numerous studies investigated agricultural processes aiming to improve material efficiency (Barros et al., 2020). Increasing phosphorus efficacy, for example, has been broadly documented as a crucial aspect of implementing circular economy principles in agriculture. Phosphorus recovery is the subject of numerous investigations concentrating on various methods. It is possible to desorb phosphorus from limestone and utilize it as a natural fertilizer for agricultural soils (Lamont et al., 2019). Due to the discharge of plant nutrients, the nitrogen and phosphorus content of agricultural fertilizers can induce negative environmental effects like the eutrophication of rivers, lakes, and other water reserves (Garcia-Nieto et al., 2016). Nonetheless, several advantages are also mentioned. Without phosphorus, plants cannot survive (Garske et al., 2020). Therefore, the environmental control of fertilizers is essential for the food supply (Ouikhalfan et al., 2022). Phosphorus, nitrogen, and potassium are critical materials for agriculture; critical and valuable is rock phosphate, which is a very scarce fossil resource and is imported from Brazil (Gameiro et al., 2019).

Using ash waste accumulated after biomass burning and supplemented with phosphorus can act as a valuable fertilizer and allow the agricultural sector to introduce circular economy principles. Waste of energy generation can be applied for land fertilization (Bekchanov & Mirzabaev, 2018). In addition, as livestock in agriculture quickens nutrient cycles (Gameiro et al., 2019), the circular economy approach must be applied to agriculture's phosphorus and nitrogen management (Papangelou et al., 2020). Though many studies deal with phosphorus-related issues, more comprehensive analyses and approaches are required to evaluate phosphorus's contribution to the circular economy. In recent years, therefore, the studies in this discipline have become more comprehensive. The theoretical and practical contributions to the subject are constantly advancing. In addition, opportunities for phosphorus-related improvements in nutrient cycles of agricultural systems are currently being investigated.

The agricultural sector's material and energy flows are significant to the circular economy. In implementation of a circular economy, principles must find methods to amplify the positive effects and mitigate the negative effects on agricultural systems. Helpful factors may impact the reutilization or conversion of waste from agricultural activities into basic materials. According to this viewpoint, the practices are designed to provide for medium- and long-term outcomes.

In agriculture, inputs and outputs of materials and energy flows indicate various options. Depending on the property, the system is more complicated. On a rural property where pigs are bred, inputs contain, e.g., energy, water, animal food, and micronutrients. Inputs for wheat, maize, and soybean plantations comprise energy and various fuels, water, seeds, fertilizers, and pesticides (Jacquet et al., 2022). There are small agricultural systems like cultivating tomatoes in a rooftop greenhouse) (Piezer et al., 2019) and large agri-food systems have different considerations and valuations of by-products (Springer & Schmitt, 2018). The input–output analysis was developed in various types of agricultural systems, providing different results. Input–output analysis for energy and materials allows us to monitor their flows (Kumar et al., 2019) and are very promising tool for potential research in the circular economy and agriculture, both delivering for sustainable development (Graedel, 2019).

Furthermore, in the past few years, there have been many studies about the significance of generating electricity from renewable sources, such as wind and solar, in agriculture. Furthermore, public sector support for agro cooperatives, awareness about waste recycling and reusing, and the possibilities of financial and environmental advantages from waste generation are necessary. As discussed in previous sections, PV panels and wind turbines comprise renewable energy, providing many advantages for rural areas. With ample open space, wind turbines can be more easily installed on rural properties. Regarding solar energy, the roofs of outbuildings, warehouses, and homes can be a foundation for solar panels. Solar energy is one of the most secure forms of energy globally. However, large-scale renewable energy implementation can harm terrestrial and aquatic systems (Ravi et al., 2016). PV agricultural systems are a new method for providing power for the agricultural sector (Xue, 2017). However, it is necessary to address the challenge of negative environmental impacts on rural areas by utilizing RESs (Bey et al., 2016; Li et al., 2017). According to Sampaio and Gonzalez (2017), it is important to increase awareness of renewable energy usage benefits in rural areas, extend R&D in the field of new technologies, and implement policies and measures that promote renewable energy generation in the agricultural sector. Furthermore, biomass, like wood, energy crops, sugar cane, etc., can produce electricity. The sugar cane can also be used to produce ethanol or biofuel produced from RESs. It is necessary to stress that biomass is the most important energy source in a circular economy regarding material and energy flows. Selecting suitable raw resources is important if the resource supply is insufficient to satisfy demand (Sherwood, 2020; Vieira et al., 2019).

Biogas is a sustainable and efficient source to produce energy from renewables. Using a biodigester system is the main way to the path of sustainable use of energy in agriculture. In addition, farm-scale energy facilities, which produce biomethane and recycle nutrients, may present future viable business opportunities in the agricultural sector (Winquist et al., 2019). This biodigester system provides good options for biogas, biomethane, and fertilizer production (Hamelin et al., 2019) using organic material previously destined for a landfill. This method permits green electricity generation. Besides that, biomethane production can provide fuel for light and heavier vehicles in the agribusiness sector and allows us to avoid GHG emissions from fossil fuels such as gasoline and diesel. Production of biogas and biomethane increases the energy supply's security (Vega-Quezada et al., 2017), reduces negative ecological impacts, and generates additional income for the agriculture sector (Lee, 2017).

Regarding biogas facilities, biodigesters are a crucial component. It is important to stress that biogas has become more economically viable due to the circular economy (Winquist et al., 2019). Opportunities for biogas exist in various sectors, including thermal, electric, and transport. Though economic factors such as initial investment and payback period still need to be fully established for biogas usage in agriculture, it is anticipated that future studies will do so. Only some studies report economic approaches in assessing biogas options based on literature review. However, Techno-Economic Analysis (TEA) is a deciding factor in selecting processes and products (Rana et al., 2020). Ferella et al. (2019) endorsed a TEA approach for evaluating existing biogas plants. They demonstrated that such plants though constructed in the past are still continuing to produce green electricity. In contrast, building new biomethane plants requires improvements in the transport sector to ensure that biomethane can contribute to developing new advanced circular economy models in agriculture. However, the nature and location of the biogas facility can strongly impact the decision to build a biogas plant. The rural producer can benefit greatly from the government's support, the decline in interest rates, and the arrangements to ensure the simplicity of investment in biogas facilities. Kwan et al. (2019) conducted a TEA to assess the recovery of food and beverage waste from sugar syrup manufacture through the integrated biorefinery facility. Another study by Karthikeyan et al. (2017) demonstrated that conventional technologies, like composting and anaerobic digestion, are necessary to ensure the benefits of food waste recycling based on TEA approach.

Creating partnerships and cooperatives between adjacent farms can deliver various ecological and financial benefits to agribusinesses (Lutz et al., 2017). The study by Kassem et al. (2020) also applied a TEA to evaluate the viability of the bioenergy generation and supply system based on a case study of 29 agricultural communities in the US. The established close relationship between farms can facilitate purchasing and selling of products and materials. Strengthening a network of rural farms and comprising cooperatives will make it easier to collect and transport agricultural waste coming from animals, vegetables, and organic matter from various farms to a farm with a biodigester. Therefore, the farm having biodigester can supply biogas to its neighbor farms. The exchange of multiple products and materials between farms can benefit agribusiness. What one organization considers as waste may be a primary material for another. Such circulation of materials and products within the cooperative can create a value-added and decrease the consumption of raw materials.

The networking within the circular economy model shows the best practices for organic waste recycling, which is a win-win opportunity for all because it reduces the negative ecological impacts of open-air disposal, recovers nutrients for agricultural production, enhances soil health, and increases the security of food supply. However, circular economy models in agriculture raise fears regarding their associated effects, such as greenhouse gas emissions, eutrophication, loss of biodiversity, and land degradation and contamination. However, the ecotoxicity effects of circular agriculture models need to be studied more (Svanstrom et al., 2017), In contrast, it is possible to measure and rectify the effects of GHG reduction, particularly in terms of changes in carbon footprint (Fantozzi & Bartocci, 2016; Schau & Fet, 2007).

According to a study by Pagotto and Halog (2016), the increase in efficacy and decrease in pollution are crucial factors in circular economy models. In several studies, life cycle analysis (LCA) was applied for the assessment of ecological impacts associated with the entire agri-food supply value chain, including work in the field, agricultural transportation, processing of agricultural products, agro-industrial supply, and consumer table. Therefore, the LCA is necessary to ensure sustainable, viable agri-food supply chains (Notarnicola et al., 2017). For instance, Noya et al. (2017) assessed the traditional linear pork processing and supply chain using LCA and concurrently compared results with alternative circular economy-based scenarios to show the possible improvements due to the application of circular economy models. The study showed that transportation and efficient use of waste and co-products resulted in the greatest economic and environmental gains. Consequently, one of the primary objectives of bioenergy in agriculture is to reduce GHG emissions and exclude fossil fuel inputs. In addition, bioenergy combined with a circular economy models produces two important outcomes: enlarged material circulation and an increase in energy efficiency. The enteric fermentation of animals (Grandl et al., 2016), agricultural waste, fertilization (Diacono et al., 2019), and feed handling (Mariantonietta et al., 2018) and other agricultural activities generate numerous GHG emissions, such as CO_2, CH_4, and N_2O.

Contrary to GHG emissions, it has been reported that eutrophication impacts are growing with the better circularity of nutrients in agri-food chains (Cobo et al., 2018). This highlights the trade-offs between the benefits and drawbacks of growing circularity in the agricultural sector (Basso et al., 2021). Phosphorus and nitrogen emissions are responsible for eutrophication's negative effects. NOx and ammonia emissions, nitrogen leaching, and runoff have been reported as sources of nitrogen-made harm

(Chojnacka et al., 2020). Furthermore, this affects the eutrophication of both terrestrial and aquatic areas (Aguilar et al., 2019).

Losses of biodiversity are an additional problem that must be taken into account in agricultural activities (Castro et al., 2019). A new branch – agroecology – has emerged as an option for biodiversity conservation Keesstra et al., 2016). A study by Mosquera-Losada et al. (2017) recommends analyzing the agricultural soil before supplementing it with various substances, like sewage sludge as agricultural fertilizer. The study showed that prior land analyses permit determining safe fertilization levels to protect biodiversity.

From a circular economy perspective, land use productivity can be important for reducing negative environmental impacts (Lord & Sakrabani, 2019). More efficient agricultural circularity practices can decrease land use by fostering the recovery of agricultural bioresources. In addition, bioeconomy and circular economy have considered residential gardens as important ways to mitigate climate change as planting citrus trees and cultivating plantations in residential gardens allow carbon sequestration. Local agricultural production can reduce reliance on products associated with distant transportation and reduce transportation-related energy consumption and GHG emissions (Tvinnereim et al., 2017).

Numerous organizations are worried about ecological problems; however, financial benefits are necessary to ensure the competitive advantage of the agricultural sector. The circular economy can generate around $8 trillion in economic potential from 2020 to 2030. The most significant outcome of the circular economy is the reuse of wastes in processing other products, as previously such wastes were only disposed of in landfills and rivers. Now these wastes can substitute non-renewable resources, generate revenues, and reduce negative environmental impacts due to circular economy practices. In addition, a few efforts have been made to evaluate agricultural systems from a broader perspective, considering their sustainability and the social dimension. Diverse approaches to LCA have been utilized for the examination of circular agricultural production systems, such as accounting for losses in agri-food systems (Vilarino et al., 2017) and closing nutrient loops. LCA has also been utilized to evaluate complete plantations (Schiipbach et al., 2020), livestock (de Boer et al., 2011), and crop cultivation (De Luca et al., 2018). Due to the need to take into account potential trade-offs within the environmental-economic-social nexus, sustainability assessments are typically complex and time-consuming approaches, despite their comprehensive nature and reliable results.

Diverse organizations have been recruited to learn and implement circular economy practices in agriculture. Electricity generation from renewables provides for reducing greenhouse gas emissions, but for a greener planet, alternative routes are required. These routes employ circular economy principles in agriculture and other sectors.

Currently, bioenergy-related regions and their proximity are prevalent. The future expansion of biomethane's use in electricity production and the supply of light and heavier vehicles from biogas-purified biomethane is illustrative. Changes and investments in technology are required, but the environmental outcomes are positive.

Regarding the situation of circular economy, therefore, the utilization of agricultural waste is of particular importance. Agricultural waste is an important resource for reducing the use of natural raw materials, producing electricity, reducing agricultural waste, and lowering food supply costs. Implementing cleaner technologies, using renewable energies, and transforming a linear economy into a circular one positively impacts the agriculture sector's competitiveness, create new jobs, generate additional income, and foster further innovations in this important sector. In agriculture, linear practices can be transitioned progressively into circular economy practices based on innovations to seize the market value while promoting the transition from linear to more circular business models. Organizations may continue operating traditional businesses while experimenting with new circular approaches, concluding the transformations then the circular business models are well-established (Barros et al., 2020).

Circular economy in agriculture supports certain United Nations Sustainable Development Goals. In rural areas, the development of a circular economy approaches and the increasing role of local farms in food production can contribute to achieving certain SDGs. The increase of food security and sustainable agro-food system development through circular economy practices can be achieved. The challenge is to double food production by 2030, creating new options for value addition and fostering the creation of new jobs in the field, as well as contributing to food reserves to reduce food price volatility. Regarding the implementation of SDG 7, from a bioeconomy standpoint, agricultural entities can produce power and biodiesel from natural raw materials, in this way contributing to certain objectives, such as considerably increasing renewable sources in the global energy balance and tripling the overall energy efficiency by 2030. Furthermore, actions to reduce climate change and its effects (Goal 13) are incorporated into rural property management practices, such as incorporating management strategies, reducing animal and agricultural waste, and promoting green power and biofuel production initiatives.

Through this literature review, it is possible to determine the most important publications in the field of agricultural circular economy. However, the development of future studies on the subject is further growing. Regarding sustainability, more efforts should be made to primarily support avoiding economic costs, creating new jobs, and generating income. Furthermore, to the circularity indicators analyzed by the Ellen MacArthur Foundation (2013), a few hints on measuring circularity can be found in the scientific literature. Farmers would be interested in a fast and simple indicator for assessing the circular economy in the agricultural sector to identify potential flows within the farm, between neighboring farms, or among agricultural cooperatives. This could lead to an increase in agricultural circular economy practices, influencing other farms. In addition, agricultural refuse could be repurposed, and additional expenses could be evaded.

The principles of circularity could be applied equally to retailers and consumers. At the retailer level, for example, such measures as donation of food surplus, collaborative partnerships waste collection, and creating incentives for customers to participate actively in such waste collection procedures, as well as price setting, to encourage competitiveness between circular products and traditional products. At the customer level, awareness is the most important issue for achieving circularity, creating initiatives for consumers to participate in circular practices and prefer products of more circular origin, and rationing the use of food products to prevent waste generation. Consequently, economic, technical, and social aspects are research priorities for the future. It is also anticipated that new circular business models will be developed to facilitate business management in the agricultural sector. To improve the management of materials and energy that circulate within and outside the entity, the instruments and indicators used to quantify and monitor the circular economy in rural areas require further development.

4.4 RENEWABLES – THE PRIMARY DRIVERS OF AGRICULTURE'S LOW-CARBON TRANSITION

Agriculture is one of the largest contributors to greenhouse gas emissions and, due to its dependence on environmental systems and limitation to altering weather conditions, is one of the industries most vulnerable to climate change impacts (Bataille et al., 2016). Nevertheless, there is no distinct decarbonization agenda or targets of GHG emission set for agriculture in EU. Even with the availability of Common Agriculture Policy (CAP) support, EU countries emphasize adaptation to climate change more than climate change mitigation actions in the agricultural sector.

So there is an immediate need for a longer-term vision outlining agriculture's role in supporting the EU's toward low-carbon transition and creating carbon neutral society by 2050. It is necessary to consider the commitments and openings provided by the EU's climate change mitigation policies and the CAP to guarantee that the appropriate incentives are established to support climate change mitigation actions and capacity development and exchange in know-how and best practices in the decarbonization of the agricultural sector among EU countries. To facilitate a more rapid transition to net-zero emissions by mid-century, this may necessitate the development of new strategies for using CAP funding and from other private investors and financial institutions, considering EU climate change mitigation targets for agriculture and shifting toward more sustainable production and consumption forms (Lehtonen et al., 2022).

OECD (2012) study titled "Linking Renewable Energy to Rural Development" identified the following critical factors for low-carbon transition in agriculture:

- Incorporation of sustainable energy strategies into the local economic expansion plans to reflect local situations and requirements.
- Integration of RESs into larger agri-food supply chains in rural areas, including forestry, animal husbandry, horticulture, food processing and manufacturing, and ecotourism.
- Limitation of the scope and duration of subsidies applied for agriculture by applying grants only to promote renewable energy projects nearing market viability.
- Refraining from imposing the development of RESs on regions not well-adapted to them.
- Concentration on mature renewable energy technologies such as biomass-derived heat, small-scale hydro, solar, and wind.
- Establishing an integrated energy system based on tiny infrastructures supporting manufacturing operations.

- Acknowledging that RESs can compete with other industries for land, water, and other inputs.
- Evaluation of potential ventures based not on short-term subsidy levels but on profitable investment and environmental criteria.
- Ensuring the local social acceptability of renewable energy projects by highlighting the benefits to local communities and by involving them in the process of renewable energy projects development through the establishment of cooperatives.

Agricultural sector is an energy-intensive industry that consumes more than 3% of the world's energy (UNECE, 2022). Agriculture utilizes fossil fuels for numerous reasons, including fertilizers, pesticides, energizing production, and agricultural operations like tillage, harvesting, fertilizing, lighting, heating, refrigeration, water pumping, and irrigation. In addition, it is projected that the future of agriculture will intensify, transforming from small to large-scale farms, which will raise energy consumption and increase energy intensity and dependence on fossil fuel supply and import (Rosa et al., 2021). Therefore, agriculture dependent on RESs is essential for achieving climate change mitigation goals, expanding the food supply system, and ensuring food security. Due to the high dependence on fossil fuels, agriculture practices based on fossil fuels are not sustainable from a long-term development perspective. It is necessary to stress that subsidized fossil fuel-based agricultural systems also dominate in countries having high energy import dependency, and this makes agricultural sector very vulnerable to high energy price fluctuations due to various global energy prices (Rosa, 2022). In addition, rural areas, especially in developing countries, lack energy security due to their limited access to affordable energy for the agricultural sector.

Energy consumption on the farm level includes using fossil fuels to power machinery operations and generate electricity necessary for irrigation and water extraction and supply. Using RESs for the electrification of farms and substituting fossil fuels and conventional power with green electricity can significantly reduce agriculture's carbon footprint (UNECE, 2022). Nevertheless, when reflecting to LCA, the change from traditional to electric machinery in agriculture reduces GHG emissions by more than 70% (Soofi et al., 2022).

In 2000 just 16% of farms were electrified, and currently, almost 30% of farms have access to electricity based on 2020 data (FAOSTAT, 2023). More than 40% of global electricity generation in 2020 was based on low-carbon fuels such as wind, solar, hydro, and nuclear (IEA, 2022). The world average carbon footprint of power generation was approximately about $450\,g\ CO_2$-eq kWh in 2020 (IEA, 2021). The wind and solar lifecycle carbon footprint is between 5 and $100\,g\ CO_2$-eq kWh (UNECE, 2022). Consequently, complete decarbonization of power used on farms will significantly reduce the carbon footprint of power generation.

Power can be produced directly on farms from abundantly available renewable energy options. One can notice that large-scale solar and wind turbines on agricultural land have become widespread, with negligible effects on cereal productivity (Sampson et al., 2020). A similar situation can be applied to adopting agri-voltaic installations, which allow increasing energy use efficiency, agricultural productivity, water savings, and low-carbon power generation (Barron-Gafford et al., 2019). Furthermore, bio-based fuels, such as ethanol, biodiesel, and biogas, are gaining popularity on farms. RESs can replace fossil fuels in conventional combustion engines without modification of the engine. Drop-in fuels can be derived from waste products, non-food crops, and algae (Kumar et al., 2020). These fuels are compatible with diesel and gasoline-powered tractors, harvesters, irrigation devices, and other agricultural machinery.

An increase in agricultural energy efficiency and energy conservation is an important driver of GHG emission reduction in the agriculture sector. Electrification of farms can increase energy use's efficacy at farms (Northrup et al., 2021). Tractors using fossil fuels can be supplanted by lighter electric vehicles, creating new opportunities for robotics and advanced automation technologies (Pearson et al., 2022). Digitalizing the agriculture sector with the help of sensors, robotics, and artificial intelligence systems can increase energy use efficiency and decrease the required energy, fertilizers, and pesticide inputs to the agricultural soils (Basso & Antle, 2020). Compared to current technologies propelled by fossil fuels, the new digitalized technologies require additional initial investment and higher costs; therefore, these are the greatest barrier to the widespread adoption of these alternative solutions in the agricultural sector. Importantly, enhanced energy efficiency can sometimes result in a rebound effect providing for a total energy consumption increase due to lower energy costs (Pellegrini & Fernández, 2018).

Also, other technologies reducing energy consumption are available like conservation tillage, organic agriculture, and crop rotation. They provide for a decrease in energy demand as well as for other inputs such as fertilizers and pesticides (Rosa et al., 2021; Nandan et al., 2021). For instance, soil cultivation is one of the most energy-intensive farm activities. Studies showed that switching to no-tillage systems can provide for a double reduction of energy demand at farms (Pearson et al., 2022). Though the no-tillage technologies' effect on greenhouse gas emission reductions depends on the geographical location of farms and no-till can increase demand for fertilizer and pesticides (Basso et al., 2021). However, reducing tilling, such as perennial cereal crops, would allow to save water and would deliver additional energy savings as well.

Agricultural irrigation is an important source of energy demand in agriculture (Lehtonen et al., 2022). Irrigation is responsible for 90% of the world's freshwater consumption and seizes more than 20% of cultivated land. To adapt to climate change, irrigating more croplands is anticipated in the future (Rosa et al., 2018, 2021). Irrigation allows to mitigation of heat and water stress on crops and reduces the impacts of climate change including air temperature extremes (Rosa, 2022). Irrigation has important environmental impacts like GHG emissions caused by fossil fuel-powered water extraction facilities (Sowby & Dicataldo, 2022). It is necessary to note that groundwater extraction requires almost 80% of the world's energy consumption; therefore, for GHG emission reduction reducing groundwater-based irrigation is crucial for climate change mitigation (Zou et al., 2015).

Generally, irrigation pumping activities contribute annually to almost 15% of global GHG emissions due to energy consumption in agriculture and is a significant GHG emission source. Approximately half of the global irrigation is executed by applying electric pumps, and the other half by employing diesel-powered pumps (Rosa & Gabrielli, 2023a). One can see that electric pumping and low-carbon electricity-based irrigation systems, like solar-powered irrigation, can significantly reduce greenhouse gas emissions from agriculture.

By introducing PV water pumping, energy consumption can be reduced by up to 95% as the carbon footprint and energy consumption of diesel water pumping is 800 g CO_2 per kWh. For solar PV water pumping, it is 45 g CO_2 per kWh (Rosa & Gabrielli, 2023a). New smart water irrigation systems, like soil moisture management systems with sensors, can significantly increase irrigation efficiency and decrease water and energy consumption. By reducing soil transpiration, such agricultural practices are beneficial: no-till farming, mulching, and agri-voltaics as they allow to keep more water in the agricultural soil, thereby diminishing the demand for water for irrigation. There is deficit irrigation, which allows cultivating crops under conditions of moderate water scarcity and can also significantly decrease the consumption of water, energy, and CO_2 emissions. Solar and wind energy can also increase drought resistance in agriculture sector as they need significantly less water to produce energy than fossil fuels (He et al., 2019).

Water management is crucial in water-scarce regions to increase water use efficiency and minimize GHG emissions from water extraction linked to irrigation systems. Though there is high potential for installing water management techniques, the lack of reliable data poses practical obstacles in this area. In this sense, the management of irrigation water plays an important role. Additional research is required to assess the benefits of water management systems and encourage producers to employ larger-scale water management practices in farms. Irrigation also has a negative impact on soil emissions of nitrous oxide and methane. Research showed that nitrous oxide emissions could grow by 50%–140% in irrigated cropping systems compared to non-irrigated ones (Trost et al., 2013). However, drip irrigation can reduce nitrous oxide by 32%–46% compared to furrow or sprinkler irrigation.

Use of fertilizers and pesticides is important for climate change mitigation due to the energy necessary to produce them. Fertilizers and pesticides are very important for increasing global food production. However, the production, transportation, and distribution of mineral fertilizers and pesticides is responsible for approximately for 42% of agricultural energy consumption. Therefore, fertilizers are generally the main contributor to agricultural energy consumption, whereas pesticides have the highest energy intensity of agricultural energy input (Lehtonen et al., 2022).

Using artificial fertilizers increases the yields of various crops by providing nutrients like nitrogen and phosphorus to the soil. The production of fertilizer is very dependent on fossil fuels and energy consumption. Producing 1 kg of nitrogen fertilizers requires almost 70 MJ of energy, and producing 1 kg of phosphate and potash fertilizers requires approximately 8 MJ and 6 MJ, correspondingly (Ouikhalfan et al., 2022). It is clear that GHG emissions due to fertilizer production depend on the type of fuel used. Notably, the energy required to fabricate 1 kg of nitrogen from the air is comparatively low and makes 1 MJ (Smith et al., 2020). However, the production of nitrogen fertilizers (such as urea) necessitates first the production of hydrogen and nitrogen, the conversion of hydrogen and nitrogen into ammonia, and the conversion of ammonia into fertilizers. So, the production of nitrogen fertilizers required to sustain the world's population is very energy-intensive. GHG emissions from nitrogen fertilizer production are higher than from phosphate (45 Mt CO_2-eq per year) and potash (34 Mt CO_2-eq per year) production. With the growth of the world population and rising demands for food and fiber, the demand for synthetic fertilizers is anticipated to increase (Rosa & Gabrieli, 2023).

So it will be essential to decarbonize the fertilizer industry to meet GHG emissions targets without compromising food security and agricultural production (Gao & Cabrera Serrenho, 2023). In phosphorus and potash fertilizer production, GHG emissions result from mining, transportation, and processing, all energy-intensive processes. The chemical and physical transformations necessary to produce these fertilizers are energy intensive. Additional GHG emissions from phosphorus fertilizer production processes are also caused by the heat produced during phosphate rock pretreatment (Ouikhalfan et al., 2022).

Utilizing low-carbon energy sources, like renewables, electrification of farms, and carbon capture and storage technologies, can allow potash and potassium fertilizer production to achieve net-zero emissions. Low-carbon heat can be produced by heat pumps or accumulated waste of heat when low-temperature heat is needed (Soofi et al., 2022). Notwithstanding the accessibility of such technologies able to decarbonize potash and phosphorus fertilizers production, these technologies' GHG emission reduction potential has not yet been quantified.

The ammonia production needed for nitrogen-based fertilizers makes up almost 2% of the world's annual energy consumption. There is high energy and carbon intensity of ammonia production if the Haber–Bosch process is used, which applies hydrogen and nitrogen as raw materials and is responsible for almost 90% of these GHG emissions (Rosa & Gabrieli, 2023).

Hydrogen production requires significant amounts of natural gas or coal. In the past 15 years, increased energy efficiency and reduced carbon intensity have significantly reduced the GHG emission intensity producing nitrogen fertilizers (IEA, 2021). Nonetheless, additional efficiency gains are anticipated to provide important energy and GHG emissions reductions (Lange, 2021). Implementing additional transformative measures in fertilizers production will be required to decarbonize the agricultural sector. Such efforts include the use of fossil fuels just together with carbon capture and storage technologies, water electrolysis powered by carbon-free power, and installation of various biochemical processes. Such advanced technologies can reduce GHG emissions associated with ammonia production by 95%. Reducing emissions by 100%, i.e., zero, would be significant in fertilizers production. Reaching net-zero GHG emissions by mitigating residual GHG emissions with CDR technologies (Rosa & Gabrieli, 2023b). Carbon capture and storage technologies can lower GHG emissions by almost 90%, while the GHG emission intensity of renewable energy technologies, such as solar PVs, exceeds zero based on the LCA approach (UNECE, 2022).

Currently, alternative technologies to the Haber–Bosch process are being developed. These technologies are based on electrochemical synthesis and plasma-activated processes and chemical reactions. Nitrogen-to-ammonia conversion processes are opposed to high-temperature and pressure processes like Haber–Bosch (Rosa & Gabrieli, 2023a). In contrast to the energy-intensive Haber–Bosch process, these technologies offer a fundamentally new method to produce ammonia from nitrogen. Solar-powered ammonia synthesis could pave the way for carbon-neutral fertilizer production and distribution systems and can reduce reliance on energy imports and disruptions of supply chains (Lange, 2021; Leger et al., 2021).

Optimizing and scaling up these fertilizers production technologies will be necessary to know whether they can provide a viable solution for agricultural sector. Biological nitrogen fixation is another potential technology that can decarbonize ammonia production and, thus, nitrogen fertilizer production. This entails nitrogen-fixing microorganisms in the soil converting atmospheric nitrogen into ammonia, which plants can use. Consequently, biological nitrogen fixation can diminish the required nitrogen fertilizer for agriculture. Only legumes such as beans and peas can support soil microorganisms that convert nitrogen to ammonia. Attempts are being made through genetic engineering to apply biological nitrogen fixation in several crops (Yan et al., 2022). Despite ongoing research and development, biological nitrogen fixation is still commercially not viable.

Use of manure as fertilizer can provide for GHG emission growth in agriculture. Manure emits twice as many greenhouse gases per unit of nitrogen as synthetic nitrogen fertilizers due to the methane emissions created during manure storage and transportation processes (Gao Cabrera Serrenho, 2023). The introduction of synthetic nitrogen fertilizers led to a terrestrial separation between crop production and livestock husbandry. This resulted in cultivating crops in the most fertile regions and relocating livestock husbandry to regions with lower land productivity, where the animal started to be imported with increased livestock production. This geographical separation of livestock and crop producers increases transportation costs and hinders the utilization of manure in agricultural production. Now reconnecting livestock and cropping systems through manure has the potential to reduce greenhouse gas emissions as manure can also be used as a source to produce biogas and generate energy (Billen et al., 2021).

Pesticides are used to secure crops from vegetation, insects, and various diseases, while fertilizers are utilized to provide nutrients for plants. The main pesticides are insecticides, herbicides, fungicides, disinfectants, and repellents. The energy intensity of pesticide production is very high, and it can release more GHG emissions per kilogram than synthetic fertilizers. Nevertheless, little attention has been paid to problems linked to the production of pesticides and related GHG emissions. Several studies tried calculating the energy consumption necessary for specific pesticide production and evaluating greenhouse gas emissions. Given production of pesticides requires almost 0.5 EJ per year of energy consumption, and that the average GHG emission factor per EJ is 69 Mt CO_2-eq (Rosa & Gabrieli, 2023a), it is possible to evaluate that GHG emissions from pesticides production make about 34 Mt CO_2-eq per year. In the meantime, decreasing efficacy and the effects of climate change are anticipated to increase the demand for pesticides, causing a vicious cycle between pesticide production which is necessary for food security, and increasing climate change.

Pesticides are based on fossil fuels, and their entire GHG emission lifecycle contributes significantly to climate change. Similar to nitrogen fertilizers, pesticides also cause GHG emissions. Some pesticides have been shown to increase nitrous oxide production in soils after application. The pesticides, like sulfuryl fluoride, are potential GHG with nearly 5,000 times higher global warming potential than CO_2 (Choudhury & Saha, 2020). Therefore, fast electrification of farms based on low-carbon energy carriers and installed CCS technologies, then fossil fuels are used, and biochemical processes can achieve significant GHG emission reductions linked to pesticide production. It is necessary to implement alternative agricultural systems, such as agroecology, which allow to decrease pesticide usage and demand. Robots were developed to precisely administer pesticides, reduce pesticide demand, and increase agricultural productivity. While genetically modified crops are frequently promoted to minimize pesticide use, scientific research indicates the opposite is true as adopting genetically modified crops has resulted in herbicide-resistant plants, requiring producers to use even more herbicides (Jacquet et al., 2022).

Developing net-zero emissions strategies requires precise estimates of GHG emissions from various agricultural activities and technologies. Recent research has quantified GHG emissions from synthetic nitrogen fertilizer production (Rosa & Gabrielli, 2023b). However, more is needed to know about GHG emission reduction strategies in the use of potash and phosphorus fertilizers, pesticides, and various farming practices like harvesting, sowing, threshing, drying, irrigation, lighting, heating, and cooling. Net-zero fertilizer production technologies preclude using fossil fuels as an energy source for electrolytic hydrogen production and biochemical processes. However, these methods have limited scope and require high investment and operational costs compared to established production methods, even though carbon taxes are increasing and putting pressure on fossil fuel consumers. It is necessary to highlight that new technologies for the production of synthetic pesticides and fertilizers have had their commercialization delayed due to the high investment costs related to their development. Currently, only 3% of the world's hydrogen is produced through water electrolysis, and only 0.01% of the world's ammonia is based on low-carbon energy use (Rosa & Gabrieli, 2023b).

Though an annual production capacity of low-carbon ammonia is anticipated to reach of 15 Mt by 2030 (Rosa & Gabrieli, 2023a), decarbonization of on-farm energy use is still a challenge. However, environmental trade-offs may exist regarding energy, land, water, and biomass nexus providing for land and water scarcity. For example, net-zero ammonia production through water electrolysis needs twenty-five times more energy, land, and water resources than traditional ammonia production methods. Consequently, the available net-zero pathways in agriculture, like using RESs, carbon capture and storage technologies, electrification of farms, and biomass usage based on circular economy principles, must be combined following the available resources. Depending on energy prices and environmental goals, production may shift from countries with many fossil fuel resources to countries with well-developed renewable energy infrastructure and abundant land and water resources (Rosa & Gabrielli, 2023b). Therefore, future energy and sustenance systems require an integrated design.

Improving fertilizer supply chain efficacy is crucial for reducing greenhouse gas emissions. Today approximately 20% of produced ammonia is utilized in food production, although 80% of it is squandered or lost because of the inefficiency of food agri-food supply systems (Rosa & Gabrielli, 2023b). Therefore, the enabling circular economy models and use of RESs in agriculture can play the crucial role in GHG emission reduction and get on the pathway of net zero carbon emission agriculture development.

Returning energy and other raw materials as agricultural inputs can reduce GHG emissions significantly especially beneficial from this point of view is biomass. In addition, today just 2% of nutrients left after consumption of agricultural products are recycled back into agricultural systems, with the remainder being discharged as air, water, and land pollutants or landfilled as municipal waste (Basso et al., 2021). Although urease and nitrification inhibitors have been utilized in the past, their applicability has been limited due to a high cost and low efficiency due to underdeveloped technologies. Additional research and product development are required to make these technologies cheaper and more affordable, to better achieve synergies between them, and ensure faster penetration of these innovative technologies in urban areas. This would allow to reduce significantly GHG emissions from agriculture.

4.5 CONCLUSIONS

Agriculture and renewable energy can provide for a successful combination, especially for remote rural areas, by ensuring the security of the energy supply, reducing energy supply dependency, ensuring climate change mitigation and saving expenses of farmers used to pay for external energy inputs.

RESs, like solar, wind, and hydro, can produce energy for free with minimal initial investment. Therefore, renewable energy can provide farmers producing energy from renewables like solar, wind biomass with a permanent source of revenue and substitute other income resources. Cultivating biomass provides more and more prospects for diversification of farm revenue, creation of new jobs, and new business opportunities in remote rural areas.

This chapter reviewed the scientific literature on circular economy practices in the agricultural sector. This study's primary outcomes are the following: (i) increasing scientific awareness of the topic; (ii) identifying potential routes for agricultural residues; and (iii) demonstrating the actual significance of the circular economy in the agricultural sector. This investigation concludes that many important scientific studies have been conducted, particularly in the last several years. The majority of investigations in this field were conducted in Europe. In Europe, circular economy practices have existed the longest, relying on medium- and long-term concrete actions. Numerous case studies and fewer reviews have been recognized. PV panels, wind turbines, biogas plants, and biomass-based plants were named four main distinct energy sources for use in agricultural environments.

REFERENCES

AL-agele, H. A., Proctor, K., Murthy, G., & Higgins, C. A. (2021). Case study of tomato (*Solanum lycopersicon* var. Legend) production and water productivity in agrivoltaic systems. *Sustainability*, *13*(5), 2850. https://doi.org/10.3390/su13052850

Aguilar, A., Twardowski, T., & Wohlgemuth, R. (2019). Bioeconomy for sustainable development. *Biotechnology Journal*, *14*(8), 1800638. https://doi.org/10.1002/biot.201800638

Alfaro, J., & Miller, S. (2014). Applying industrial symbiosis to smallholder farms: Modeling a case study in Liberia, West Africa. *Journal of Industrial Ecology*, *18*(1), 145–54. https://doi.org/10.1111/jiec.12077

Amjith, L., & Bavanish, B. (2022). A review on biomass and wind as renewable energy for sustainable environment. *Chemosphere*, *293*, 133579.

Amoah-Antwi, C., Kwiatkowska-Malina, J., Thornton, S.F., Fenton, O., Malina, G., & Szara, E. (2020). Restoration of soil quality using biochar and brown coal waste: A review. *Science of the Total Environment*, *722*, 137852. https://doi.org/10.1016/j.scitotenv.2020.137852

Asnaz, M. S. K., & Dolcek, A. O. (2021). Comparative performance study of different types of solar dryers towards sustainable agriculture. *Energy Reports*, *7*, 6107–6108.

Awasthi, M. K., Sarsaiya, S., Wainaina, S., Rajendran, K., Kumar, S., Quan, W., Duan, Y., Awasthi, S. K., Chen, H., Pandey, A., Zhang, Z., Jain, A., & Taherzadeh, M. J. (2019). A critical review of organic manure biorefinery models toward sustainable circular bioeconomy: Technological challenges, advancements, innovations, and future perspectives. *Renewable and Sustainable Energy Reviews*, *111*, 115–131. https://doi.org/10.1016/j.rser.2019.05.017

Balafoutis, A. T., Evert, F. K. V., & Fountas, S. (2020). Smart farming technology trends: Economic and environmental effects, labor impact, and adoption readiness. *Agronomy*, *10*, 743.

Balussou, D., McKenna, R., Most, D, & Fichtner, W. (2018). A model-based analysis of the future capacity expansion for German biogas plants under different legal frameworks. *Renewable and Sustainable Energy Reviews*, *96*, 119–131. https://doi.org/10.1016/j.rser.2018.07.041

Banerjee, S., Das, R., & Bhattacharjee, C. (2021). Biofuel cells for water desalination. In: *Biofuel cells* (pp. 345–375). Wiley. Available at: https://doi.org/10.1002/9781119725008.ch13

Banos, R., Manzano-Agugliaro, F., Montoya, F. G., Gil, C., Alcayde, A., & Gomez, J. (2011). Optimization methods applied to renewable and sustainable energy: A review. *Renewable and Sustainable Energy Review*, *15*, 1753–1766. https://doi.org/10.1016/j.rser.2010.12.008

Barron-Gafford, G. A., Pavao-Zuckerman, M. A., Minor, R. L., Sutter, L. F., Barnett-Moreno, I., Blackett, D. T., Thompson, M., Dimond, K., Gerlak, A. K., Nabhan, G. P., & Macknick, J. E. (2019). Agrivoltaics provide mutual benefits across the food-energy-water nexus in drylands. *Nature Sustainability*, *2*, 848–855.

Barros, M. V., Salvador, R., de Francisco, A. C., Piekarski, C. M. (2020). Mapping of research lines on circular economy practices in agriculture: From waste to energy. *Renewable and Sustainable Energy Reviews*, *131*, 109958. https://doi.org/10.1016/j.rser.2020.109958

Basso, B., & Antle, J. (2020). Digital agriculture to design sustainable agricultural systems. *Nature Sustainability*, *3*(4), 254–256.

Basso, B., Jones, J. W., Antle, J., Martinez-Feria, R. A., & Verma, B. (2021). Enabling circularity in grain production systems with novel technologies and policy. *Agricultural Systems*, *193*, 103244.

Bataille, C., Waisman, H., Colombier, M., Segafredo, L., & Williams, J. (2016). The deep decarbonization pathways project (DDPP): Insights and emerging issues. *Climate Policy*, *16*, S1–S6.

Bekchanov, M, & Mirzabaev, A. (2018). Circular economy of composting in Sri Lanka: Opportunities and challenges for reducing waste related pollution and improving soil health. *Journal of Cleaner Production*, *202*, 1107–1119. https://doi.org/10.1016/j.jclepro.2018.08.186

Bey, M., Hamidat, A., Benyoucef, B., Nacer, T. (2016). Viability study of the use of grid connected photovoltaic system in agriculture: Case of Algerian dairy farms. Renew Sustain Energy Rev 63, 333–3345. https://doi.org/10.1016/j.rser.2016.05.066.

Billen, G., Aguilera, E., Einarsson, R., Gamier, J., Gingrich, S., Grizzetti, B., Lassaletta, L., Le Noe, J., & Sanz-Cobena, A. (2021). Reshaping the European agro-food system and closing its nitrogen cycle: The potential of combining dietary change, agroecology, and circularity. *One Earth, 4*, 839–850.

Bluemling, B., & Wang, F. (2018). An institutional approach to manure recycling: Conduit brokerage in Sichuan Province, China. *Resources, Conservation & Recycling, 139*, 396–406. https://doi.org/10.1016/j.resconrec.2018.08.001

Boh, M. Y., & Clark, O. G. (2020). Nitrogen and phosphorus flows in Ontario's food systems. *Resources, Conservation & Recycling, 154*, 104639. https://doi.org/10.1016/j.resconrec.2019.104639

Bordoloi, S. (2021). Simulation and analysis of green house based agri-voltaic system using energy 3D software. *Journal of Engineering Technology, 10*, 6–11.

Cardoso, J. S., Silva, V., Chavando, J. A. M., Eusébio, D., Hall, M. J., & Costa, M. (2021). Small-scale biomass gasification for green ammonia production in Portugal: A techno-economic study. *Energy & Fuels, 35*, 13847–13862.

Carroquino, J., Bernal-Agustm, J.-L., & Dufo-Lopez, R. (2019). Standalone renewable energy and hydrogen in an agricultural context: A demonstrative case. *Sustainability, 11*, 951.

Castro, A. J., Lopez-Rodriguez, M. D., Giagnocavo, C., Gimenez, M., Cèspedes, L., La Calle, A., Gallardo, M., Pumares, P., Cabello, J., Rodriguez, E., & Uclès, D. (2019). Six collective challenges for sustainability of Almeria greenhouse horticulture. *International Journal of Environmental Research and Public Health, 16*(21), 4097. https://doi.org/10.3390/ijerph16214097

Chojnacka, K., Moustakas, K., & Witek-Krowiak, A. (2020). Bio-based fertilizers: A practical approach towards circular economy. *Bioresource Technology, 295*, 122223. https://doi.org/10.1016/j.biortech.2019.122223

Choudhury, P. P., & Saha, S. (2020). Dynamics of pesticides under changing climatic scenario. *Environmental Monitoring and Assessment, 192*, 1–3.

Caruso, M. C., Braghieri, A., Capece, A., Napolitano, F., Romano, P., Galgano, F., & Altieri, G., Genovese, F. (2019). Recent updates on the use of agro-food waste for biogas production. *Applied Sciences, 9*(6), 1217. https://doi.org/10.3390/app9061217

Chel, A., Tiwari, G. N., & Chandra, A. (2009). Sizing and cost estimation methodology for stand-alone residential PV power system. *International Journal of Agile Systems and Management, 4*, 21–40.

Chel, K. (2011). Renewable energy for sustainable agriculture. *Agronomy for Sustainable Development, 31*(1), 91–118. https://hal.science/hal-00930477

Chen, L., Cong, R. G., Shu, B., & Mi, Z. F. (2017). A sustainable biogas model in China: The case study of Beijing Deqingyuan biogas project. *Renewable and Sustainable Energy Reviews, 78*, 773–779. https://doi.org/10.1016/j.rser.2017.05.027

Chertow, M. R. (2000). Industrial symbiosis: Literature and taxonomy. *Annual Review of Energy and the Environment, 25*(1), 313–337.

Chiaramonti, D., & Panoutsou, C. (2019). Policy measures for sustainable sunflower cropping in EU-MED marginal lands amended by biochar: Case study in Tuscany, Italy. *Biomass and Bioenergy, 126*, 199–210. https://doi.org/10.1016/j.biombioe.2019.04.021

Cobo, S., Dominguez-Ramos, A., & Irabien, A. (2018). Trade-offs between nutrient circularity and environmental impacts in the management of organic waste. *Environmental Science &Technology, 52*(19),10923–10933. https://doi.org/10.1021/acs.est.8b01590

de Boer, U., Cederberg, C., Eady, S., Gollnow, S., Kristensen, T., Macleod, M., Meul, M., Nemecek, T., Phong, L.T., Thoma, G., Van Der Werf, H. M. G., Williams, A. G., & Zonderland- Thomassen, M. A. (2011). Greenhouse gas mitigation in animal production: Towards an integrated life cycle sustainability assessment. *Current Opinion in Environmental Sustainability, 3*(5), 423–31. https://doi.org/10.1016/j.cosust.2011.08.007

De Luca, A.I., Falcone, G., Stillitano T., Iofrida, N., Strano, A., & Gulisano, G. (2018). Evaluation of sustainable innovations in olive growing systems: A life cycle sustainability Assessment case study in southern Italy. *Journal of Cleaner Production, 171*, 1187–1202. https://doi.org/10.1016/j.jclepro.2017.10.119

Diacono, M., Persiani, A., Testani, E., Montemurro, F., Ciaccia, C. (2019). Recycling agricultural wastes and by-products in organic farming: Biofertilizer production, yield performance and carbon footprint analysis. *Sustainability, 11*(14), 3824. https://doi.org/10.3390/su11143824

Donner, M., Gohier, R., & de Vries, H. (2020). A new circular business model typology for creating value from agro-waste. *Science of the Total Environment, 716*, 137065. https://doi.org/10.1016/j.scitotenv.2020.137065

Dumont, B., Fortun-Lamothe, L., Jouven, M., Thomas, M., & Tichit, M. (2013). Prospects from agroecology and industrial ecology for animal production in the 21st century. *Animal, 7*(6), 1028–1043. https://doi.org/10.1017/S1751731112002418

Egea, F. J., Torrente, R. G., & Aguilar, A. (2018). An efficient agro-industrial complex in Almeria (Spain): Towards an integrated and sustainable bioeconomy model. *New Biotechnology, 40*, 103–112. https://doi.org/10.1016/j.nbt.2017.06.009

Eker, B. (2005) Solar powered water pumping systems. *Trakia Journal of Sciences, 3*, 7–11.

Elfasakhany, A. (2020). Dual and ternary biofuel blends for desalination process: Emissions and heat recovered assessment. *Energies, 14*, 61.

Elkhalifa, S., Al-Ansari, T., Mackey, H. R., & McKay, G. (2019). Food waste to biochars through pyrolysis: A review. *Resources, Conservation and Recycling, 144*, 310–320. https://doi.org/10.1016/j.resconrec.2019.01.024

Fantozzi, F., & Bartocci, P. (2016). Carbon footprint as a tool to limit greenhouse gas emissions. *Greenhouse Gases, 285*. https://doi.org/10.5772/62281

FAOSTAT (2023). Climate change: Agrifood systems emissions. Available at: www.fao.org/faostat/en/#data (Accessed June 2023).

Faridi, H., Arabhosseini, A, Zarei, G., & Okos, M. (2019). Utilization of soil temperature modeling to check the possibility of Earth-air heat exchanger for agricultural building. *Iranica Journal of Energy & Environment, 10*, 260–268.

Ferella, F., Cucchiella, F., D'Adamo, I., & Gallucci, K. (2019). A techno-economic assessment of biogas upgrading in a developed market. *Journal of Cleaner Production, 210*, 945–957. https://doi.org/10.1016/j.jclepro.2018.11.073

Ferjani, A. I., Jeguirim, M., Jellali, S., Limousy, L., Courson, C., Akrout, H., Thevenin, N., Ruidavets, L., Muller, A., & Bennici, S. (2019). The use of exhausted grape marc to produce biofuels and biofertilizers: Effect of pyrolysis temperatures on biochars properties. *Renewable and Sustainable Energy Reviews, 107*, 425–433. https://doi.org/10.1016/j.rser.2019.03.034

Fernandez-Mena, H., Nesme, T., & Pellerin, S. (2016). Towards an agro-industrial ecology: A review of nutrient flow modelling and assessment tools in agro-food systems at the local scale. *Science of the Total Environment, 543*, 467–479. https://doi.org/10.1016/j.scitotenv.2015.11.032

Filho, L. W. (2018). Bioeconomy meets the circular economy: The RESYNTEX and FORCE projects. In: *Towards a sustainable bioeconomy: Principles, challenges and perspectives* (p. 567–575). Cham: Springer. https://doi.org/10.1007/978-3-319-73028-8_29

Frey, G. W., & Linke, D. M. (2002). Hydropower as a renewable and sustainable energy resource meeting global energy challenges in a reasonable way. *Energy Policy, 30*, 141261–141265.

Gao, Y., & Cabrera Serrenho, A. (2023). Greenhouse gas emissions from nitrogen fertilizers could be reduced by up to one-fifth of current levels by 2050 with combined interventions. *Nature Food, 4*, 170–178.

Garcia-Nieto, P. J., Garcia-Gonzalo, E., Fernandez, J. A., & Muniz, C. D. (2016). Using evolutionary multivariate adaptive regression splines approach to evaluate the eutrophication in the Pozon de la Dolores lake (Northern Spain). *Ecological Engineering*, *94*, 36–51. https://doi.org/10.1016/j.ecoleng.2016.05.047

Garske, B., Stubenrauch, J., & Ekardt, F. (2020). Sustainable phosphorus management in European agricultural and environmental law. *Review of European, Comparative & International Environmental Law*, *29*, 107–117. https://doi.org/10.1111/reel.12318

Gameiro, A., Bonaudo, T., & Tichit, M. (2019). Nitrogen, phosphorus and potassium accounts in the Brazilian livestock agro-industrial system. *Regional Environmental Change*, *19*(3), 893–905. https://doi.org/10.1007/s10113-018-1451-2

Gao, M., Wang, D., Wang, H., Wang, X., & Feng, Y. (2019). Biogas potential, utilization and countermeasures in agricultural provinces: A case study of biogas development in Henan Province, China. *Renewable and Sustainable Energy Reviews*, *99*, 191–200. https://doi.org/10.1016/j.rser.2018.10.005

Ghobadpour, A., Boulon, L., Mousazadeh, H., Malvajerdi, A. S., & Rafiee, S. (2019). State of the art of autonomous agricultural off-road vehicles driven by renewable energy systems. *Energy Procedia*, *162*, 4–13.

Gonocruz, R.A., Nakamura, R., Yoshino, K., Homma, M., Doi, T., Yoshida, Y., & Tani, A. (2021). Analysis of the rice yield under an agrivoltaic system: A case study in Japan. *Environments*, *8*, 65.

Gonzalez-de-Soto, M., Emmi, L., & Gonzalez-de-Santos, R. (2019). Hybrid-powered autonomous robots for reducing both fuel consumption and pollution in precision agriculture tasks. *Agricultural Robots-Fundamentals and Applications*. https://doi.org/10.5772/intechopen.79875

Gorjian, S., Calise, F., Kant, K., Ahamed, M. S., Copertaro, B., Najafi, G., Zhang, X., Aghaei, M., & Shamshiri, R. R. (2021). A review on opportunities for implementation of solar energy technologies in agricultural greenhouses. *Journal of Cleaner Production*, *285*, 124807.

Gorjian, S., Ebadi, H., Trommsdorff, M., Sharon H., Demant, M., & Schindele, S. (2021). The advent of modern solar-powered electric agricultural machinery: A solution for sustainable farm operations. *Journal of Cleaner Production*, *292*, 126030.

Gorjian, S., Fakhraei, O., Gorjian, A., Sharafkhani, A., & Aziznejad, A. (2022). Sustainable food and agriculture: Employment of renewable energy technologies. *Current Robotics Reports*, *3*, 153–163. https://doi.org/10.1007/s43154-022-00080-x

Gourdo, L., Fatnassi, H., Tiskatine, R., Wifaya, A., Demrati, H., Aha- roune, A., & Bouirden, L. (2019). Solar energy storing rock-bed to heat an agricultural greenhouse. *Energy*, *169*, 206–212.

Graedel, T. E. (2019). Material flow analysis from origin to evolution. *Environmental Science & Technology*, *53*(21), 12188–12196. https://doi.org/10.1021/acs.est.9b03413

Grandl, F., Luzi, S. P., Furger, M., Zeitz, J. O., Leiber, F., Ortmann, S., & Schwarm, A. (2016). Biological implications of longevity in dairy cows: 1. Changes in feed intake, feeding behavior, and digestion with age. *Journal of Dairy Science*, *99*(5), 3457–3471. https://doi.org/10.3168/jds.2015-10261

Grant, G. B., Seager, T. P., Massard, G., & Nies, L. (2010). Information and communication technology for industrial symbiosis. *Journal of Industrial Ecology*, *14*(5), 740–753. https://doi.org/10.1111/j.1530-9290.2010.00273.x

Grimm, D., & Wosten, H. A. B. (2018). Mushroom cultivation in the circular economy. *Applied Microbiology and Biotechnology*, *102*, 7795–7803. https://doi.org/10.1007/s00253-018-9226-8

Guo, S., Zhang, Y., Qu, H., Li, M., Zhang, S., Yang, J., Zhang, X., & Tan, S. C. (2022). Repurposing face mask waste to construct floating photo-thermal evaporator for autonomous solar ocean farming. *EcoMat*, *4*, 1–10.

Hamelin, L., Borzęcka, M., Kozak, M., & Pudelko, R. (2019). A spatial approach to bio-economy: Quantifying the residual biomass potential in the EU-27. *Renewable and Sustainable Energy Reviews*, *100*, 127. https://doi.org/10.1016/j.rser.2018.10.017

He, X., Feng, K., Li, X., Craft, A. B., Wada, Y., Burek, P, Wood, E. F., & Sheffield, J. (2019). Solar and wind energy enhances drought resilience and groundwater sustainability. *Nature Communications*, *10*, 4893.

Herczeg, G., Akkerman, R., & Hauschild, M. Z. (2018). Supply chain collaboration in industrial symbiosis networks. *Journal of Cleaner Production*, *171*, 1058–1067. https://doi.org/10.1016/j.jclepro.2017.10.046

Hidalgo, D., Martm-Marroquin, J. M., & Corona, F. (2019). A multi-waste management concept as a basis towards a circular economy model. *Renewable and Sustainable Energy Reviews*, *111*, 481–489. https://doi.org/10.1016/j.rser.2019.05.048

Horton, P., Long, S. P., Smith, P., Banwart, S. A., & Beerling, D. J. (2021). Technologies to deliver food and climate security through agriculture. *Nature Plants*, *7*, 250–255.

IEA (International Energy Agency). (2011). Deploying renewables 2011: Best and future policy practice. Paris: IEA. Available at: https://www.iea.org/reports/deploying-renewables-2011-best-and-future-policy-practice (Accessed June 2023).

IEA (International Energy Agency). (2021). Net zero by 2050. Available at: https://www.iea.org/reports/net-zero-by-2050 (Accessed June 2023).

IEA (International Energy Agency). (2022). *Renewable electricity*. Paris: IEA. Available at: https://www.iea.org/reports/renewables-2022 (Accessed June 2023).

IEA Bioenergy. (2023). How bioenergy contributes to a sustainable future. Available at: http://www.ieabioenergyreview.org/ (Accessed June 2023).

Ihoume, I., Tadili, R., Arbaoui, N., Bazgaou, A., Idrissi, A., Benchrifa, M., & Fatnassi, H. (2022). Performance study of a sustainable solar heating system based on a copper coil water to air heat exchanger for greenhouse heating. *Solar Energy*, *232*, 128–138.

IlenMacArthurFoundation(2013).TowardsthecirculareconomyVol. 1:aneconomicandbusiness rationaleforanacceleratedtransition, Availableat:https://www.ellenmacarthurfoundation.org/towards-the-circular-economy-vol-1-an-economic-and-business-rationale-for-an

Ingrao, C., Faccilongo, N., Di Gioia, L., & Messineo, A. (2018). Food waste recovery into energy in a circular economy perspective: A comprehensive review of aspects related to plant operation and environmental assessment. *Journal of Cleaner Production*, *84*, 869–892. https://doi.org/10.1016/j.jclepro.2018.02.267

IRENA. (2021). World energy transitions outlook: 1.5 degrees pathway. International Renewable Energy Agency. Available at: https://irena.org/publications/2021/March/World-Energy-Transitions-Outlook (Accessed June 2023).

Jacquet, F., Jeuffroy, M. H., Jouan, J., Le Cadre, E., Litrico, I., Malausa, T., Reboud, X., & Huyghe, C. (2022). Pesticide-free agriculture as a new paradigm for research. *Agronomy for Sustainable Development*, *42*, 8.

Jeswani, H. K., Chilvers, A., & Azapagic, A. (2020). Environmental sustainability of biofuels: A review. *Proceedings of the Royal Society A*, *476*(2243), 20200351.

Karthikeyan, O. P., Mehariya, S., & Wong, J. W. C. (2017). Bio-refining of food waste for fuel and value products. *Energy Procedia*, *136*, 14–21. https://doi.org/10.1016/j.egypro.2017.10.253

Kassem, N., Sills, D., Posmanik, R., Blair, C., & Tester, J. W. (2020). Combining anaerobic digestion and hydrothermal liquefaction in the conversion of dairy waste into energy: A techno economic model for New York state. *Waste Management*, *103*, 228–239. https://doi.org/10.1016/j.wasman.2019.12.029

Keesstra, S. D., Bouma, J., Wallinga, J., Tittonell, P., Smith, P., Cerda, A., Montanarella, L., Quinton, J. N., Pachepsky, Y., Van Der Putten, W. H., & Bardgett, R. D. (2016). The significance of soils and soil science towards realization of the United Nations Sustainable Development Goals. *Soils*, *2*, 111–128. https://doi.org/10.5194/soil-2-111-2016

Khan, Z. A., Imran, M., Altamimi, A., Diemuodeke, O. E., & Abdelatif, A. O. (2022). Assessment of wind and solar hybrid energy for agricultural applications in Sudan. *Energies*, *15*, 5. https://doi.org/10.3390/en15010005

Kizito, S., Luo, H., Lu, J., Bah, H., Dong, R., & Wu, S. (2019). Role of nutrient-enriched biochar as a soil amendment during maize growth: Exploring practical alternatives to recycle agricultural residuals and to reduce chemical fertilizer demand. *Sustainability*, *11*(11), 3211. https://doi.org/10.3390/su11113211

Kougias, P. G., & Angelidaki, I. (2018). Biogas and its opportunities – A review. *Frontiers of Environmental Science & Engineering*, *12*, 1–12. https://doi.org/10.1007/s11783-018-1037-8

Kristensen, D. K., Kjeldsen, C., & Thorsoe, M. H. (2016). Enabling sustainable agro-food futures: Exploring fault lines and synergies between the integrated territorial paradigm, rural eco-economy and circular economy. *Journal of Agricultural and Environmental Ethics*, *29*(5), 749–765. https://doi.org/10.1007/s10806-016-9632-9

Kumar, A., Purohit, P., Rana, S., & Kandpal, T. C. (2002). An approach to the estimation of the value of agricultural residues used as biofuels. *Biomass & Bioenergy*, *22*, 195–203.

Kumar, M., Haillot, D., & Gibout, S. (2022). Survey and evaluation of solar technologies for agricultural greenhouse application. *Solar Energy*, *232*(12), 18–34.

Kumar, R., Strezov, V., Weldekidan, H., He, J., Singh, S., Kan, T., & Dastjerdi, B. (2020) Lignocellulose biomass pyrolysis for bio-oil production: A review of biomass pre-treatment methods for production of drop-in fuels. *Renewable and Sustainable Energy Reviews*, *123*, 109763.

Kumar, S. S., Kumar, V., Kumar, R., Malyan, S. K., & Pugazhendhi, A. (2019). Microbial fuel cells as a sustainable platform technology for bioenergy, biosensing, environmental monitoring, and other low power device applications. *Fuel*, *255*, 115682.

Kwan, T. H., Ong, K. L., Haque, M. A., Kulkarni, S., & Lin, C. S. K. (2019). Biorefinery of food and beverage waste valorisation for sugar syrups production: Techno-economic assessment. *Process Safety and Environmental Protection*, *121*, 194–208. https://doi.org/10.1016/j.psep.2018.10.018

Kwon, H., Liu, X., Xu, H., & Wang, M. (2021). Greenhouse gas mitigation strategies and opportunities for agriculture. *Agronomy Journal*, *113*, 4639–4647.

Lamont, K., Pensini, E., Daguppati, P., Rudra, R., van de Vegte, J., & Levangie, J. (2019). Natural reusable calcium-rich adsorbent for the removal of phosphorus from water: Proof of concept of a circular economy. *Canadian Journal of Civil Engineering*, *46*(5), 458–461. https://doi.org/10.1139/cjce-2018-0512

Lange, J. P. (2021). Towards circular carbo-chemicals-the metamorphosis of petrochemicals. *Energy & Environmental Science*, *14*, 4358–4376.

Lee, D. H. (2017). Econometric assessment of bioenergy development. *International Journal of Hydrogen Energy*, *42*(45), 27701–27717. https://doi.org/10.1016/j.ijhydene.2017.08.055

Leger, D., Matassa, S., Noor, E., Shepon, A., Milo, R., & Bar-Even, A. (2021). Photovoltaic-driven microbial protein production can use land and sunlight more efficiently than conventional crops. *Proceedings of the National Academy of Sciences of the United States of America*, *118*, e2015025118.

Lehtonen, H., Huan-Niemi, E., & Niemi, J. (2022). The transition of agriculture to low carbon pathways with regional distributive impacts. *Environmental Innovation and Societal Transitions*, *44*, 1–13, https://doi.org/10.1016/j.eist.2022.05.002

Li, G., Jin, Y., Akram, M. W., & Chen, X. (2017). Research and current status of the solar photovoltaic water pumping system – A review. *Renewable and Sustainable Energy Reviews*, *79*, 440–458. https://doi.org/10.1016/j.rser.2017.05.055.h

Ling, C., Wang, Y., Min, C., & Zhang, Y. (2018). Economic evaluation of reverse osmosis desalination system coupled with tidal energy. *Frontiers in Energy*, *12*, 297–304.

Lord, R., & Sakrabani, R. (2019). Ten-year legacy of organic carbon in non-agricultural (brownfield) soils restored using green waste compost exceeds 4 per mille per annum: Benefits and trade-offs of a circular economy approach. *Science of the Total Environment, 686*, 1057–1068. https://doi.org/10.1016/j.scitotenv.2019.05.174

Lutz, J., Smetschka, B., & Grima, N. (2017). Farmer cooperation as a means for creating local food systems-potentials and challenges. *Sustainability, 9*(6), 925. https://doi.org/10.3390/su9060925

Lyytimaki, J., Nygrėn, N. A., Pulkka, A., & Rantala, S. (2018). Energy transition looming behind the headlines? Newspaper coverage of biogas production in Finland. *Energy, Sustainability and Society, 8*(1), 15. https://doi.org/10.1186/s13705-018-0158-z

Maap, O., & Grundmann, P. (2018). Governing transactions and interdependences between linked value chains in a circular economy: The case of wastewater reuse in braunschweig (Germany). *Sustainability, 10*(4), 1125. https://doi.org/10.3390/SU10041125

Mariantonietta, F., Alessia, S., Francesco, C., & Giustina, P. (2018). GHG and cattle farming: CO-assessing the emissions and economic performances in Italy. *Journal of Cleaner Production, 172*, 3704–3012. https://doi.org/10.1016/jjclepro.2017.07.167

Mosquera-Losada, M. R., Santiago-Freijanes, J. J., Rigueiro-Rodríguez, A., Rodríguez-Rigueiro, F. J., Arias Martínez, D., Pantera, A., Ferreiro-Domínguez, N. (2020). *The importance of agroforestry systems in supporting biodiversity conservation and agricultural production: a European perspective*. Reconciling agricultural production with biodiversity conservation. Cambridge: Burleigh Dodds Science Publishing Limited, p. 282.

Melian-Martel, N., del Rfo-Gamero, B., & Schallenberg-Rocirfguez, J. (2021). Water cycle driven only by wind energy surplus: Towards 100% renewable energy islands. *Desalination, 515*, 115216.

Mohan, S. V., Nikhil, G. N., Chiranjeevi, P., Reddy, C. N., Rohit, M. V., Kumar, A. N., & Sarkar, O. (2016). Waste biorefinery models towards sustainable circular bioeconomy: Critical review and future perspectives. *Bioresource Technology, 215*, 2–12. https://doi.org/10.1016/j.biortech.2016.03.130

Molina-Moreno, V., Leyva-Diaz, J. C., Llorens-Montes, F. J., & Cortes-Garcia, F. J. (2017). Design of indicators of circular economy as instruments for the evaluation of sustainability and efficiency in wastewater from pig farming industry. *Water, 9*(9), 653. https://doi.org/10.3390/w9090653

Morseletto, P. (2020). Restorative and regenerative: Exploring the concepts in the circular economy. *Journal of Industrial Ecology, 24*(2), 763–773. https://doi.org/10.1111/jiec.12987

Mosquera-Losada, M. R., Amador-Garcia, A., Rigueiro-Rodriguez, A., Ferreiro-Dominguez, N. (2019). Circular economy: Using lime stabilized bio-waste based fertilisers to improve soil fertility in acidic grasslands. *Catena, 179*, 119–128. https://doi.org/10.1016/j.catena.2019.04.008

Murray, A., Skene, K., & Haynes, K. (2017). The circular economy: An interdisciplinary exploration of the concept and application in a global context. *Journal of Business Ethics, 140*(3), 369–380. https://doi.org/10.1007/s10551-015-2693-2

Nandan, R., Poonia, S. P., Singh, S. S., Nath, C. P., Kumar, V., Malik, R. K., McDonald, A., & Hazra, K. K. (2021). Potential of conservation agriculture modules for energy conservation and sustainability of rice-based production systems of Indo-Gangetic Plain region. *Environmental Science and Pollution Research, 28*(1), 246–261. https://doi.org/10.1007/s11356-020-10395-x. Epub 2020 Aug 18. PMID: 32808133; PMCID: PMC7782432.

Nielsen, P. H. (2017). Microbial biotechnology and circular economy in wastewater treatment. *Microbial Biotechnology, 10*(5), 1102–1105. https://doi.org/10.1111/1751-7915.12821

Niutanen, V., & Korhonen, J. (2003). Industrial ecology flows of agriculture and food industry in Finland: Utilizing by-products and wastes. *International Journal of Sustainable Development & World Ecology, 10*(2), 133–147. https://doi.org/10.1080/13504500309469792

Northrup, D. L., Basso, B., Wang, M. Q., Morgan, C. L., & Benfey, P. N. (2021). Novel technologies for emission reduction complement conservation agriculture to achieve negative emissions from row-crop production. *Proceedings of the National Academy of Sciences, 118*, e2022666118.

Notarnicola, B., Sala, S., Anton, A., McLaren, S. J., Saouter, E., & Sonesson, U.(2017). The role of life cycle assessment in supporting sustainable agri-food systems: A review of the challenges. *Journal of Cleaner Production, 140*, 399–409. https://doi.org/10.1016/j.jclepro.2016.06.071

Noya, I., Aldea, X., Gonzalez-Garcia, S., Gasol, C. M., Moreira, M. T., Amores, M. J., Marin, D., & Boschmonart-Rives, J. (2017). Environmental assessment of the entire pork value chain in Catalonia – A strategy to work towards circular economy. *Science of the Total Environment, 589*, 122–129. https://doi.org/10.1016/j.scitotenv.2017.02.186

OECD. (2012). *Linking energy to rural development, OECD Green growth studies*. Paris: OECD Publishing. Available at: https://doi.org/10.1787/9789264180444-en (Accessed June 2023).

Omer, A. M. (2008). Energy, environment and sustainable development. *Renewable and Sustainable Energy Reviews, 12*, 2265–2300. https://doi.org/10.1016/j.rser.2007.05.001

Ometto, A. R., Ramos, P. A. R., & Lombardi, G. (2007). The benefits of a Brazilian agro-industrial symbiosis system and the strategies to make it happen. *Journal of Cleaner Production, 15*(13–14), 1253–1258. https://doi.org/10.1016/j.jclepro.2006.07.021

Ouikhalfan, M., Lakbita, O., Delhali, A., Assen, A. H., & Belmabkhout, Y. (2022). Toward net-zero emission fertilizers industry: Greenhouse gas emission analyses and decarbonization solutions. *Energy Fuels, 8*, 4198–4223.

Pagotto, M., & Halog, A. (2016). Towards a circular economy in Australian agri-food industry: An application of input-output oriented approaches for analyzing resource efficiency and competitiveness potential. *Journal of Industrial Ecology, 20*(5), 1176–1186. https://doi.org/10.1111/jiec.12373

Papangelou, A., Achten, W. M., & Mathijs, E. (2020). Phosphorus and energy flows through the food system of Brussels Capital Region. *Resources, Conservation & Recycling, 156*, 104687. https://doi.org/10.1016/j.resconrec.2020.104687

Parmar, R., Banerjee, C., & Tripathi, A. K. (2021). Performance analysis of cost effective portable solar photovoltaic water pumping system. *Current Photovoltaic Research, 9*, 51–58.

Patricio, J., Axelsson, L., Blomė, S., & Rosado, L. (2018). Enabling industrial symbiosis collaborations between SMEs from a regional perspective. *Journal of Cleaner Production, 202*, 1120–1130. https://doi.org/10.1016/j.jclepro.2018.07.230

Pearson, S., Camacho-Villa, T. C., Valluru, R., Gaju, O., Rai, M. C., Gould, I., Brewer, S., & Sklar, E. (2022). Robotics and autonomous systems for net zero agriculture. *Current Robotics Reports, 3*, 57–64.

Pellegrini, P., & Fernández, R. J. (2018). Crop intensification, land use, and on-farm energy-use efficiency during the worldwide spread of the green revolution. *Proceedings of the National Academy of Sciences, 115*, 2335–2340.

Piezer, K., Petit-Boix, A., Sanjuan-Delmas, D., Briese, E., Celik, I., Rieradevall, J., Gabarrell, X., Josa, A., & Apul, D. (2019). Ecological network analysis of growing tomatoes in an urban rooftop greenhouse. *Science of the Total Environment, 651*, 1495–504. https://doi.org/10.1016/j.scitotenv.2018.09.293

Quaglia, G., Visconte, C., Scimmi, L.S., Melchiorre, M., Cavallone, P., & Pastorelli, S. (2020). Design of a UGV powered by solar energy for precision agriculture. *Robotics, 9*(1), 13.

Ramos-Suarez, J. L., Ritter, A., Gonzalez, J. M., & Pèrez, A. (2019). Biogas from animal manure: A sustainable energy opportunity in the Canary Islands. *Renewable and Sustainable Energy Reviews, 104*, 137–150. https://doi.org/10.1016/j.rser.2019.01.025

Rana, M. S., Bhushan, S., Prajapati, S. K., & Kavitha, S. (2020). Techno-economic analysis and environmental aspects of food waste management. In: *Food waste to valuable resources* (pp. 325–342). Academic Press. https://doi.org/10.1016/B978-0-12-818353-3.00015-8

Ravi, S., Macknick, J., Lobell, D., Field, C., Ganesan, K., Jain, R., Elchinger, M., & Stoltenberg, B. (2016). Colocation opportunities for large solar infrastructures and agriculture in drylands. *Applied Energy, 165*, 383–392. https://doi.org/10.1016/j.apenergy.2015.12.078

Riaz, M. H., Imran, H., Alam, H., Alam, M. A., & Butt, N. Z. (2022). Crop-specific optimization of bifacial PV arrays for agrivoltaic food-energy production: The light-productivity-factor approach. *IEEE Journal of Photovoltaics, 12*, 572–580.

Reisinger, A., Clark, H., Cowie, A. L., Emmet-Booth, J., Gonzalez Fischer, C., Herrero, M., Howden, M., & Leahy, S. (2021). How necessary and feasible are reductions of methan emissions from livestock to support stringent temperature goals? *Philosophical Transactions of the Royal Society A, 379*(2210), 20200452.

Rosa, L. (2022). Adapting agriculture to climate change via sustainable irrigation: Biophysical potentials and feedbacks. *Environmental Research Letters, 17*, 063008.

Rosa, L., & Gabrielli, P. (2023a). Achieving net-zero emissions in agriculture: A review. *Environment Research Letters, 18*, 063002. https://doi.org/10.1088/1748-9326/acd5e8

Rosa, L., & Gabrielli, P. (2023b). Energy and food security implications of transitioning synthetic nitrogen fertilizers to net-zero emissions. *Environmental Research Letters, 18*, 014008

Rosa, L., Rulli, M. C., Ali, S., Chiarelli, D. D., DelFAngelo, J., Mueller, N. D., Scheidel, A., Siciliano, G., & D'Odorico, P. (2021). Energy implications of the 21st century agrarian transition. *Nature Communications, 12*, 2319.

Rosa, L., Rulli, M. C., Davis, K. F., Chiarelli, D. D., Passera, C., & D'Odorico, P. (2018). Closing the yield gap while ensuring water sustainability. *Environmental Research Letters, 13*, 104002.

Salvador, R., Barros, M. V., da Luz, L. M., Piekarski, C. M., & de Francisco, A. C. (2019). Circular business models: Current aspects that influence implementation and unaddressed subjects. *Journal of Cleaner Production, 250*, 119555. https://doi.org/10.1016/j.jclepro.2019.119555

Sampaio, P. G. V., & Gonzalez, M. O. A. (2017). Photovoltaic solar energy: Conceptual framework. *Renewable and Sustainable Energy Reviews, 74*, 590–601. https://doi.org/10.1016/j.rser.2017.02.081

Sampson, G. S., Perry, E. D., & Tayler, M. R. (2020). The on-farm and near-farm effects of wind turbines on agricultural land values. *Journal of Agricultural and Resources Economics, 45*, 410–427.

Sanchez-Molina, J. A., Reinoso, J. V., Aciėn, F. G., Rodriguez, F., & Lopez, J. C. (2014). Development of a biomass-based system for nocturnal temperature and diurnal CO2 concentration control in greenhouses. *Biomass & Bioenergy, 67*, 60–71.

Santos, V. E. N., & Magrini, A. (2018). Biorefining and industrial symbiosis: A proposal for regional development in Brazil. *Journal of Cleaner Production, 177*, 19–33. https://doi.org/10.1016/j.jclepro.2017.12.107

Schau, E. M., & Fet, A. M. (2008). LCA studies of food products as background for environmental product declarations. *The International Journal of Life Cycle Assessment, 13*(3), 255–264. https://doi.org/10.1065/lca2007.12.372

Sethi, C. K., Acharya, S. K., Patnaik, P. P., & Behera, A. (2021). A review on solar drying of marine applications. In: *Current advances in mechanical engineering: Select proceedings of ICRAMERD 2020* (pp. 271–281). Springer Nature. Available at: https://link.springer.com/10.1007/978-981-33-4795-3_26 (Accessed June 2023).

Sherwood, J. (2020). The significance of biomass in a circular economy. *Bioresource Technology, 300*, 122755. https://doi.org/10.1016/j.biortech.2020.122755

Schüpbach, B., Roesch, A, Herzog, F., Szerencsits, E., & Walter, T. (2020). Development and application of indicators for visual landscape quality to include in life cycle sustainability assessment of Swiss agricultural farms. *Ecological Indicators*, *110*, 105788. https://doi.org/10.1016/j.ecolind.2019.105788

Smith, C., Hill, A. K., & Torrente-Murciano, L. (2020). Current and future role of Haber-Bosch ammonia in a carbon-free energy landscape. *Energy & Environmental Science*, *13*, 331–344.

Soofi, A. F., Manshadi, S. D., & Saucedo, A. (2022). Farm electrification: A road-map to decarbonize the agriculture sector. *The Electricity Journal*, *35*, 107076.

Sowby, R. B., & Dicataldo, E. (2022). The energy footprint of US irrigation: A first estimate from open data. *Energy Nexus*, *6*, 100066.

Springer, N. P., & Schmitt, J. (2018). The price of byproducts: Distinguishing co-products from waste using the rectangular choice-of-technologies model. *Resources, Conservation & Recycling*, *138*, 231–237. https://doi.org/10.1016/j.resconrec.2018.07.034

Svanstrom, M., Heimersson, S., Peters, G., Harder, R., Fons, D., Finnson, A., & Olsson, J. (2017). Life cycle assessment of sludge management with phosphorus utilisation and improved hygienisation in Sweden. *Water Science and Technology*, *75*(9), 2013–2024. https://doi.org/10.2166/wst.2017.073

Tan, R. R. (2019). Data challenges in optimizing biochar-based carbon sequestration. *Renewable and Sustainable Energy Reviews*, *104*, 174–177. https://doi.org/10.1016/j.rser.2019.01.032

Therond, O., Duru, M., Roger-Estrade, J., & Richard, G. (2017). A new analytical framework of farming system and agriculture model diversities. A review. *Agronomy for Sustainable Development*, *37*(3), 21. https://doi.org/10.1007/s13593-017-0429-7

Toop, T. A., Ward, S., Oldfield, T., Hull, M., Kirby, M. E., & Theodorou, M. K. (2017). Agro cycle-developing a circular economy in agriculture. *Energy Procedia*, *123*, 76–80. https://doi.org/10.1016/j.egypro.2017.07.269

Trokanas, N., Cecelja, F., & Raafat, T. (2014). Semantic input/output matching for waste processing in industrial symbiosis. *Computers & Chemical Engineering*, *66*, 259–268. https://doi.org/10.1016/j.compchemeng.2014.02.010

Trommsdorff, M., Kang, J., Reise, C., Schindele, S., Bopp, G., Ehmann, A., Weselek, A., Hogy, P., & Obergfell, T. (2021). Combining food and energy production: Design of an agrivoltaic system applied in arable and vegetable farming in Germany. *Renewable and Sustainable Energy Reviews*, *140*, 110694.

Trost, B., Prochnow, A., Drastig, K., Meyer-Aurich, A., Ellmer, F., & Baumecker, M. (2013). Irrigation, soil organic carbon and N2O emissions. A review. *Agronomy for Sustainable Development*, *33*, 733–749.

Tvinnereim, E., Liu, X., & Jamelske, E. M. (2017). Public perceptions of air pollution and climate change: Different manifestations, similar causes, and concerns. *Climatic Change*, *140*(3–4), 399–412. https://doi.org/10.1007/s10584-016-1871-2

United Nations Economic Commission for Europe (UNECE). (2022). *Carbon neutrality in the UNECE region: Integrated life-cycle assessment of electricity sources*. Geneva: United Nations. Available at: https://unece.org/sites/default/files/2022-04/LCA_3_FINAL%20March%202022.pdf (Accessed June 2023).

Vaneeckhaute, C., Styles, D., Prade, T., Adams, P., Thelin, G., Rodhe, L., Gunnarsson, I., & D'Hertefeldt, T. (2018). Closing nutrient loops through decentralized anaerobic digestion of organic residues in agricultural regions: A multi-dimensional sustainability assessment. *Resources, Conservation & Recycling*, *136*, 110–117. https://doi.org/10.1016/j.resconrec.2018.03.027

Vega-Quezada, C., Blanco, M., & Romero, H. (2017). Synergies between agriculture and bioenergy in Latin American countries: A circular economy strategy for bioenergy production in Ecuador. *New Biotechnology*, *39*, 81–89. https://doi.org/10.1016/j.nbt.2016.06.730

Vieira, S., Barros, M. V., Sydney, A. C. N., Piekarski, C. M., de Francisco, A. C., de Souza Vandenberghe, L. P., & Sydney, E. B. (2019). Sustainability of sugarcane lignocellulosic biomass pre treatment for the production of bioethanol. *Bioresource Technology*, *299*, 122635. https://doi.org/10.1016/j.biortech.2019.122635

Vilarino, M. V., Franco, C., & Quarrington, C. (2017). Food loss and waste reduction as an integral part of a circular economy. *Frontiers in Environmental Science*, *5*, 21. https://doi.org/10.3389/fenvs.2017.00021

Vourdoubas, J. (2020). Possibilities of using small wind turbines for electricity generation in agricultural greenhouses in the Island of Crete, Greece. *International Journal of Agriculture and Environmental Research*, *6*, 746–761.

Wainaina, S., Awasthi, M. K., Sarsaiya, S., Chen, H., Singh, E., Kumar, A., & Kumar, S. (2020). Resource recovery and circular economy from organic solid waste using aerobic and anaerobic digestion technologies. *Bioresource Technology*, *301*, 122778. https://doi.org/10.1016/j.biortech.2020.122778

Winquist, E., Rikkonen, P., Pyysiainen, J., & Varho, V. (2019). Is biogas an energy or a sustainability product? – Business opportunities in the Finnish biogas branch. *Journal of Cleaner Production*, *233*, 1344–1354. https://doi.org/10.1016/j.jclepro.2019.06.181

Xu, X., Ma, Z., Chen, Y., Gu, X., Liu, Q., Wang, Y., Sun, M., & Chang, D. (2018). Circular economy pattern of livestock manure management in Longyou, China. *Journal of Material Cycles and Waste Management*, *20*(2), 1050–1062. https://doi.org/10.1007/s10163-017-0667-4

Xue, J. (2017). Photovoltaic agriculture – New opportunity for photovoltaic applications in China. *Renewable and Sustainable Energy Reviews*, *73*, 1–9. https://doi.org/10.1016/j.rser.2017.01.098

Yadav, S., Panda, S. K., Tiwari, G. N., Al-Helal, I. M., & Hachem-Vermette, C. (2022). Periodic theory of greenhouse integrated semi-transparent photovoltaic thermal (GiSPVT) system integrated with earth air heat exchanger (EAHE). *Renewable Energy*, *184*, 45–55.

Yahya, M., Fahmi, H., & Hasibuan, R. (2022). Experimental performance analysis of a pilot-scale biomass-assisted recirculating mixed-plow dryer for drying paddy. *International Journal of Food Science*, *2022*, 1–15.

Yan, D., Tajima, H., Cline, L. C., Fong, R. Y., Ottaviani, J. I., Shapiro, H. Y., & Blumwald, E. (2022). Genetic modification of flavone biosynthesis in rice enhances biofilm formation of soil diazotrophic bacteria and biological nitrogen fixation. *Plant Biotechnology Journal*, *20*, 2135–2148.

Zabaniotou, A. (2018). Redesigning a bioenergy sector in EU in the transition to circular waste-based bioeconomy – A multidisciplinary review. *Journal of Cleaner Production*, *177*, 197–206. https://doi.org/10.1016/j.jclepro.2017.12.172

Zhu, Q., Lowe, E. A., Wei, Y. A., & Barnes, D. (2007). Industrial symbiosis in China: A case study of the Guitang Group. *Journal of Industrial Ecology*, *11*(1), 31–42. https://doi.org/10.1162/jiec.2007.929

Zou, X., Li, Y. E., Li, K., Cremades, R., Gao, Q., Wan, Y., & Qin, X. (2015). Greenhouse gas emissions from agricultural irrigation in China. *Mitigation and Adaptation Strategies for Global Change*, *20*, 295–315.

5 The Role of Agricultural Business in Low-Carbon Transition

Dalia Streimikiene

5.1 GREEN GROWTH AND AGRICULTURAL BUSINESS

As the global community confronts the urgent challenges posed by climate change, the imperative for sustainable practices across industries becomes increasingly evident. Among these, the agricultural sector stands at the forefront of the battle against greenhouse gas emissions and environmental degradation. Agricultural businesses play a pivotal role in the low-carbon transition, with their decisions and actions shaping the trajectory of sustainable development.

This comprehensive teaching material explores the multifaceted role of agricultural business in the low-carbon transition, encompassing four key subtopics. We delve into the concept of green growth and its integration into agricultural practices, highlighting how businesses can navigate the evolving landscape of sustainability while promoting economic prosperity.

The second subtopic delves into the pivotal role of agricultural businesses in the low-carbon energy transition. We investigate the challenges and opportunities faced by these businesses as they adopt cleaner and renewable energy sources, reducing their carbon footprint and contributing to a greener future.

Our exploration continues with an examination of green agricultural initiatives (GAI); as agricultural businesses increasingly recognize the significance of eco-friendly practices in their operations. We analyze the innovative strategies and technologies that can be implemented to achieve sustainable agricultural production while safeguarding the environment.

Furthermore, the material delves into the critical aspect of corporate social responsibility (CSR) in agriculture. With heightened consumer awareness and growing demands for ethical practices, businesses must adopt a holistic approach to address social and environmental concerns. We explore how agricultural companies can embrace CSR principles to create positive impacts on local communities, foster biodiversity conservation, and strengthen their commitment to sustainability.

This educational material is meticulously designed to provide definitions, in-depth explanations, and real-world examples, ensuring a comprehensive understanding of the role of agricultural business in the low-carbon transition. Whether you are an educator seeking to enrich your students' knowledge or a student aspiring to grasp the significance of sustainable practices in agriculture, this resource serves as an invaluable guide in navigating the path toward a greener and more sustainable future.

Green growth is a concept that seeks to reconcile economic growth with environmental sustainability, emphasizing the importance of decoupling economic activities from excessive resource consumption and environmental degradation. In the context of the agricultural sector, green growth represents a paradigm shift toward sustainable farming practices that promote biodiversity, conserve natural resources, and reduce greenhouse gas emissions. Agricultural businesses play a central role in driving this transition, as they are key stakeholders responsible for producing the world's food and fiber while minimizing their ecological footprint.

Green growth in agriculture refers to the adoption of practices and technologies that foster sustainable agricultural production while mitigating negative environmental impacts. It involves employing innovative methods that maintain or enhance the productivity and profitability of agricultural activities while conserving natural resources and ecosystems.

Green growth denotes a sustainable and environmentally responsible approach to economic development and progress. It aims to foster economic growth and prosperity while simultaneously reducing the negative impacts on the environment, preserving natural resources, and promoting social inclusivity. This concept recognizes that traditional economic growth models, which solely focus on increasing gross domestic product (GDP) and maximizing profit, can lead to ecological degradation, resource depletion, and social inequality in the long term.

The main principles of green growth encompass the following aspects (Hallegatte et al., 2011; Toman, 2012; Capasso et al., 2019; Hao et al., 2020; Abad-Segura et al., 2020; Hysa et al., 2020; Moranta et al., 2021; Liu et al., 2021; O'Neill, 2020; Hermelin & Andersson, 2018; Taghizadeh-Hesary & Yoshino, 2020):

1. **Environmental Sustainability**: Green growth emphasizes the need to preserve and enhance the natural environment for current and future generations. It promotes the responsible use of resources, such as water, energy, and land, to ensure that ecosystems can regenerate and continue providing vital services.
2. **Low-Carbon Economy**: One of the central pillars of green growth is the transition to a low-carbon economy. This involves reducing greenhouse gas emissions by promoting energy efficiency, adopting renewable energy sources, and implementing technologies that produce fewer carbon emissions.
3. **Circular Economy**: Green growth encourages the adoption of a circular economy, where products, materials, and resources are reused, refurbished, or recycled to minimize waste generation and reduce the overall environmental impact.
4. **Biodiversity Conservation**: Protecting biodiversity is crucial for maintaining ecological balance and essential ecosystem services. Green growth strategies incorporate measures to safeguard natural habitats, preserve endangered species, and promote sustainable agriculture and forestry practices.

5. **Social Inclusivity and Equity**: Green growth aims to ensure that economic benefits are distributed fairly and equitably among different sections of society. It seeks to reduce poverty, provide decent work opportunities, and enhance social welfare while transitioning to sustainable economic practices.
6. **Technological Innovation**: Embracing technological advancements is fundamental to green growth. Innovation and research play a pivotal role in developing cleaner technologies, more efficient processes, and sustainable solutions across various sectors.
7. **Long-Term Planning and Policy Frameworks**: Successful green growth necessitates the development and implementation of long-term strategies and policies that support sustainable practices. Governments, businesses, and other stakeholders collaborate to establish frameworks that align economic objectives with environmental and social goals.
8. **Green Finance and Investment**: Mobilizing financial resources to support green initiatives is crucial for implementing green growth strategies. This includes attracting investments in renewable energy, sustainable infrastructure, green businesses, and nature-based solutions.

In the context of agricultural business, green growth implies promoting sustainable agricultural practices that minimize environmental degradation, improve resource efficiency, and ensure food security. This involves adopting eco-friendly farming methods, reducing the use of harmful chemicals, optimizing water usage, and promoting biodiversity conservation in agricultural landscapes.

Green growth represents a holistic and forward-looking approach to economic development that harmonizes economic progress with environmental protection and social well-being. By integrating sustainability principles into policies, businesses, and everyday practices, societies can strive for a more resilient, equitable, and environmentally conscious future.

The global agricultural sector faces numerous challenges, including population growth, climate change, and resource scarcity. Green growth is essential to address these challenges, as it offers a pathway to meet the increasing demand for food and other agricultural products while safeguarding the environment for future generations.

The necessity of green growth in agriculture stems from the urgent need to address the challenges posed by traditional agricultural practices to the environment, society, and the economy. Agriculture is a vital sector that provides food, fiber, and fuel to sustain human life and economic activities. However, conventional agricultural methods have often led to negative consequences (Figure 5.1) which the application of green growth principles in agriculture helps to solve.

Intensive tillage and the use of heavy machinery can lead to soil compaction and erosion, reducing soil fertility and productivity. Prolonged soil degradation can result in decreased crop yields and increased vulnerability to droughts and floods (Alam, 2014). Excessive use of chemical fertilizers and pesticides in traditional agriculture can lead to runoff into water bodies, causing water pollution (Braden & Shortle, 2013). This contamination can harm aquatic ecosystems, impact drinking water

KEY NEGATIVE CONSEQUENCES OF TRADITIONAL AGRICULTURE

- Soil Degradation
- Water Pollution
- Biodiversity Loss
- Greenhouse Gas Emissions
- Deforestation
- Water Scarcity
- Health Risks
- Loss of Traditional Knowledge
- Dependency on External Inputs
- Farmer Debt and Financial Stress
- Food Safety Concerns

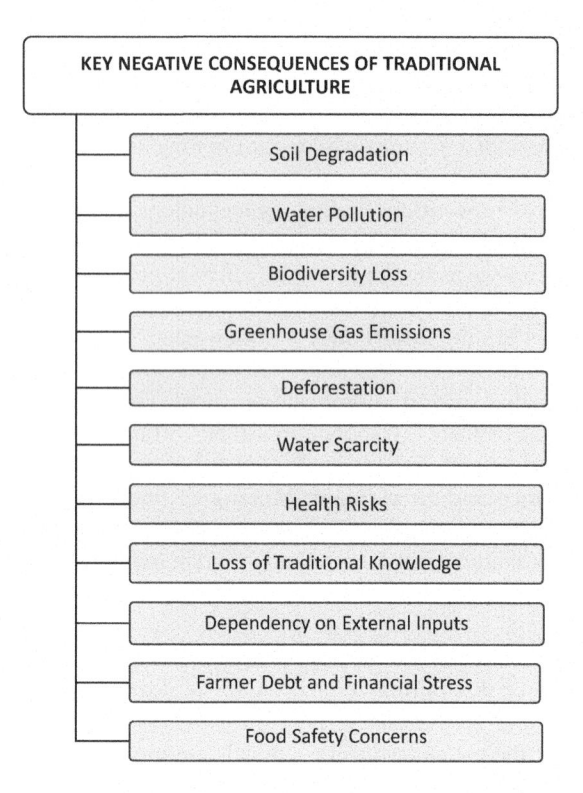

FIGURE 5.1 Negative consequences of traditional agriculture. (Created by authors.)

sources, and disrupt the balance of aquatic life (Stoyanova & Harizanova, 2019). Traditional agriculture often involves the clearing of land for monoculture crops, leading to the destruction of natural habitats and loss of biodiversity. This reduction in biodiversity can negatively affect ecosystem resilience and disrupt ecosystem services, such as pollination and natural pest control (Eriksson, 2021; Chaudhary et al., 2016). Conventional agriculture contributes to greenhouse gas emissions through activities such as methane release from livestock, nitrous oxide emissions from fertilizers, and carbon dioxide emissions from land-use changes and fossil fuel use (Gołasa et al., 2021). These emissions contribute to global climate change and its associated impacts. Expanding agricultural land often leads to deforestation, especially in regions with high rates of forest conversion for agriculture. Deforestation contributes to carbon emissions, loss of biodiversity, and disruptions to local and global climate patterns (Abman et al., 2020). In some regions, traditional agricultural practices consume vast amounts of water, leading to water scarcity and competition for water resources between agriculture, industry, and communities (Rosegrant et al., 2009). Unsustainable water use can lead to depletion of aquifers and rivers, affecting ecosystems and rural livelihoods. Pesticides and chemical fertilizers used in traditional agriculture can pose health risks to farmers, farmworkers, and consumers. Exposure to these chemicals can lead to acute and chronic health issues, affecting respiratory, neurological, and reproductive systems. Intensive modern agriculture practices may lead to the abandonment of traditional and indigenous agricultural knowledge and practices, which are often adapted to local environments and more sustainable. Traditional agriculture's heavy reliance on chemical inputs and fossil fuels can create economic vulnerabilities for farmers, as they become dependent on external suppliers and market fluctuations. Addressing these negative consequences of traditional agricultural methods and transitioning to more sustainable and regenerative practices is crucial for ensuring long-term food security, environmental preservation, and sustainable development.

Adopting green growth principles in agriculture is essential for the following reasons (Shen et al., 2021; Devi et al., 2023; Pogonyshev et al., 2021; Koohafkan et al., 2012; Shen et al., 2020; Evans & Lawson, 2020; Nwachukwu, 2023; Ji & Hoti, 2021; Li et al. 2021; Eyhorn et al., 2019):

- **Environmental Preservation**: Conventional agriculture has been associated with deforestation, soil erosion, water pollution from chemical fertilizers and pesticides, and biodiversity loss. Green growth in agriculture seeks to reverse these trends by promoting sustainable land management practices, conservation of natural habitats, and reduced use of harmful chemicals, thereby preserving the environment and its ecosystems.
- **Sustainable Resource Management**: Conventional agriculture often relies heavily on non-renewable resources, such as fossil fuels, and depletes essential resources like water and fertile soil. Green growth in agriculture focuses on sustainable resource management, optimizing water usage, adopting renewable energy sources, and promoting soil conservation to ensure the long-term viability of agricultural practices.
- **Climate Change Mitigation**: Agriculture is a significant contributor to greenhouse gas emissions, mainly through methane from livestock and nitrous oxide from fertilizers. By adopting climate-smart agricultural practices, such as agroforestry, sustainable livestock management, and reduced tillage, the sector can play a role in mitigating climate change and reducing its carbon footprint.
- **Biodiversity Conservation**: Traditional agriculture often leads to the loss of biodiversity due to monoculture crops and habitat destruction. Green growth in agriculture incorporates practices that promote biodiversity conservation, such as crop rotation, polyculture, and maintaining natural habitats on farms, which enhance ecosystem resilience and contribute to sustainable food production.
- **Resource Efficiency**: Green growth in agriculture emphasizes resource efficiency, aiming to produce more food with fewer inputs. This involves optimizing water usage, promoting efficient irrigation systems, and adopting precision agriculture techniques to minimize waste and maximize productivity.
- **Adaptation to Climate Change**: Climate change poses challenges to agricultural productivity due to shifting weather patterns and increased frequency of extreme events. Green growth strategies focus on building resilient agricultural systems that can withstand climate variability and adapt to changing conditions.

- **Sustainable Food Production**: With a growing global population, ensuring food security becomes paramount. Green growth in agriculture promotes sustainable food production that meets the nutritional needs of the population without depleting natural resources or compromising the ability of future generations to meet their own needs.
- **Consumer Demand**: Increasingly, consumers are seeking environmentally friendly and sustainable products, including food. Green growth in agriculture responds to this demand, enhancing market access for farmers who adopt sustainable practices and cater to environmentally conscious consumers.
- **Economic Benefits**: Green growth in agriculture can lead to economic benefits for farmers and rural communities. By improving resource efficiency and reducing waste, farmers can cut production costs and increase profitability. Additionally, embracing sustainable practices can open up new market opportunities for environmentally conscious consumers.
- **Social Inclusivity**: Green growth in agriculture also considers the social dimension, aiming to improve livelihoods and working conditions for farmers and agricultural laborers. By promoting sustainable practices and providing support for smallholder farmers, it can contribute to poverty reduction and rural development.
- **Global Sustainability Goals**: Green growth in agriculture aligns with international sustainability goals, such as the United Nations' Sustainable Development Goals (SDGs) and the Paris Agreement on climate change. Implementing sustainable agricultural practices is crucial for achieving these global objectives.

Green growth in agriculture is not only necessary but also imperative to ensure a sustainable and secure future for our planet and its inhabitants. By embracing environmentally friendly and socially inclusive practices, the agricultural sector can play a pivotal role in addressing pressing global challenges, such as climate change, resource depletion, and food insecurity. Policymakers, farmers, businesses, and consumers all have a role to play in promoting green growth in agriculture and creating a resilient and prosperous future for all.

Green growth in agriculture aligns with the triple bottom line (TBL) principle, focusing on three core dimensions: economic prosperity, social well-being, and environmental stewardship. Agricultural businesses aiming for green growth must balance these dimensions to create a sustainable and resilient food system.

The TBL approach in agriculture is a framework that encourages businesses and farmers to consider not only economic performance but also social and environmental aspects of their operations (Singh & Srivastava, 2022; Arora et al., 2016). The TBL approach aims to create a balance between economic prosperity, social well-being, and environmental sustainability, ensuring that agricultural practices contribute positively to all three dimensions. The three pillars of the TBL are often referred to as "Profit, People, and Planet" (Price & Hacker, 2009; Svensson et al., 2018; Zanin et al., 2020):

- **Profit**: The economic dimension of the TBL focuses on financial performance and profitability. It involves making sound financial decisions, optimizing resource use, and seeking economic growth and efficiency in

The Role of Agricultural Business in Low-Carbon Transition

agricultural operations. Farmers and agricultural businesses aim to generate profits, create value, and contribute to economic development while ensuring the long-term viability and success of their ventures.

- **People**: The social dimension of the TBL emphasizes the well-being of people involved in agriculture, including farmers, farmworkers, rural communities, and consumers. It considers fair labor practices, community development, social inclusivity, and the overall quality of life of those impacted by agricultural activities. The people aspect also includes considerations for food safety, access to nutritious food, and the health and safety of agricultural workers.
- **Planet**: The environmental dimension of the TBL focuses on minimizing negative environmental impacts and promoting sustainable practices in agriculture. This includes reducing greenhouse gas emissions, conserving water resources, promoting biodiversity, adopting regenerative farming techniques, and mitigating pollution and waste. The planet aspect seeks to ensure that agricultural practices do not deplete natural resources or harm ecosystems, and instead, contribute to environmental preservation and resilience.

The TBL approach recognizes that a narrow focus on economic profit alone is insufficient for long-term success and sustainability. By integrating social and environmental considerations into decision-making processes, the TBL framework encourages more responsible and holistic practices in agriculture (Loviscek, 2021; Zegar & Wrzaszcz, 2017; Wang et al., 2009).

Some examples of applying the TBL approach in agriculture include (Figure 5.2) (Arora et al., 2016; Zanin et al., 2020; Nogueira et al., 2020; Mili & Loukil, 2023):

- **Sustainable Farming Practices**: Farmers adopt conservation agriculture, organic farming, or agroforestry methods that promote soil health, reduce chemical inputs, and enhance biodiversity while maintaining economic productivity.
- **Fair Trade and Ethical Sourcing**: Agricultural businesses and consumers support fair trade practices that ensure farmers receive fair compensation for their products and adhere to ethical labor and social standards.

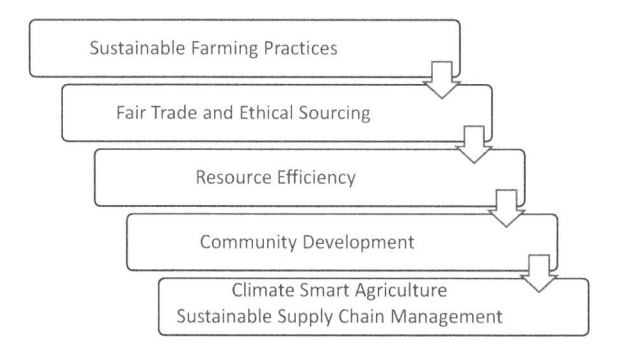

FIGURE 5.2 Examples of applying the triple bottom line approach in agriculture. (Created by authors.)

- **Resource Efficiency**: Implementing precision agriculture technologies and water-efficient irrigation systems helps optimize resource use, reducing costs and environmental impact.
- **Community Development**: Agricultural businesses engage with local communities, support rural development projects, and contribute to initiatives that enhance the well-being of people living in and around farming areas.
- **Climate Smart Agriculture**: Farmers adopt climate-resilient practices to adapt to changing weather patterns and contribute to climate change mitigation by sequestering carbon in soils or using renewable energy.
- **Sustainable Supply Chain Management**: Agricultural companies incorporate sustainability criteria when selecting suppliers and partners, ensuring that their entire supply chain aligns with social and environmental values.

The TBL approach in agriculture acknowledges that the success of the agricultural sector is intertwined with the well-being of people and the health of the planet. By adopting a balanced and comprehensive approach that considers economic, social, and environmental factors, agriculture can become more sustainable, resilient, and equitable for current and future generations.

The importance of green growth strategies in agricultural business lies in their ability to create a harmonious balance between economic growth, social well-being, and environmental protection. By implementing sustainable practices and responsible approaches, agricultural businesses can play a crucial role in addressing global challenges while securing a more sustainable and prosperous future for themselves and the planet.

Implementing green growth strategies in agricultural businesses involves adopting sustainable practices and integrating environmentally friendly principles into their operations. The main strategies for green growth in agricultural businesses (European Commission, 2023; Yıldırım & Yılmaz, 2023; Wezel et al., 2013; Amede et al., 2023; Velasco-Muñoz et al., 2022; Smidt et al., 2016; Kumar et al., 2020; Finco et al., 2023; Morgan et al., 2010; Nair et al., 2010; Tavares et al., 2019) are as follows:

- **Sustainable Farming Practices**: Adopting sustainable farming practices is at the core of green growth in agriculture. This includes practices such as conservation agriculture, organic farming, agroforestry, crop rotation, and integrated pest management (IPM). These methods promote soil health, reduce the use of synthetic chemicals, and enhance biodiversity while maintaining or even increasing agricultural productivity.
- **Water Management**: Implementing efficient water management practices is crucial for green growth. This involves using precision irrigation techniques, rainwater harvesting, and drip irrigation systems to optimize water usage, reduce waste, and mitigate water scarcity.
- **Renewable Energy**: Transitioning to renewable energy sources, such as solar, wind, or bioenergy, for powering farm operations and facilities can significantly reduce the carbon footprint of agricultural businesses. Expanding the strategy of renewable energy in agricultural businesses

involves integrating clean and sustainable energy sources to power farm operations and reduce the carbon footprint of the agricultural sector. Renewable energy solutions not only contribute to green growth but also offer several benefits for farmers and the environment.

Solar panels can be installed on farms to harness the power of the sun and convert it into electricity. Solar energy is particularly useful for powering farm buildings, irrigation systems, and other electrical equipment. It can reduce electricity costs, provide energy independence, and lower greenhouse gas emissions.

Wind turbines can be installed in suitable locations on farms to generate electricity from wind power. Wind energy can complement solar energy and provide a stable power supply even during cloudy days. It is especially relevant for farms located in windy regions.

Biogas can be produced from organic materials like manure, crop residues, and food waste through anaerobic digestion. The biogas can be used as a renewable fuel for electricity generation or as a replacement for fossil fuels in heating and cooking.

Farms with access to running water or streams may consider small-scale hydroelectric power systems to generate electricity. Hydroelectric power is a reliable and consistent renewable energy source, especially for farms located near water bodies.

Embracing renewable energy demonstrates an agricultural business's commitment to sustainability and environmental responsibility, which can enhance brand reputation and attract environmentally conscious consumers. By integrating renewable energy solutions, agricultural businesses can not only enhance their environmental performance but also improve their economic efficiency and contribute to a greener, more sustainable future for agriculture and society as a whole.

- Employing precision agriculture technologies, data analytics, and IoT (Internet Efficient Resource Use of Things) devices can help optimize the use of inputs like fertilizers, pesticides, and water, leading to cost savings and reduced environmental impacts.
- **Carbon Sequestration**: Implementing practices that enhance carbon sequestration in soils, such as cover cropping and agroforestry, can contribute to mitigating climate change while improving soil health.
- **Waste Management and Recycling**: Implementing waste management practices, such as composting agricultural residues and recycling packaging materials, can reduce waste generation and contribute to a circular economy.
- **Biodiversity Conservation**: Preserving and promoting biodiversity on agricultural lands through measures like maintaining hedgerows, wildlife corridors, and native plant species can improve ecosystem services and enhance pollination and natural pest control.
- **Sustainable Livestock Management**: Implementing practices that prioritize animal welfare, reduce methane emissions, and optimize feed efficiency in livestock operations can make the livestock sector more sustainable.
- **Value-Added Products**: Developing and marketing value-added agricultural products, such as organic or sustainably produced goods, can create new market opportunities and attract environmentally conscious consumers.

- **Certification and Standards**: Obtaining eco-certifications, such as organic or fair-trade labels, can enhance the credibility of agricultural businesses committed to green growth and sustainable practices.
- **Partnerships and Collaboration**: Collaborating with NGOs, research institutions, and government agencies can foster knowledge exchange and support the implementation of sustainable practices.
- **Consumer Education and Awareness**: Engaging with consumers and raising awareness about the importance of sustainable agriculture can help drive demand for green products and support the adoption of sustainable practices throughout the supply chain.
- **Financial Incentives**: Governments and financial institutions can provide incentives, grants, or loans to support agricultural businesses' transition to sustainable practices and green technologies.
- **Continuous Improvement**: Encouraging a culture of continuous improvement within the agricultural business, where sustainable practices are regularly evaluated and optimized, ensures ongoing progress toward green growth goals.
- **Precision Agriculture**: The implementation of precision agriculture technologies allows farmers to optimize resource use and minimize waste. By using sensors, GPS, and data analytics, agricultural businesses can precisely apply inputs like fertilizers and water, enhancing efficiency and reducing environmental pollution.
- **Climate-Smart Agriculture**: Climate-smart agriculture focuses on adapting to climate change while mitigating its effects. Techniques such as drought-resistant crop varieties, climate-resilient livestock management, and climate-smart irrigation contribute to increased agricultural productivity in the face of changing weather patterns.

By implementing these strategies, agricultural businesses can contribute to green growth, improving their environmental performance, reducing resource consumption, and creating more sustainable and resilient farming systems. Green growth not only benefits the environment but can also lead to increased efficiency, cost savings, and improved market competitiveness in the long run.

There are some Green Business Models in agriculture (Figures 5.3 and 5.4).

The papers propose that there are multiple successful environmentally friendly business models in the field of agriculture. Saenchaiyathon and Wongthongchai (2021) discovered that adopting a green operational approach can result in enhanced efficiency, improved environmental impact, and better economic performance for small and medium-sized agri-food businesses. Donner et al. (2021) identified key factors that determine the success and risks of eco-innovative business models focused on a circular economy by utilizing agricultural waste or by-products. Rezaee et al. (2020) devised a green entrepreneurship model specific to the agricultural sector, incorporating elements such as scientific capacity, strategic positioning, and policy development. Pan (2021) observed that the presence of green intellectual capital, green innovation, and environmental leadership can provide agricultural companies with a sustainable competitive advantage. Overall, the collective findings suggest

The Role of Agricultural Business in Low-Carbon Transition

GREEN BUSINESS MODELS IN AGRICULTURE (1)		
	Organic Farming	Organic farming is a well-known green business model that relies on natural processes and avoids the use of synthetic chemicals, pesticides, and genetically modified organisms. Organic farmers focus on soil health, biodiversity conservation, and sustainable farming practices, providing consumers with food produced in an environmentally friendly manner.
	Agroforestry	Agroforestry integrates trees and crops on the same piece of land, promoting biodiversity, soil conservation, and climate change mitigation. This model allows farmers to diversify income streams by harvesting both agricultural products and tree products like fruits, nuts, or timber.
	Community Supported Agriculture (CSA)	CSA is a direct marketing model where consumers become shareholders in a farm's produce, sharing the risks and rewards of farming. By fostering direct relationships between farmers and consumers, CSA encourages local, seasonal, and sustainable food production.
	Precision Agriculture	Precision agriculture uses advanced technologies, such as GPS, sensors, and data analytics, to optimize the use of inputs like water, fertilizers, and pesticides. This model enhances resource efficiency, reduces waste, and minimizes environmental impacts.
	Vertical Farming	Vertical farming involves growing crops in stacked layers or vertically inclined surfaces, often indoors or in urban environments. This approach maximizes land use, reduces water consumption, and minimizes transportation emissions associated with long-distance food supply chains.
	Sustainable Aquaculture	Sustainable aquaculture practices prioritize fish health and welfare while minimizing environmental impacts. This includes adopting responsible feed sources, reducing water pollution, and using recirculating systems to conserve water.

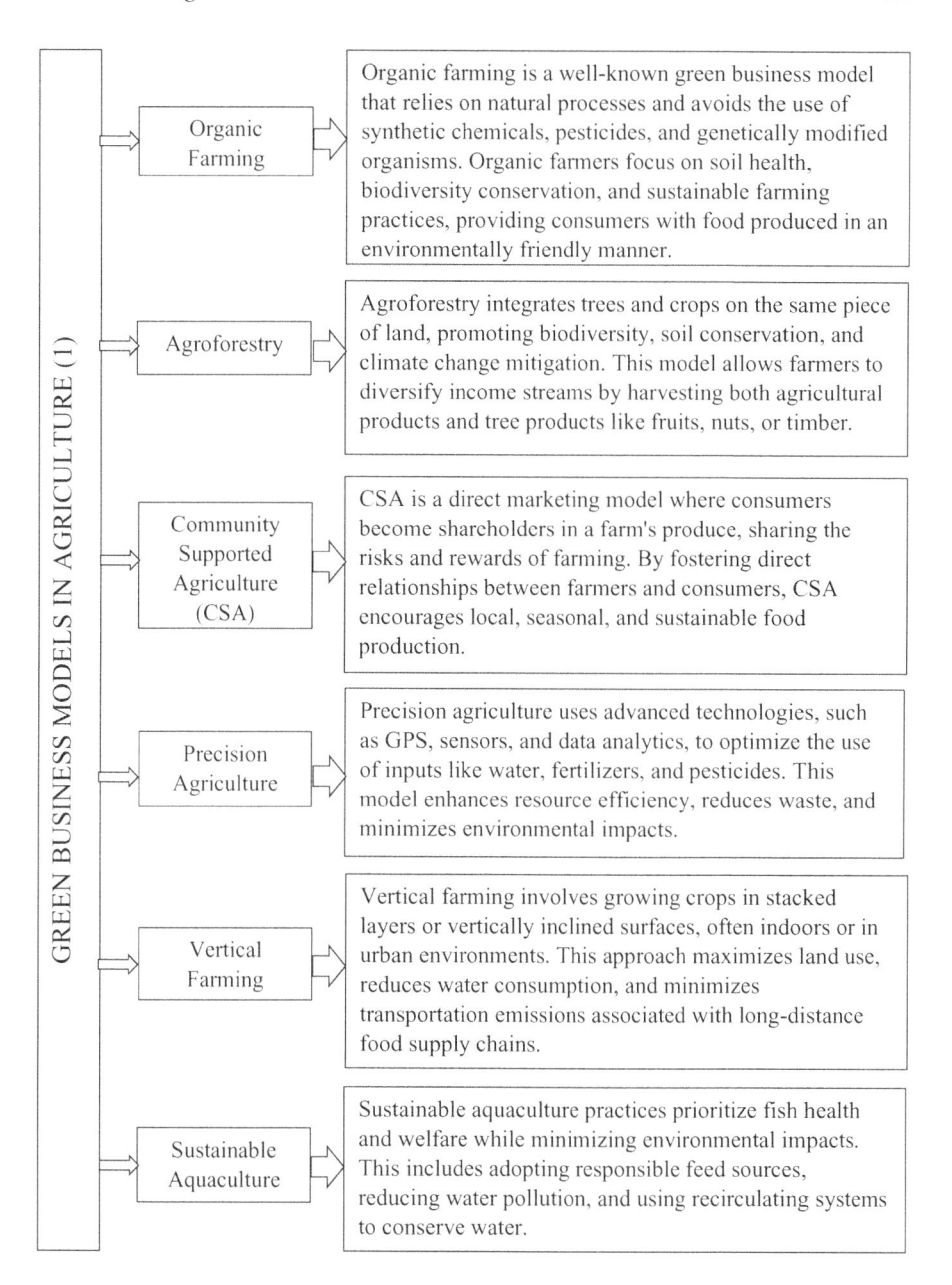

FIGURE 5.3 Green business models in agriculture (1). (Created by authors.)

that prosperous green business models in agriculture encompass the adoption of eco-friendly strategies, the integration of innovative technologies, and the consideration of environmental and social aspects while catering to market demands.

Green operation strategies and social enterprise business models have proven to be effective in supporting small-scale farmers. Tinsley and Agapitova (2018) emphasize

GREEN BUSINESS MODELS IN AGRICULTURE (2)

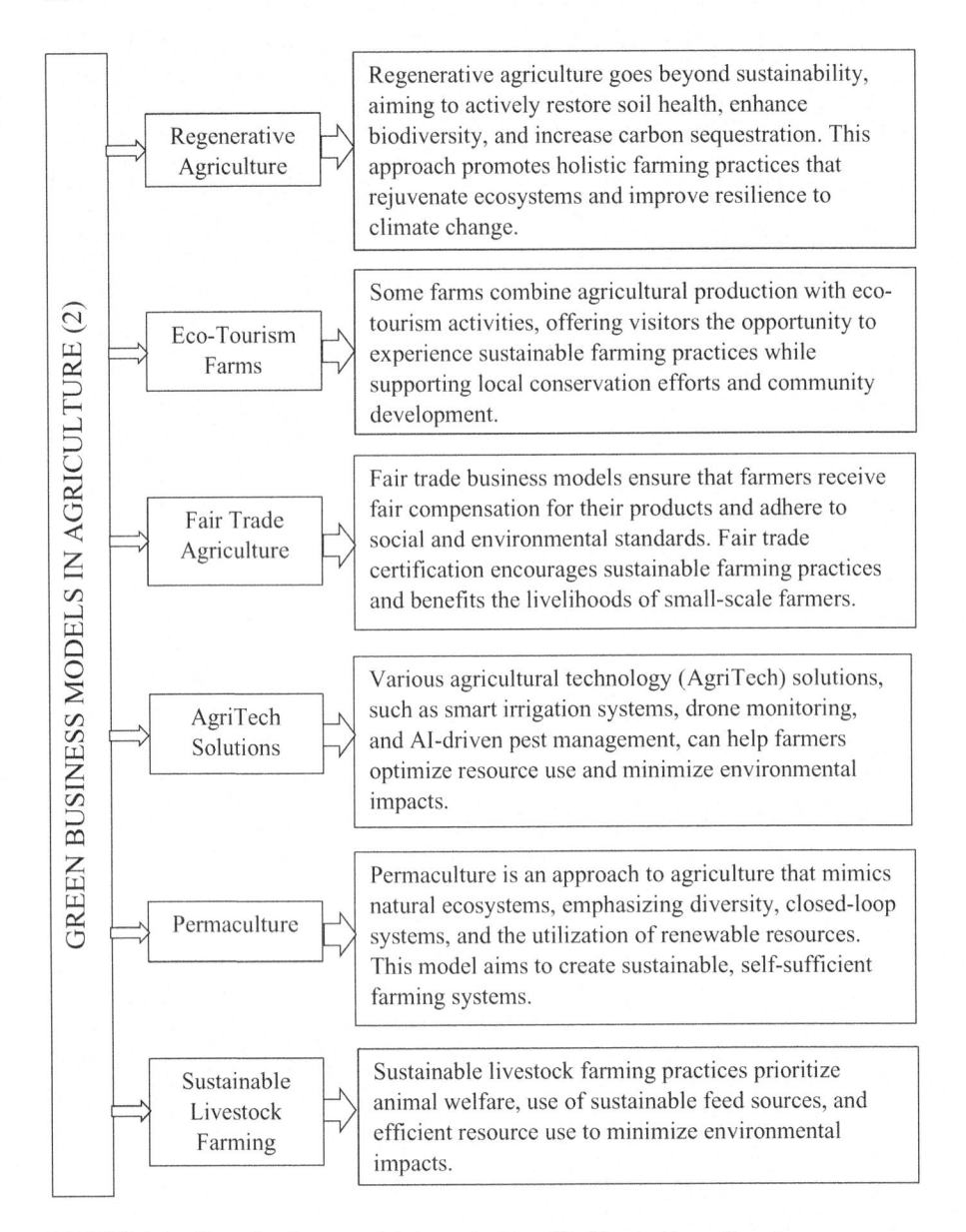

Regenerative Agriculture	Regenerative agriculture goes beyond sustainability, aiming to actively restore soil health, enhance biodiversity, and increase carbon sequestration. This approach promotes holistic farming practices that rejuvenate ecosystems and improve resilience to climate change.
Eco-Tourism Farms	Some farms combine agricultural production with eco-tourism activities, offering visitors the opportunity to experience sustainable farming practices while supporting local conservation efforts and community development.
Fair Trade Agriculture	Fair trade business models ensure that farmers receive fair compensation for their products and adhere to social and environmental standards. Fair trade certification encourages sustainable farming practices and benefits the livelihoods of small-scale farmers.
AgriTech Solutions	Various agricultural technology (AgriTech) solutions, such as smart irrigation systems, drone monitoring, and AI-driven pest management, can help farmers optimize resource use and minimize environmental impacts.
Permaculture	Permaculture is an approach to agriculture that mimics natural ecosystems, emphasizing diversity, closed-loop systems, and the utilization of renewable resources. This model aims to create sustainable, self-sufficient farming systems.
Sustainable Livestock Farming	Sustainable livestock farming practices prioritize animal welfare, use of sustainable feed sources, and efficient resource use to minimize environmental impacts.

FIGURE 5.4 Green business models in agriculture (2). (Created by authors.)

the achievements of social enterprises in tackling the challenges faced by smallholder farmers. The study identified more than 100 social enterprises employing nine distinct business models that span the entire agriculture value chain. Saenchaiyathon and Wongthongchai (2021) further corroborate the benefits of green operation strategies, demonstrating that they lead to improved efficiency, environmental impact, and economic performance for agri-food small and medium-sized enterprises (SMEs). In addition, Zhu and Chen (2022) highlight the variation in preferences among small-scale farmers for green agriculture policy incentives. Farmers' age, education level, family size, and involvement in agricultural organizations significantly influence their preferences. The collective findings suggest that successful green business models for small-scale farmers should address specific pain points along the value chain, integrate green operation strategies, and tailor policy incentives according to farmers' characteristics and preferences.

Effective green business models for large-scale farmers vary depending on the unique circumstances and challenges faced by each country. Barth and Melin (2018) introduce a Lean Implementation Framework designed to enhance production, profitability, and environmental sustainability among small and mid-size farms in Sweden. Sher et al. (2019) identify obstacles hindering the adoption of green entrepreneurial farming in Pakistan, including barriers related to training and development, as well as the limited involvement of the government in supporting such initiatives. Medina et al. (2015) examine how European farmers are adapting to the Common Agricultural Policy (CAP) and its impact on farm pathway dependence, revealing that intensive and extensive farmers employ distinct coping strategies. Rezaee et al. (2020) present a green entrepreneurship model tailored to the agricultural sector in Kermanshah County, Iran, which is developed through expert interviews and encompasses causal conditions, interventional variables, strategies, and consequences for the region. Successful green business models for large-scale farmers necessitate a customized approach that considers the specific challenges and opportunities unique to each country or region.

Adopting green business models in agriculture not only benefits the environment and society but can also enhance the resilience and profitability of agricultural operations in the long term. These models offer opportunities for farmers and agribusinesses to respond to consumer demands for sustainable products while contributing to the global effort to address environmental challenges and promote a more sustainable future.

To promote green growth in agriculture, policymakers can implement a range of policies and incentives that encourage sustainable practices and support the adoption of environmentally friendly technologies. These policies aim to create an enabling environment for agricultural businesses to transition toward greener and more sustainable practices.

Some key policies and incentives for green growth in agriculture are presented in Figure 5.5.

Governments can provide financial incentives such as *subsidies and grants* to support farmers in adopting sustainable practices and technologies (Heyl et al., 2022). These financial incentives can help offset the initial costs of transitioning to green practices, making them more accessible to farmers. The research papers propose that governments have employed subsidies as a means to encourage environmentally friendly development in agriculture; however, the effectiveness of these policies varies. Akkaya et al. (2019) discovered that subsidy policies outperform hybrid policies in terms of achieving greater social welfare. Yet, policies involving zero-expenditure, where taxes imposed on non-adopters are used to fund subsidies for adopters, result in a decline in social welfare. Pan (2022) investigated the influence of grain subsidy policies on farmers' green production behavior in China and found

POLICY AND INCENTIVES FOR GREEN GROWTH IN AGRICULTURE
- Subsidies and Grants
- Tax Incentives
- Eco-Certification and Labeling
- Research and Development Funding
- Education and Training
- Regulatory Measures
- Market Access and Support
- Land-Use Planning
- Climate Change Adaptation and Resilience
- Green Financing
- Public-Private Partnerships
- Sustainable Agricultural Practices Incentives

FIGURE 5.5 Policy and incentives for green growth in agriculture. (Created by authors.)

that providing grain subsidies directly to farmland operators often leads to excessive fertilizer application behavior. On the other hand, Feng et al. (2021) observed that policy-driven afforestation has played a significant role in large-scale greening in China, while policies linked to agricultural intensification have contributed to improved vegetation cover.

Overall, the papers indicate that subsidies can be effective in promoting environmentally conscious growth in agriculture, but the specific design and implementation of these policies can have varying impacts on environmental outcomes.

Tax incentives, such as tax breaks or credits, can be provided to agricultural businesses that demonstrate a commitment to environmentally sustainable and green growth practices. These incentives serve to reduce the overall tax burden, encouraging more businesses to invest in sustainable approaches. The research papers present diverse findings on the effectiveness of tax incentives in promoting green growth within the agriculture sector. Tax policy tools, including tax incentives, have a positive impact on environmental sustainability in agriculture. Guo et al. (2022) highlighted the efficacy of digital inclusive finance, which may incorporate tax incentives, in effectively fostering green development in agriculture in China. However, Wang (2022) found that subsidies can negate the positive effects of tax incentives on green innovation in Chinese listed companies. Broader importance of reducing greenhouse gas emissions in agriculture does not directly address the effectiveness of tax incentives. Tax incentives can indeed be effective in promoting green growth in agriculture, but their success may be contingent on specific contextual factors, such as financing constraints and the presence of subsidies.

Governments can establish *eco-certification programs or labeling schemes* that recognize and reward farmers and agricultural products adhering to sustainable standards. These certifications can help consumers make informed choices and create a market demand for sustainable products.

Investing in research and development related to sustainable agriculture can drive innovation and the development of green technologies. Governments can allocate funds for research projects that focus on improving agricultural productivity while minimizing environmental impact. Policy measures can be directed toward providing education and training to farmers on sustainable practices and techniques. This can help raise awareness about the benefits of green growth and build capacity within the agricultural sector. Governments can introduce regulations that set environmental standards for agricultural practices, limiting the use of harmful chemicals and promoting sustainable land management. These regulations can drive the adoption of green practices and ensure compliance across the industry. Policymakers can facilitate market access for green products by promoting sustainable public procurement, supporting farmers' access to green markets, and facilitating trade of sustainable agricultural products. Effective land-use planning can help designate areas for sustainable agriculture, protect natural habitats, and preserve agricultural land from urbanization and conversion. Policies that support climate change adaptation and resilience in agriculture can help farmers cope with the impacts of climate change and encourage the adoption of climate-smart practices. Governments and financial institutions can establish specialized green financing mechanisms that offer preferential loan terms or lower interest rates for investments in sustainable agriculture. Encouraging public–private partnerships can facilitate knowledge sharing, technology transfer, and the implementation of sustainable agriculture projects. Governments can provide additional incentives for practices like agroforestry, conservation agriculture, and organic farming, which have significant environmental benefits.

5.2 BUSINESS IN LOW-CARBON ENERGY TRANSITION

The world is witnessing a seismic shift toward renewable energy sources as a response to the pressing challenges posed by climate change and the need for a low-carbon future. The transition to low-carbon energy is not only driven by environmental concerns but also presents a significant economic opportunity for businesses. Embracing renewable energy and transitioning away from fossil fuels can lead to enhanced competitiveness, reduced operational costs, and positive brand positioning. In this teaching material, we explore the crucial role of businesses in the low-carbon energy transition and how they can contribute to a cleaner, more sustainable future.

5.2.1 Definition of Low-Carbon Energy

Low-carbon energy refers to energy sources that produce minimal greenhouse gas emissions during their lifecycle, including production, distribution, and consumption. These energy sources, predominantly renewables, such as solar, wind, hydro, geothermal, and biomass, play a vital role in reducing carbon emissions and mitigating climate change. The transition toward low-carbon energy is a critical aspect of combating climate change and reducing the overall carbon footprint of human activities.

The primary goal of low-carbon energy is to limit the release of greenhouse gases into the atmosphere, which are major contributors to global warming and climate change. The main greenhouse gases include carbon dioxide (CO_2), methane (CH_4), nitrous oxide (N_2O), and fluorinated gases. By transitioning to low-carbon energy sources, we can mitigate the adverse effects of climate change, reduce air pollution, and promote sustainable development.

Here are some key low-carbon energy sources and technologies (Gul & Chaudhry, 2022; Jianqiang et al., 2011; Ge et al., 2022; Azar et al., 2010; Nurdiawati & Urban, 2021):

- **Renewable Energy**: Renewable energy sources harness natural processes that replenish themselves over time and have minimal or no emissions during operation. Common examples include:
  a. *Solar Energy*. Capturing sunlight and converting it into electricity using photovoltaic cells or solar thermal systems.
  b. *Wind Energy*. Utilizing wind turbines to convert kinetic energy from the wind into electricity.
  c. *Hydropower*. Generating electricity from the flow of water in rivers or from stored water in reservoirs.
  d. *Geothermal Energy*. Tapping into the heat from the Earth's core to produce electricity or for direct heating purposes.
  e. *Biomass Energy*. Using organic materials such as wood, agricultural residues, or waste to produce heat or electricity.
- **Nuclear Energy**: Nuclear power generates electricity through controlled nuclear reactions that release heat. While nuclear power plants produce low-carbon emissions during operation, they have concerns related to safety, radioactive waste disposal, and non-proliferation of nuclear weapons.
- **Carbon Capture and Storage (CCS)**: CCS technologies capture carbon dioxide emissions from industrial processes or power plants and store them deep underground in geological formations to prevent their release into the atmosphere.

CCS holds significant prospects in the global efforts to combat climate change and achieve a low-carbon future. While the technology is still in its early stages of deployment, it offers several potential benefits and opportunities (Araújo & Medeiros, 2017; Haszeldine, 2009; Boot-Handford et al., 2014; Bobicki et al., 2012; Theuer & Olarte, 2023; Grove & Peridas, 2023):

1. **Greenhouse Gas Emissions Reduction**: CCS provides a means to capture carbon dioxide emissions from industrial processes and power plants, preventing them from being released into the atmosphere. This can significantly reduce greenhouse gas emissions and help countries and industries meet their climate targets.

2. **Compatibility with Existing Infrastructure**: CCS can be retrofitted to existing fossil fuel-based power plants and industrial facilities, allowing them to continue operation while reducing their carbon footprint. This feature is especially valuable in regions heavily reliant on fossil fuels for energy generation.
3. **Base Load Electricity Generation**: Unlike some renewable energy sources, CCS-equipped power plants can provide stable, continuous electricity generation, making them suitable for meeting base load demand and ensuring grid stability.
4. **Decarbonizing Hard-to-Abate Sectors**: Industries with high carbon emissions that are challenging to decarbonize, such as cement, steel, and chemical production, can benefit from CCS implementation, enabling them to continue operations while reducing their environmental impact.
5. **Carbon Neutrality for Bioenergy**: Biomass-based power generation with CCS, known as Bioenergy with Carbon Capture and Storage (BECCS), can achieve carbon neutrality or even create net-negative emissions, as the captured CO_2 is effectively removed from the atmosphere.
6. **Job Creation and Economic Opportunities**: The deployment of CCS technologies can create jobs in research, development, construction, and maintenance of the infrastructure, fostering economic growth and technological innovation.
7. **Bridge to a Low-Carbon Future**: While renewable energy sources are being rapidly adopted, complete decarbonization will take time. CCS can act as a transitional technology, helping bridge the gap between fossil fuels and a fully renewable energy system.
8. **International Collaboration**: CCS can foster global collaboration in climate change mitigation efforts. Countries with abundant geological storage capacities can collaborate with nations seeking to implement CCS technologies but lack suitable storage options.

Despite its promising potential, there are also challenges and concerns associated with CCS (Corsten et al., 2013). The current cost of CCS implementation is relatively high compared to conventional energy sources, limiting its widespread adoption (Rubin et al., 2015). However, ongoing research and development are expected to drive down costs in the future. Ensuring the safe and permanent storage of captured carbon dioxide underground is a critical aspect of CCS. Leakage or accidental release of stored CO_2 could have adverse environmental and health impacts. Some communities may have concerns about the safety and environmental implications of CCS, which could lead to opposition and hinder project development. Identifying suitable geological formations for long-term CO_2 storage can be challenging, and the available capacity may not be sufficient to accommodate all emissions.

- **Energy Efficiency**: Enhancing energy efficiency in buildings, transportation, and industries reduces energy consumption and, consequently, the associated carbon emissions.

The research papers present varied findings regarding the impact of energy efficiency on the cost of energy. While energy-efficient appliances reduce the marginal price of the services they provide, the actual energy savings tend to be smaller than what engineering projections suggest. Jacobsen (2015) conducted a study and found no evidence indicating that electricity prices influence consumers' inclination to choose high-efficiency appliances. On the other hand, Fan et al. (2017) discovered a positive correlation between energy efficiency and the financial performance of energy-intensive firms in China. Allan et al. (2007) found that enhancing energy efficiency can result in rebound effects of approximately 30%–50%, without any backfire.

Energy efficiency offers a wide range of benefits for consumers, both in terms of cost savings and improved quality of life. One of the most immediate and tangible benefits of energy efficiency is lower energy bills (Figus et al., 2016; Freire-González et al., 2017). By using energy-efficient appliances, lighting, and heating/cooling systems, consumers can significantly decrease their monthly energy expenses, leaving more money in their pockets. Energy-efficient improvements, such as better insulation, high-efficiency windows, and well-sealed doors, can help maintain a more consistent indoor temperature. This contributes to a more comfortable and pleasant living environment throughout the year.

Energy efficiency offers numerous benefits for businesses across various sectors (Gennitsaris et al., 2023; Agrawal et al., 2023). Implementing energy-efficient practices and technologies can lead to financial savings, operational improvements, and enhanced environmental stewardship. One of the most significant benefits for businesses is cost savings. Energy-efficient measures reduce energy consumption, leading to lower utility bills and operational expenses. These savings can directly contribute to increased profits and improved financial stability. Businesses that prioritize energy efficiency can gain a competitive edge in the market. Consumers and investors are increasingly valuing environmentally responsible practices, which can attract more customers and positively influence brand perception.

- **Hydrogen**: Producing hydrogen using renewable electricity through a process called electrolysis can create a low-carbon energy carrier that can be used in various sectors, including transportation and industrial processes. Hydrogen is considered a versatile and promising low-carbon energy source that has the potential to play a significant role in the transition to a more sustainable and clean energy future. It can be produced through various methods and utilized in different sectors to reduce greenhouse gas emissions and promote energy diversification.

Hydrogen can be burned in conventional gas turbines to generate electricity, with water vapor as the only emission. It can also be used in conjunction with natural gas to reduce emissions in existing power plants. Hydrogen is a crucial feedstock in industries such as ammonia production, petroleum refining, and steel manufacturing. Replacing fossil fuel-derived hydrogen with low-carbon hydrogen can reduce emissions from these processes. However, it faces certain challenges (Zhu & Wei, 2022; Görlach et al., 2023; Gorji, 2023; EFI-KAPSARC Workshop Report, 2023):

1. **Production Efficiency**: The production of hydrogen requires energy, and the efficiency of different production methods varies. Green hydrogen production, while carbon-neutral, can be energy-intensive if not powered by renewable sources.
2. **Infrastructure Development**: Establishing a hydrogen infrastructure, including production facilities, transportation, storage, and distribution networks, requires substantial investment and planning.
3. **Cost**: The cost of hydrogen production, especially green hydrogen, is currently higher compared to conventional fuels. However, as technology advances and economies of scale are achieved, costs are expected to decrease.
4. **Storage and Transport**: Hydrogen has a low energy density per unit volume, which presents challenges in storage and transport. Compressed or liquefied hydrogen must be handled safely and efficiently.
5. **Scaling Up**: The widespread adoption of hydrogen technologies requires policy support, international collaboration, and investment to scale up production and utilization.

Hydrogen, when produced and utilized with a focus on low-carbon methods, has the potential to contribute significantly to global efforts to mitigate climate change, reduce emissions, and create a more sustainable energy landscape.

- **Advanced Nuclear**: Ongoing research and development are focused on advanced nuclear technologies that aim to improve safety, waste management, and efficiency while reducing greenhouse gas emissions.

Advanced nuclear technologies represent a new generation of nuclear power concepts that aim to address various challenges associated with traditional nuclear reactors. These technologies offer the potential for enhanced safety, improved efficiency, reduced waste generation, and expanded fuel options (Cao & Chen, 2022; Qureshi, 2015).

The transition to low-carbon energy involves not only adopting cleaner energy sources but also implementing policies, regulations, and incentives to accelerate the deployment of these technologies. Governments, businesses, and individuals all play essential roles in driving the low-carbon energy transition to address the challenges posed by climate change and work toward a more sustainable future.

The combustion of fossil fuels for energy production is a primary driver of global greenhouse gas emissions. To limit global warming to manageable levels, businesses must transition toward low-carbon energy alternatives to curb their carbon footprint and contribute to the goals outlined in international climate agreements.

The imperative of low-carbon energy transition is driven by the urgent need to mitigate the impacts of climate change, ensure energy security, and foster sustainable development on a global scale. This transition is essential to address the challenges posed by increasing greenhouse gas emissions, environmental degradation, and the finite nature of fossil fuel resources.

Shifting toward energy sources with lower carbon emissions carries substantial economic and societal consequences. The varying governance models might result in distinct low-carbon futures, and at present, civil society's influence on this transition is restricted. King (2018) demonstrates that transitioning to a low-carbon state is likely to cause a decrease in net energy per capita, potentially impacting current ways of life, unless substantial enhancements in end-use efficiency are realized. Foran (2011) proposes that a transition to renewable energy with restrained growth, coupled with reductions in personal consumption and the establishment of a generational sovereign wealth fund, could lead to a reduction of up to 70% in CO_2 emissions in Australia. Finally, Scheffran and Froese (2016) underscores the significance of creating conducive environments that facilitate the spread of low-carbon technologies, investments, and social learning, all of which are pivotal in supporting the shift toward low-carbon societies.

Businesses can reap numerous opportunities and benefits by embracing sustainable practices, including the transition to low-carbon energy. Incorporating environmentally friendly strategies not only aligns with societal expectations but also yields positive economic outcomes and competitive advantages. There are some key opportunities and benefits for businesses, which are presented in Figures 5.6 and 5.7.

Incorporating low-carbon energy transition into a business strategy is not just an ethical choice; it's also a strategic one that enhances competitiveness, profitability, and long-term viability. By embracing sustainable practices and adopting cleaner energy sources, businesses can position themselves for success in a changing economic and environmental landscape.

While transitioning to low-carbon energy sources is critical for addressing climate change and achieving sustainable development, there are several potential risks and challenges that businesses, governments, and communities need to consider. These risks can impact various aspects of the transition process and require careful planning and mitigation strategies. There are some potential risks associated with transitioning to low-carbon energy sources (Nazir et al., 2020; Malecha, 2022; Findlay, 2020; Schmidt, 2014; International Energy Agency, 2023; IAEA, 2021; World Economic Forum, 2023; Zhou et al., 2023; Serafeim, 2022; Gibon et al., 2017; Qadrdan et al., 2019; Lempert & Trujillo, 2018; Sovacool et al., 2019; Mauleón, 2020; Heal, 2020; Papadis & Tsatsaronis, 2020; Wu et al., 2018; Fragkos et al., 2021; Basok & Baseyev, 2022; Nurdiawati & Urban, 2021; Kyriakopoulos, 2021):

1. **Technology Risks**

    Some low-carbon energy technologies, especially advanced ones, may still be in the experimental stage, posing uncertainties regarding their performance, reliability, and long-term viability. Investing heavily in specific low-carbon technologies could lead to "lock-in" if better alternatives emerge, potentially resulting in stranded assets. Technology risks associated with transitioning to low-carbon energy sources encompass uncertainties related to the adoption, performance, and integration of new and

Opportunities and Benefits for Businesses (1)

Cost Savings

Energy Efficiency. Implementing energy-efficient technologies and practices reduces energy consumption and lowers operational costs over time.

Resource Optimization. Efficient use of resources, such as water and raw materials, leads to cost savings and reduces waste disposal expenses.

Competitive Advantage

Market Differentiation. Businesses that prioritize sustainability and low carbon energy transition can stand out in the market, attracting environmentally conscious consumers and investors.

Brand Reputation. Demonstrating a commitment to sustainability enhances brand reputation and can foster customer loyalty.

Innovation and Technological Leadership

Research and Development. Pursuing sustainable practices often requires innovative solutions. Investing in research and development can lead to new products, services and technologies.

Adoption of Clean Technologies. Embracing low carbon energy sources and technologies positions businesses as leaders in a rapidly evolving and environmentally conscious marketplace.

Regulatory Compliance and Risk Management

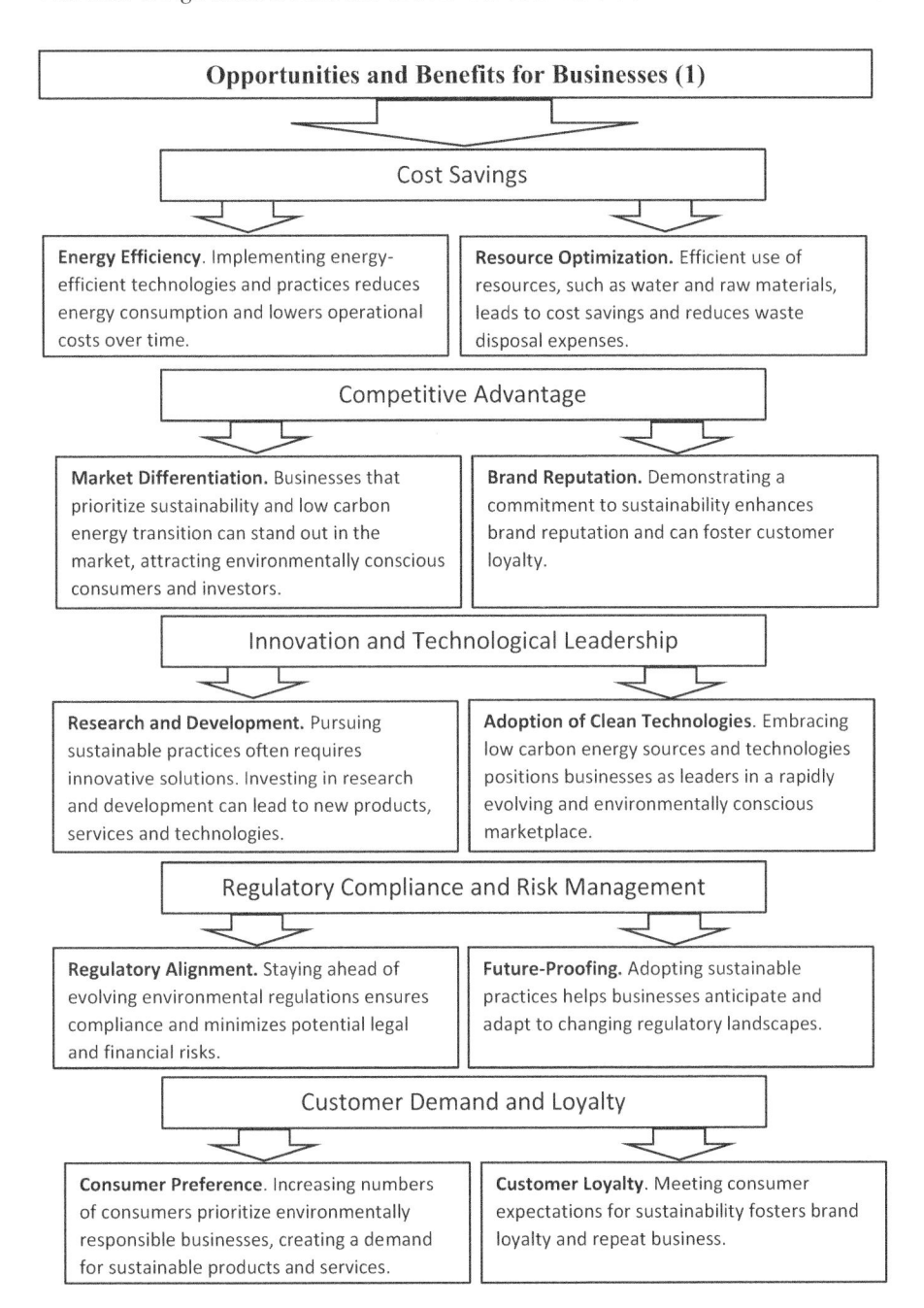

Regulatory Alignment. Staying ahead of evolving environmental regulations ensures compliance and minimizes potential legal and financial risks.

Future-Proofing. Adopting sustainable practices helps businesses anticipate and adapt to changing regulatory landscapes.

Customer Demand and Loyalty

Consumer Preference. Increasing numbers of consumers prioritize environmentally responsible businesses, creating a demand for sustainable products and services.

Customer Loyalty. Meeting consumer expectations for sustainability fosters brand loyalty and repeat business.

FIGURE 5.6 Businesses opportunities and benefits by embracing sustainable practices, including the transition to low-carbon energy (1). (Created by authors.)

Opportunities and Benefits for Businesses (2)

Investor Attraction

ESG Considerations. Environmental, Social, and Governance (ESG) factors are becoming integral to investment decisions. Businesses with strong sustainability practices are more appealing to socially responsible investors.

Risk Mitigation

Supply Chain Resilience. Implementing sustainable practices in supply chains reduces risks related to resource scarcity, regulatory changes, and disruptions.

Employee Engagement and Productivity

Talent Attraction. Employees are drawn to companies with strong sustainability values, enhancing recruitment efforts.

Morale and Retention. Engaged employees thrive in purpose-driven workplaces, leading to higher job satisfaction and lower turnover rates.

Long-Term Viability

Resilience. Embracing sustainable practices enhances a business's adaptability to changing market dynamics and consumer preferences.

Access to Resources. As environmental challenges intensify, businesses that prioritize sustainability are more likely to have continued access to vital resources.

Public Relations Opportunities

CSR Initiatives. Demonstrating commitment to social responsibility and environmental stewardship fosters positive public relations and media coverage.

International Market Access

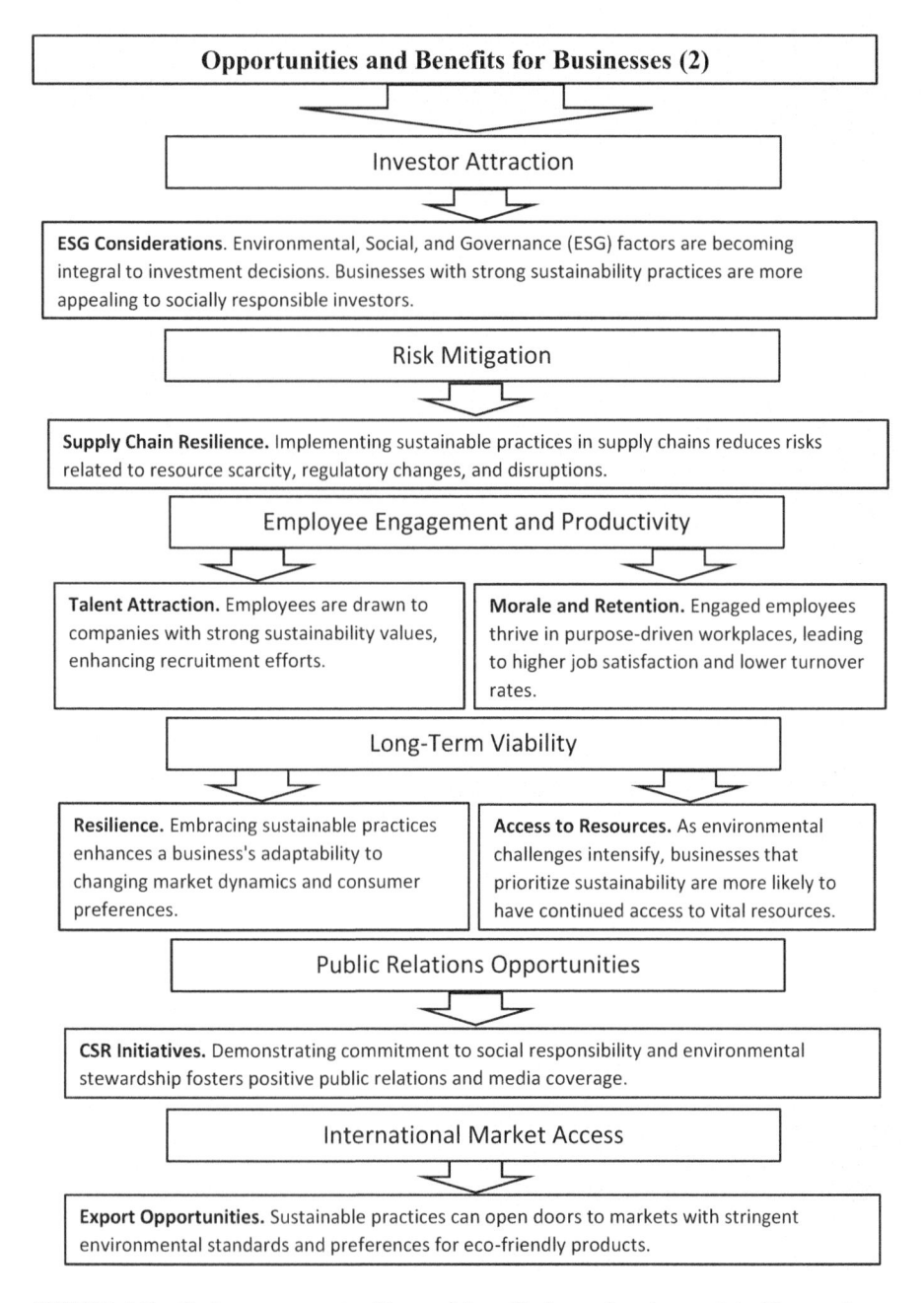

Export Opportunities. Sustainable practices can open doors to markets with stringent environmental standards and preferences for eco-friendly products.

FIGURE 5.7 Businesses opportunities and benefits by embracing sustainable practices, including the transition to low-carbon energy (2). (Created by authors.)

evolving technologies. These risks can impact the successful implementation of low-carbon energy projects and require careful consideration and mitigation strategies. Some of technology risks in the context of transitioning to low-carbon energy sources are presented in Figure 5.8.

The Role of Agricultural Business in Low-Carbon Transition 215

```
TECHNOLOGY RISKS IN THE CONTEXT OF
TRANSITIONING TO LOW CARBON ENERGY SOURCES
   ├── Emerging Technology Uncertainty
   ├── Performance Reliability
   ├── Technological Lock-In
   ├── Compatibility and Integration
   ├── Supply Chain Vulnerabilities
   ├── Regulatory and Standards Challenges
   ├── Cybersecurity Concerns
   ├── Technological Learning Curves
   └── Economic Viability
```

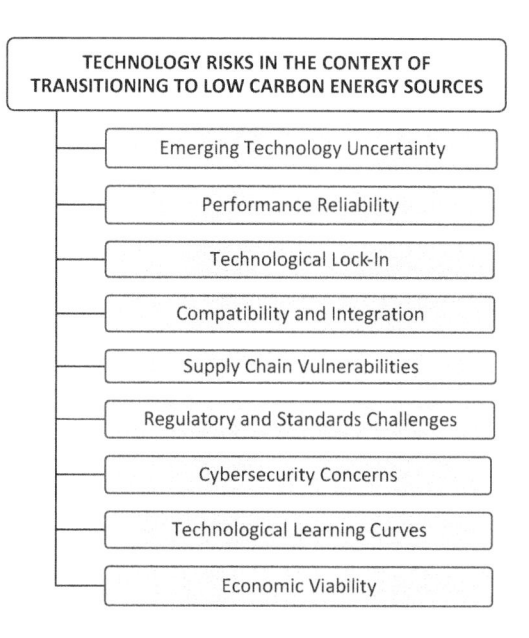

FIGURE 5.8 Technology risks in the context of transitioning to low-carbon energy sources. (Created by authors.)

Emerging Technology Uncertainty. Adopting nascent or unproven technologies carries the risk of technical failures, performance shortcomings, or unexpected challenges. Businesses and governments investing in emerging technologies may face uncertainty regarding their long-term viability and commercial feasibility.

Performance Reliability. Low-carbon energy technologies, especially renewable sources like solar and wind, can be influenced by weather conditions, leading to intermittency and variability in energy production. Energy storage solutions, such as batteries, must demonstrate reliability and long-term performance to ensure stable energy supply.

Technological Lock-In. Heavy investments in specific low-carbon energy technologies could lead to "lock-in," where businesses become dependent on certain technologies that might become obsolete or outperformed by newer alternatives. Rapid technological advancements could render existing infrastructure and technologies less competitive.

Compatibility and Integration. Integrating diverse low-carbon energy sources into existing energy grids may require complex engineering solutions to ensure compatibility and efficient distribution. Smart grid technology and energy management systems are necessary for managing the intermittent nature of renewable energy sources and optimizing energy use.

Supply Chain Vulnerabilities. Dependence on specific critical materials for manufacturing low-carbon technologies, such as rare earth elements for certain renewable energy technologies, can create supply chain vulnerabilities. Shortages or disruptions in the supply of these materials could impact the production and deployment of low-carbon energy systems.

Regulatory and Standards Challenges. Rapidly evolving technological landscapes may lead to a lack of standardized regulations and standards, creating uncertainty for businesses and investors. Navigating complex regulatory frameworks could pose challenges in gaining necessary approvals and certifications for new technologies.

Cybersecurity Concerns. As low-carbon energy systems become more interconnected and reliant on digital infrastructure, they become susceptible to cybersecurity threats and attacks. Ensuring robust cybersecurity measures is crucial to protect critical infrastructure and data.

Technological Learning Curves. Transitioning to new technologies may require a learning curve for operators and maintenance staff, leading to potential operational challenges and downtime. Developing a skilled workforce familiar with advanced technologies is essential for their successful adoption.

Economic Viability. The economic viability of certain low-carbon energy technologies, such as advanced nuclear or CCS, could be affected by high development and operational costs. Uncertainties in cost projections and return on investment could impact the willingness of businesses and investors to adopt these technologies.

Addressing technology risks involves robust research and development efforts, pilot projects, collaboration between academia and industry, and continuous monitoring and adaptation to emerging trends. Governments, businesses, and research institutions should work together to ensure that new technologies are thoroughly tested, scalable, and capable of meeting the demands of a low-carbon energy future while minimizing potential technological risks.

2. **Economic Risks**

The upfront costs of transitioning to low-carbon energy sources can be significant, and businesses may face financial challenges in adopting new technologies or retrofitting existing infrastructure. Rapid changes in technology and policy can lead to market uncertainties, affecting investment decisions and returns.

3. **Infrastructure and Supply Chain Risks**

Transitioning to certain low-carbon technologies (e.g., electric vehicles) may require upgrading infrastructure (e.g., charging stations), which could face delays or logistical challenges. New materials or components needed for low-carbon technologies may be subject to supply chain disruptions or shortages.

4. **Intermittency and Reliability Risks**

Renewable energy sources like solar and wind can be intermittent, leading to challenges in maintaining consistent energy supply. Implementing efficient energy storage solutions to address intermittency requires investment and the development of appropriate technologies.

5. **Transitioning Workforce Risks**

The shift away from traditional energy sectors may result in job losses in those industries, requiring strategies for workforce transition and reskilling.

The workforce may lack the necessary skills to operate and maintain new low-carbon energy technologies, necessitating training and education programs.

6. **Policy and Regulatory Risks**

    Changes in government policies, subsidies, or incentives can impact the viability and economics of low-carbon energy projects. Unclear or evolving regulations can create uncertainty and hinder investment decisions.

7. **Environmental Risks**

    The development of renewable energy projects can lead to land use conflicts, habitat disruption, and unintended environmental consequences. Scaling up certain technologies, such as lithium-ion batteries, may lead to increased demand for specific materials, potentially causing resource scarcity and environmental impacts.

8. **Social and Community Risks**

    Local communities may have concerns about the visual impact, noise, or other effects of renewable energy installations. The transition could inadvertently leave marginalized communities without access to affordable clean energy technologies.

9. **Energy Security Risks**

    Overreliance on intermittent renewable sources could impact grid stability and security, requiring robust energy storage and management solutions.

    Addressing these risks requires a holistic approach that involves collaboration among stakeholders, clear policy frameworks, investment in research and development, and comprehensive risk assessment and mitigation strategies. While the transition to low-carbon energy sources offers significant benefits, it's important to manage and minimize potential risks to ensure a smooth and successful transition process.

    The research papers propose that effective strategies for transitioning to energy sources with lower carbon emissions encompass several approaches. These include heightened investments in research and development (Jianqiang et al., 2011), expediting the industrialization of low-carbon energy technologies (Jianqiang et al., 2011), reinforcing government policy guidance efforts (Jianqiang et al., 2011), and advocating for the adoption of renewable energy sources, energy storage solutions, and energy-efficient technologies (Zhao 2020). However, achieving widespread adoption of low-carbon vehicles and fuels will be demanding and necessitate robust sales growth by no later than 2025. Furthermore, curtailing personal consumption and channeling saved consumption opportunities toward an intergenerational sovereign wealth fund can serve as a buffer for society and future generations against disruptions while also reducing carbon emissions (Foran, 2011).

Strategies for Low-carbon Energy Transition:

- Adopting Renewable Energy Sources

    Adopting renewable energy sources is a pivotal strategy for achieving low-carbon energy transition due to its numerous environmental, economic, and social benefits. Renewable energy technologies harness naturally replenishing sources of energy, such as sunlight, wind, water, and geothermal heat, to generate electricity and heat without emitting greenhouse gases. The implications of a renewable energy strategy for the transition to low-carbon energy are presented in Figure 5.9.

Renewable energy sources produce little to no direct greenhouse gas emissions during operation, helping to significantly mitigate climate change. Unlike fossil fuels, renewables do not release carbon dioxide, methane, or other pollutants that contribute to global warming. Replacing fossil fuel-based power generation with renewables reduces air pollutants like sulfur dioxide, nitrogen oxides, and particulate matter, which have adverse effects on human health and the environment. Renewable energy sources are abundant and widely distributed, reducing dependence on finite and often imported fossil fuels. This enhances energy security by diversifying energy sources and reducing exposure to geopolitical risks.

The renewable energy sector generates significant employment opportunities in manufacturing, installation, maintenance, and research. It stimulates economic growth by fostering innovation and attracting investments. Unlike fossil fuel prices, which can be volatile and subject

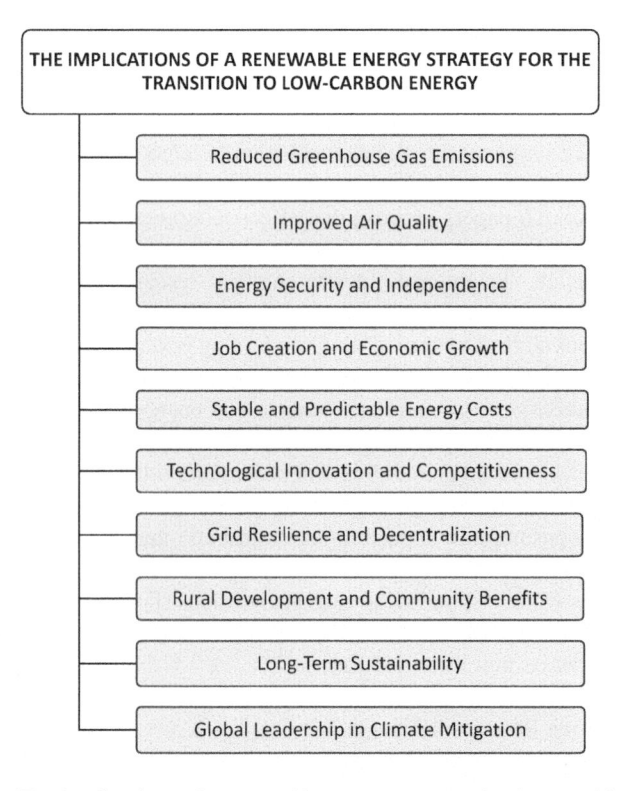

THE IMPLICATIONS OF A RENEWABLE ENERGY STRATEGY FOR THE TRANSITION TO LOW-CARBON ENERGY

- Reduced Greenhouse Gas Emissions
- Improved Air Quality
- Energy Security and Independence
- Job Creation and Economic Growth
- Stable and Predictable Energy Costs
- Technological Innovation and Competitiveness
- Grid Resilience and Decentralization
- Rural Development and Community Benefits
- Long-Term Sustainability
- Global Leadership in Climate Mitigation

FIGURE 5.9 The implications of a renewable energy strategy for the transition to low-carbon energy.

to geopolitical tensions, the cost of renewable energy tends to be more stable over time, providing energy cost predictability for businesses and consumers. The adoption of renewable energy technologies fosters technological innovation, encouraging research and development in areas such as solar panels, wind turbines, energy storage, and grid integration.

Distributed renewable energy systems, such as rooftop solar panels and community wind projects, contribute to grid resilience by reducing the risk of centralized system failures. Renewable energy projects often benefit rural communities by providing income from land lease agreements and job opportunities, revitalizing local economies. Renewable energy sources are inherently sustainable as they do not deplete finite resources. They have the potential to provide energy for generations to come without environmental degradation. Transitioning to renewable energy positions countries and businesses as leaders in climate change mitigation, aligning with international agreements and demonstrating commitment to a sustainable future. The reduction of air pollutants and greenhouse gas emissions from renewable energy adoption leads to improved public health outcomes, fewer respiratory illnesses, and better environmental quality.

- Investing in Energy Storage

    Energy storage solutions, such as batteries, enable businesses to store excess energy generated from renewable sources and use it during peak demand periods or when renewable generation is low. This helps ensure a consistent energy supply and maximizes the utilization of renewable energy.

- Implementing Energy Management Systems

    Energy management systems enable businesses to monitor and optimize their energy consumption. Through smart technology and data analytics, businesses can identify inefficiencies, manage peak demand, and make informed decisions to reduce energy waste.

5.3 GREEN AGRICULTURAL INITIATIVES

GAIs are at the forefront of transforming conventional farming practices into sustainable and environmentally friendly endeavors. As the global population continues to grow, the demand for food and agricultural products places increasing pressure on natural resources and ecosystems. Green initiatives in agriculture aim to strike a balance between meeting this demand and preserving the planet's biodiversity and ecological health. In this teaching material, we explore the diverse range of GAIs and the transformative impact they can have on the agricultural sector.

In recent years, the global agricultural sector has been facing a growing need for sustainable and environmentally friendly practices. GAIs have emerged as a response to the challenges posed by climate change, resource depletion, and the need to ensure food security for a growing population.

GAI encompass a wide array of practices, technologies, and policies designed to promote sustainable farming and reduce the ecological footprint of agricultural operations. These initiatives focus on improving resource efficiency, conserving biodiversity, minimizing chemical inputs, and adopting climate-resilient approaches.

Researchers often describe GAIs as a set of practices that integrate ecological principles into farming systems. This involves diversifying crops, promoting soil health through cover cropping and reduced tillage, and implementing IPM techniques. The goal is to create resilient agricultural systems capable of withstanding environmental changes.

The research papers present diverse viewpoints regarding the characterization of GAIs. Burch et al. (2001) delve into the utilization of "green" terminology in discussions about agri-food, using examples from the organic farming sectors in Australia and New Zealand. Koohafkan et al. (2012) introduce ten attributes outlining the essence of green agriculture, underlining the necessity for flexible and contextually suitable sustainability and resiliency principles and boundaries. Sdrolia and Zarotiadis (2018) tackle the absence of a universally accepted description for green products and introduce a comprehensive definition that encompasses the entire life cycle of a product. Lastly, Bachynskyi (2021) examines the incentive mechanisms used within EU nations to stimulate environmentally conscious endeavors in agriculture, suggesting the establishment of an information system to facilitate the integration of green initiatives in Ukrainian practices. In essence, GAIs involve ecologically sound practices that are both sustainable and resilient, adapting to local contexts and emphasizing the importance of clear definitions and incentive tools to support their successful implementation.

Different countries conceptualize and implement green agriculture initiatives differently. Bachynskyi (2021) examines the assortment of incentive mechanisms adopted by EU member states to invigorate environmentally conscious endeavors within agriculture, encompassing regulatory, informative, and economic instruments. Magaudda et al. (2020) underscore the necessity for guidelines that facilitate regional strategies and efficient allocation of funding for greening policies within Italy's framework of the CAP. Nwachukwu (2023) places significant emphasis on the imperative of sustainable advancements in green technology for food production, a vital component in achieving both environmentally friendly agriculture and ensuring food security. Mouysset (2014) juxtaposes diverse optimal public strategies for agricultural policies within France, drawing attention to the trade-offs intertwined with policy execution, social acceptance, technical complexities, and equitable performance distribution among regions. In essence, GAIs encompass a comprehensive process where nature conservation activities within enterprises are systematically organized through coordinated efforts of agricultural producers, governmental structures, administrative bodies, and united territorial communities. This orchestration involves the utilization of a range of tools and measures to stimulate ecologically responsible activities.

Here are a few examples of how different countries might define GAIs:

- **United States**: In the United States, GAIs refer to efforts aimed at promoting sustainable farming practices that reduce the environmental impact of agriculture. These initiatives may include programs that encourage the adoption of practices such as conservation tillage, cover cropping, precision agriculture, IPM, and organic farming. The U.S. Department of Agriculture (USDA) defines GAIs as strategies that improve soil health, reduce water and energy use, minimize chemical inputs, and enhance biodiversity on farmland.

The objectives of GAIs in the United States aim to encourage profitable and environmentally sustainable farming practices. Hrubovcak et al. (1999) underscore that adopting green technologies like conservation tillage, IPM, improved nutrient management, and precision agriculture can guide agriculture toward a more sustainable trajectory. Nevertheless, the lack of markets for the environmental benefits tied to these technologies can hinder their advancement. Nwachukwu (2023) underscores the significance of sustainable advancements in green technology within food production, driven by renewable energy and organic fertilizers, to ensure food security for the growing global population. Nistor (2015) emphasizes the necessity of closely aligning agricultural policies with environmental policies to achieve sustainable economic growth. Lastly, Bachynskyi (2021) suggests that successful green initiatives in agriculture demand collaborative efforts among the government, society, united territorial communities, and agricultural businesses, and highlights that incentive mechanisms like ecological taxes, quotas, funds, subsidies, and tax benefits can incentivize enterprises to embrace environmentally conscious practices.

- **European Union**: In the European Union (EU), GAIs are often associated with the CAP and its efforts to support environmentally friendly farming practices. GAIs in the EU involve measures to promote agroecological approaches, agroforestry, sustainable water management, and the preservation of High Nature Value farmlands. These initiatives are designed to align agricultural production with broader environmental and climate goals, emphasizing the importance of biodiversity conservation and carbon sequestration.

  Primary aims of the European Union's GAIs encompass the promotion of sustainable farming practices and the alleviation of environmental strains. Kowalska and Bieniek (2022) discover that the European Commission is targeting 25% organic land usage in agriculture by 2030, yet existing incentives fall short of meeting this objective. Szabo et al. (2022) propose that the European Green Agreement might yield both positive and negative effects on crop production, though directed assistance could facilitate the realization of sustainability goals. Blake (2020) underscores the imperative to optimize resource utilization and adopt prudent approaches to bolster sustainable agriculture. Pe'er et al. (2020) contend that the (CAP is inadequately addressing sustainability concerns and urges immediate intervention to achieve sustainable food production, biodiversity preservation, and climate mitigation. In essence, the papers collectively convey that while the EU's GAIs aim to foster sustainability and alleviate environmental strains, achieving these objectives necessitates more targeted support and expedited action.

- **China**: In China, GAIs are part of the country's commitment to sustainable development and ecological civilization. Chinese GAIs encompass strategies to enhance soil fertility, improve water management, reduce pollution from agricultural runoff, and encourage the adoption of precision farming technologies. These initiatives aim to increase agricultural productivity while minimizing negative environmental impacts, in line with China's goals of environmental protection and rural revitalization.

  The research collectively implies that China's initiatives for sustainable agriculture aim to transform the existing resource-intensive and environmentally costly farming into a eco-friendlier agriculture and rural environment characterized by heightened productivity, efficient resource utilization, and reduced environmental impact (Shen et al., 2020). The objectives of China's sustainable agriculture endeavors encompass the enhancement of supply capacity, resource efficiency, environmental quality, ecosystem preservation, and the well-being of farmers (Liu et al., 2020). The studies also address the hurdles and possible remedies for realizing green agricultural progress in China, including interdisciplinary innovations, comprehensive enhancements along the entire food chain, and region-specific strategies (Shen et al., 2020). The authors of Qiu et al. (2020) propose that China can derive insights from the European Union's strategies geared toward greening agriculture, particularly concerning agricultural inputs, land usage, and the monitoring of greening outcomes. Lastly, Lu et al. (2020) scrutinize the transition of both China and the UK from narrow farm output policies to broader green development approaches, highlighting the significant mutual benefits attainable through enhanced collaboration in problem identification, technological and managerial responses, and policy formulation.

- **India**: In India, GAIs focus on sustainable practices that increase agricultural productivity and resilience while addressing challenges related to water scarcity, soil degradation, and climate change. These initiatives involve the promotion of organic farming, integrated nutrient management, watershed development, and the use of drought-resistant crop varieties. India's National Mission for Sustainable Agriculture (NMSA) reflects the country's emphasis on improving the livelihoods of farmers through ecologically responsible practices.

  Conventional agricultural methods in India possess the capacity to counteract the adverse impacts of climate change and enhance food security (Patel et al., 2020). However, the Green Revolution has contributed to environmental deterioration, and ensuring sustainability remains a significant hurdle for Indian agriculture (Puttaswamaiah et al., 2006). Strategies have been enacted to foster sustainable farming practices, including the adoption of organic cultivation; nevertheless, barriers involving economic feasibility, institutional backing, and market entry persist (Puttaswamaiah et al., 2006). The implementation of sustainable agricultural practices is imperative to elevate agricultural output, enrich soil health, attain socio-economic equilibrium, as well as to safeguard natural resources and the environment (Singh et al., 2019; Prasad, 2006). Yet, India encounters ongoing obstacles in realizing sustainable agriculture, encompassing the need to rectify gender disparities and regional disparities, in addition to enhancing the financial viability of Indian agriculture (Prasad, 2006). On the whole, the studies propose that while India has experienced successes and holds potential for eco-friendly agricultural initiatives, overcoming challenges is essential to achieving sustainable agriculture.

- **Brazil**: In Brazil, GAIs center around the conservation of the Amazon rainforest and other vital ecosystems. Efforts are made to promote sustainable land use practices, discourage deforestation, and support agroforestry models that integrate crops, livestock, and trees. GAIs in Brazil aim to balance agricultural expansion with environmental conservation, often in the context of combating illegal logging and promoting responsible land tenure.

    Brazil has effectively implemented successful eco-friendly agricultural initiatives. Embrapa, a public research entity, has exerted a significant influence on Brazil's agricultural transformation (Schmidt & Correa, 2014). The National School Meal Program (PNAE) and the Food Procurement Program (PAA) have embedded agrobiodiversity within their procurement schemes, fostering the enrichment of local production systems through increased biodiversity (Resque et al., 2019). The proposal for the integration of crops, livestock, and forestry (ICLF) aligns with the principles of sustainable agriculture and stands as a pivotal approach to enhancing production sustainably in Brazil (Reis, 2016). Lastly, Sills and Caviglia-Harris (2015) discovered that participation in an internationally supported farmer association that advocated for sustainable agricultural practices led to more diversified production systems, including greater allocation of land to agroforestry.

These examples illustrate that the definition of GAIs can vary from country to country based on local agricultural practices, environmental challenges, and policy priorities. However, the common thread among these definitions is the emphasis on sustainable farming practices that minimize negative environmental impacts while ensuring food security and the well-being of rural communities.

5.3.1 THE IMPORTANCE OF GREEN AGRICULTURAL INITIATIVES

Agriculture is an essential industry, providing food, fiber, and livelihoods for billions of people. However, conventional agricultural practices have often led to deforestation, soil degradation, water pollution, and greenhouse gas emissions. GAIs offer a pathway to address these challenges and create a more sustainable and resilient food system.

GAIs are of paramount importance for several critical reasons, encompassing environmental, economic, and social aspects. These initiatives focus on promoting sustainable and eco-friendly practices within the agricultural sector to address the challenges posed by climate change, resource depletion, and food security concerns.

Here are some key reasons highlighting the significance of GAIs (Shen et al., 2020; Wang, 2022; Solazzo et al., 2016; Li et al., 2022; Guyomard et al., 2023; Helga et al., 2023):

- **Climate Change Mitigation**: Agriculture is a significant contributor to greenhouse gas emissions, primarily through activities like livestock production, fertilizer use, and land conversion. GAIs encourage practices that reduce carbon emissions and enhance carbon sequestration, helping to mitigate the impacts of climate change.
- **Preservation of Biodiversity**: Conventional agricultural practices often involve monocultures and the use of synthetic pesticides and fertilizers, leading to biodiversity loss and soil degradation. Green initiatives promote crop diversification, agroforestry, and IPM, helping to maintain ecosystem balance and support various species.
- **Water Resource Management**: Agriculture accounts for a substantial portion of global water use. Green initiatives promote efficient water management techniques such as drip irrigation, rainwater harvesting, and soil moisture monitoring. These practices reduce water wastage and increase resilience against drought conditions.
- **Soil Health and Conservation**: Soil degradation, erosion, and nutrient depletion are common issues associated with intensive agriculture. Green initiatives encourage practices like cover cropping, reduced tillage, and organic farming, which enhance soil health, structure, and fertility, leading to higher agricultural productivity and reduced environmental impact.
- **Food Security and Resilience**: As the global population grows, ensuring food security becomes a critical challenge. GAIs focus on sustainable farming practices that maintain and enhance soil fertility, reduce reliance on chemical inputs, and increase crop diversity, thereby contributing to long-term food security.

Figure 5.10 provides a comprehensive picture of how organic agriculture initiatives contribute to food security and resilience (FAO, 2017; Helga et al., 2023; Scialabba, 2013; Halberg & Müller, 2013; Ward & Reynolds, 2013; Badgley et al., 2007; Meemken & Qaim, 2018):

- *Diverse Crop Production.* Green initiatives encourage the cultivation of a diverse range of crops, including traditional and locally adapted varieties. This diversification reduces the risk of crop failure due to pests, diseases, or adverse weather conditions. If one crop fails, others can still thrive, ensuring a consistent food supply.
- *Climate-Resilient Crops.* With climate change leading to unpredictable weather patterns, green initiatives promote the adoption of climate-resilient crop varieties. These crops are specifically bred to withstand extreme temperatures, droughts, floods, and other climatic challenges, ensuring that farmers can still harvest even under adverse conditions.
- *Conservation Agriculture.* Green agricultural practices like reduced tillage and cover cropping help conserve soil moisture and prevent erosion. This is especially important in regions prone to droughts or heavy rainfall. Conserved soil moisture supports crop growth during dry spells, while erosion prevention maintains fertile topsoil for sustainable farming.

The Role of Agricultural Business in Low-Carbon Transition 225

- Diverse Crop Production
- Diverse Crop Production
- Conservation Agriculture
- Water Management
- Crop Rotation and Intercropping
- Agroforestry
- Local Food Systems
- Seed Banks and Genetic Diversity
- Capacity Building and Knowledge Sharing
- Community Resilience
- Reduced Dependence on Imports
- Long-Term Sustainability

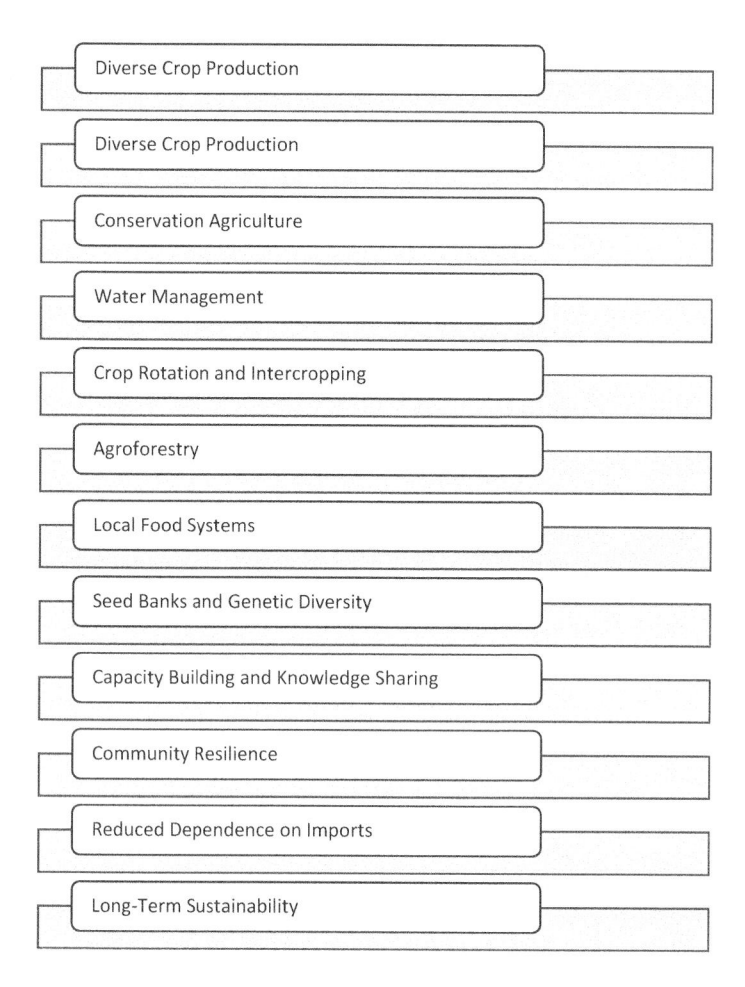

FIGURE 5.10 Impact of organic agriculture initiatives on food security and resilience. (Created by authors.)

- *Water Management.* Efficient water management techniques, such as drip irrigation and rainwater harvesting, are promoted by green initiatives. These practices optimize water use, particularly in water-scarce areas, ensuring that crops receive adequate moisture while minimizing water wastage.
- *Crop Rotation and Intercropping.* Green initiatives encourage farmers to practice crop rotation and intercropping. Crop rotation replenishes soil nutrients by alternating different types of crops, while intercropping helps optimize space and resource utilization. Both strategies contribute to increased yields and overall food production.
- *Agroforestry.* Integrating trees and shrubs within agricultural landscapes enhances biodiversity, provides shade, and improves soil fertility. Agroforestry systems are more resilient against climate extremes and

often yield a mix of food, timber, and other useful products, ensuring a diverse and stable food supply.
- *Local Food Systems.* Green initiatives support the development of local food systems, reducing dependence on distant sources and increasing food security. Local food production, distribution, and consumption networks can quickly adapt to disruptions in global supply chains, ensuring that communities have access to fresh and nutritious food.
- *Seed Banks and Genetic Diversity.* Preserving diverse crop varieties in seed banks safeguards genetic resources that can be crucial in adapting to changing conditions. These seeds are often better suited to local climates and can be used to develop new resilient varieties.
- *Capacity Building and Knowledge Sharing.* Green initiatives involve training farmers in sustainable practices, empowering them with the skills to adapt to changing conditions. Knowledge sharing through workshops, extension services, and community networks enhances farmers' ability to respond to challenges effectively.
- *Community Resilience.* GAIs foster stronger community bonds and cooperative networks. When communities work together to implement sustainable practices, share resources, and support each other during tough times, their collective resilience improves.
- *Reduced Dependence on Imports.* By promoting local food production, green initiatives reduce a community's reliance on imported food, which can be vulnerable to disruptions. This strengthens local economies and ensures a consistent food supply, even when external sources are compromised.
- *Long-Term Sustainability.* Green initiatives prioritize practices that maintain ecosystem health, soil fertility, and natural resources. This focus on sustainability ensures that farming can continue for generations without depleting the land's capacity to produce food.

In essence, GAIs contribute significantly to food security and resilience by fostering adaptive practices, diversifying agricultural systems, conserving resources, and enhancing local community capacities. By striking a balance between ecological considerations, economic viability, and social well-being, these initiatives create a more robust and secure food future.

- **Rural Livelihoods and Economic Growth**: Many rural communities depend on agriculture for their livelihoods. Green initiatives can improve the income and quality of life of smallholder farmers by promoting sustainable practices that lead to increased yields, reduced input costs, and access to premium markets for organic and sustainably grown products.

GAIs have a profound impact on rural livelihoods and economic growth by promoting sustainable farming practices, improving income opportunities, and fostering local economic development. These initiatives prioritize environmental conservation, social equity, and economic viability, creating a positive cycle that benefits both farmers and

The Role of Agricultural Business in Low-Carbon Transition 227

THE BENEFITS OF GREEN AGRICULTURAL INITIATIVES FOR RURAL LIVELIHOODS AND THE ECONOMY

- Increased Agricultural Productivity
- Reduced Input Costs
- Access to Premium Markets
- Diversification of Income Sources
- Improved Food Security at the Local Level
- Capacity Building and Skill Development
- Tourism and Ecotourism
- Rural Entrepreneurship
- Infrastructure Development
- Natural Resource Management
- Community-Based Economic Development
- Gender Empowerment
- Strengthening Resilience

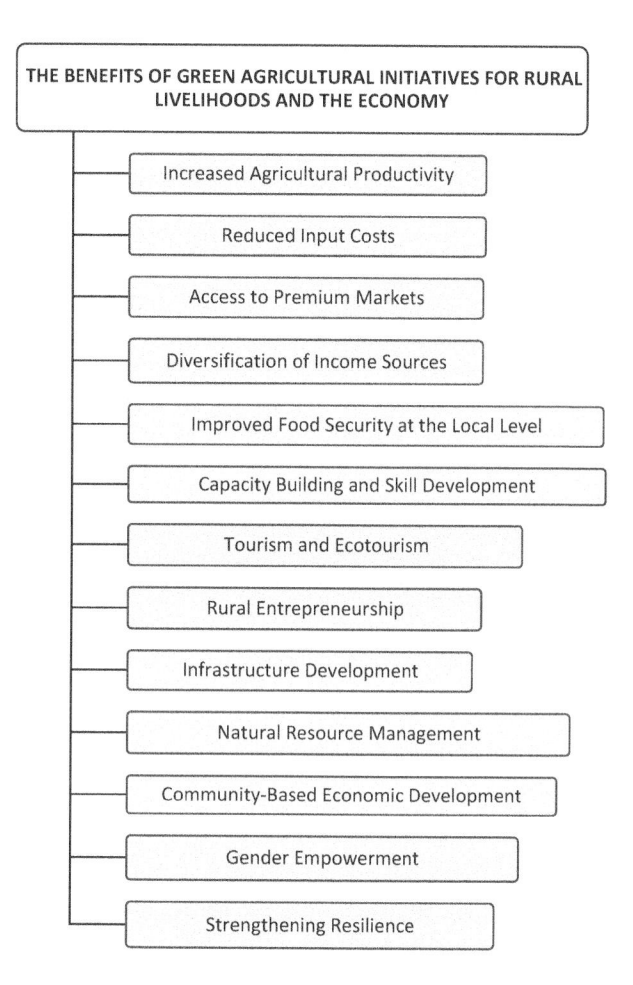

FIGURE 5.11 The benefits of green agricultural initiatives for rural livelihoods and the economy. (Created by authors.)

the broader community. Figure 5.11 shows how green agriculture initiatives contribute to rural livelihoods and economic growth (Hsu & Liao, 2018; Unay-Gailhard & Bojnec, 2019; Wang, 2022; Pan et al., 2021; Dang et al., 2020; Uchida et al., 2008; Geoghegan et al., 2023).

Green initiatives often emphasize techniques that enhance soil fertility, reduce pest pressures, and optimize resource use. This leads to increased yields and improved crop quality, which in turn translates to higher incomes for farmers. By promoting organic farming practices and reducing reliance on expensive synthetic inputs such as pesticides and fertilizers, green initiatives help farmers cut down on production costs. This allows them to maintain profitability even in the face of market fluctuations.

Many consumers and markets are increasingly valuing environmentally friendly and organic products. GAIs enable farmers to tap into these premium markets, commanding higher prices for their produce and boosting their income.

Green initiatives often encourage diversification beyond traditional crops, including livestock rearing, agroforestry, and value-added products like processed foods, crafts, and herbal medicines. This diversification spreads risk and provides additional income streams. When communities have access to diverse and locally grown food sources, their food security improves. This not only ensures a stable food supply for households but also reduces their vulnerability to price fluctuations in global markets.

Green initiatives provide training and technical support to farmers, enhancing their skills and knowledge. This empowers farmers to adopt innovative and sustainable practices, contributing to higher productivity and profitability. Initiatives that encourage value addition and processing of agricultural products create opportunities for rural entrepreneurship. Farmers can establish small-scale businesses such as food processing units, packaging facilities, and craft workshops, generating jobs and income for themselves and others.

Green initiatives often prioritize community involvement and cooperation. This can lead to the formation of cooperatives, community-supported agriculture schemes, and farmer networks, which collectively contribute to rural economic development. Green practices, such as agroforestry and sustainable water management, ensure the long-term health of natural resources. This is essential for maintaining agricultural productivity, preserving water sources, and preventing soil degradation, all of which underpin economic growth.

Some green initiatives focus on promoting agrotourism or ecotourism, allowing visitors to experience sustainable farming practices and enjoy the natural beauty of rural areas. This generates additional income for local communities and stimulates rural economies.

As rural economies grow, there is often increased demand for infrastructure improvements, such as better roads, transportation systems, and communication networks. Economic growth driven by GAIs can stimulate investments in these areas. Green initiatives can also empower women in rural communities. By offering training, access to resources, and income-generating opportunities, these initiatives can help overcome gender inequalities and contribute to women's economic empowerment.

By adopting sustainable practices that protect natural resources, farmers are better equipped to withstand the impacts of climate change, market shocks, and other disruptions. This resilience contributes to the long-term stability of rural economies. GAIs have the potential to transform rural livelihoods and drive economic growth by enhancing agricultural productivity, fostering diversification, promoting entrepreneurship, and ensuring the sustainable use of natural resources. These initiatives create a synergy between environmental stewardship and economic prosperity, leading to improved well-being for rural communities and the broader society.

- **Reduced Chemical Use and Health Benefits**: Conventional agriculture often relies heavily on synthetic pesticides and fertilizers, which can have negative impacts on both human health and the environment. Green initiatives prioritize the use of natural and less toxic alternatives, promoting safer working conditions for farmers and healthier food for consumers.

- **Carbon Sequestration and Bioenergy**: Certain green agricultural practices, such as agroforestry and cover cropping, enhance carbon sequestration in soils and vegetation. Additionally, crops like switchgrass and algae can be cultivated for bioenergy production, contributing to renewable energy sources.
- **Adaptation to Changing Conditions**: GAIs emphasize adaptive strategies to cope with changing climatic conditions, such as the development of drought-resistant crop varieties, flood management techniques, and resilient farming systems.
- **Public Awareness and Education**: By promoting green initiatives, society can become more aware of the importance of sustainable agriculture and make informed choices about food consumption, supporting a shift toward more eco-friendly practices.

In conclusion, GAIs are critical for achieving a more sustainable and resilient food system that balances environmental preservation, economic viability, and social well-being. Implementing these initiatives can lead to a healthier planet, enhanced food security, and improved livelihoods for both farmers and communities.

Several key GAIs have been developed to promote sustainable and environmentally friendly practices within the agricultural sector. These initiatives aim to address various challenges such as climate change, resource depletion, and food security concerns. The most commonly used and discussed GAIs (Aulakh et al., 2022; Reganold & Wachter, 2016; Research Institute of Organic Agriculture, 2023; Murthy et al., 2017; Cardinael et al., 2020; Thiesmeier & Zander, 2023; Ghosh et al., 2019; Stagnari et al., 2009; Chaudhary et al., 2022; Norton et al., 2019; Adetunji et al., 2020; Stein et al., 2023; Yu et al., 2022; Zegada-Lizarazu & Monti, 2011; Stoicea et al., 2023; Ertop et al., 2023; Velasco Muñoz et al., 2023; Yang et al., 2023; Purohit et al., 2017; Koutsos & Menexes, 2019; Sahu et al., 2019; Kumar et al., 2020; Hristov et al., 2020; Rotz, 2020; Paul, 2018; Samoggia et al., 2019; Hirschfeld & van Acker, 2021; Dang et al., 2020; Battaglia et al., 2023; Wambugu et al., 2023; Soini Coe & Coe, 2023) are as follows:

- **Organic Farming**: Organic farming is a key GAI that prioritizes sustainable and environmentally friendly practices. It involves cultivating crops and raising livestock without the use of synthetic pesticides, fertilizers, genetically modified organisms (GMOs), and other chemicals. Instead, organic farming relies on natural methods to maintain soil fertility, control pests, and promote ecosystem health. The main forms and components of organic farming are presented in Figure 5.12.

Organic farming systems can vary based on climate, geography, and cultural practices, but the core principles remain consistent: promoting soil health, minimizing environmental impact, and providing consumers with safe and nutritious food.

- **Agroforestry**: Agroforestry involves integrating trees or shrubs into agricultural landscapes. This initiative improves soil fertility, provides shade for crops and livestock, sequesters carbon, and enhances biodiversity. It also offers additional products like timber, fruits, and medicinal plants.

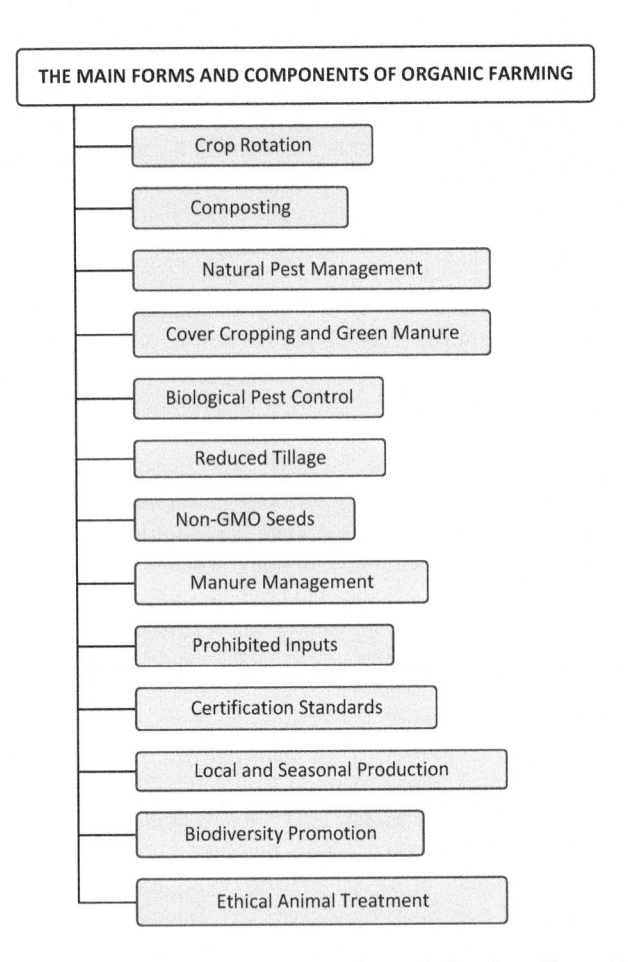

```
THE MAIN FORMS AND COMPONENTS OF ORGANIC FARMING
    ├── Crop Rotation
    ├── Composting
    ├── Natural Pest Management
    ├── Cover Cropping and Green Manure
    ├── Biological Pest Control
    ├── Reduced Tillage
    ├── Non-GMO Seeds
    ├── Manure Management
    ├── Prohibited Inputs
    ├── Certification Standards
    ├── Local and Seasonal Production
    ├── Biodiversity Promotion
    └── Ethical Animal Treatment
```

FIGURE 5.12 The main forms and components of organic farming. (Created by authors.)

- **Conservation Agriculture**: Conservation agriculture emphasizes minimal soil disturbance, crop residue retention, and diversified crop rotations. These practices help prevent erosion, maintain soil moisture, and improve soil structure while reducing the need for tillage. Conservation agriculture emphasizes a holistic approach that aims to balance agricultural production with environmental preservation. Its practices contribute to improving soil health, mitigating climate change, conserving water resources, and ensuring sustainable and resilient food production systems.
- **Integrated Pest Management (IPM)**: IPM involves using a combination of biological, cultural, physical, and chemical methods to manage pests and diseases. This approach minimizes the use of synthetic pesticides, reduces environmental impact, and encourages natural predators. IPM is a comprehensive approach to managing pests and diseases in agriculture while minimizing the use of synthetic chemicals. It involves a range of stakeholders who collaborate to implement various strategies to control pests and maintain crop health.

The main actors in the IPM process are farmers who monitor their crops for pest and disease outbreaks, implement preventive measures, and, based on their observations, make decisions about pest management strategies. The latter are joined by agricultural scientists who provide valuable information on pest biology, behavior, and management strategies. They study the local ecosystem, identify pest problems, and recommend appropriate IPM methods that are effective and environmentally friendly. Government agencies, Agricultural and Environmental Regulatory Agencies, are also involved in this process and play an important role in promoting IPM through policy development, regulations, and incentives. They can also fund research, education, and the implementation of IPM practices. Non-governmental organizations (NGOs) focused on agriculture, environmental protection, and sustainable development often support IPM initiatives. They provide resources, training, and awareness to promote IPM adoption among farmers.

Biological control specialists study and present the natural enemies of pests (predators, parasites, and pathogens). They help farmers create balanced ecosystems that support manageable levels of pest populations. Agronomists advise on crop management practices that promote healthy plants because strong and healthy crops are more resistant to pests and diseases.

Although IPM focuses on reducing the use of chemicals, agrochemical companies can provide more selective and less harmful products that comply with IPM principles.

We can also see an important role for consumers and markets, as the demand for sustainably produced food can encourage farmers to adopt IPM practices. Consumers who understand the benefits of IPM may prefer products grown using these methods.

- **Cover Cropping and Green Manure**: Cover crops are planted between main crops to protect the soil from erosion, improve soil fertility, and enhance nutrient cycling. Green manure crops are intentionally grown and then incorporated into the soil to enrich it with organic matter.
- **Crop Rotation**: Crop rotation involves planting different crops in a specific sequence over time. This practice helps break pest and disease cycles, enhances soil health, and reduces the need for chemical inputs.
- **Rainwater Harvesting**: Rainwater harvesting captures and stores rainwater for irrigation and other agricultural uses. This initiative conserves water resources and ensures a stable supply of water for crops, especially in arid and semi-arid regions.

Rainwater harvesting is practiced in various countries around the world, but its prevalence and popularity often depend on local climatic conditions, water availability, and cultural practices. Rainwater harvesting has been widely adopted in India, especially in regions with limited access to freshwater sources. The government has promoted rainwater harvesting systems for both rural and urban areas to address water scarcity issues. Due to its arid and semi-arid climate, Australia has embraced rainwater harvesting as a key strategy for water conservation. Many Australian households, particularly in rural and remote areas, use rainwater tanks to supplement their water supply. Rainwater harvesting is common in Brazil, particularly in the northeastern region where water scarcity is a significant challenge. The government has promoted rainwater harvesting systems in both urban and rural areas to enhance water security. Rainwater harvesting is practiced in various regions of China, particularly in areas prone to drought. The Chinese government has supported rainwater harvesting projects as part of its efforts to manage water resources more sustainably.

It's important to note that rainwater harvesting can be found in many other countries as well, and its popularity often depends on local context, government policies, and community initiatives to address water scarcity and ensure sustainable water management.

- **Drip Irrigation**: Drip irrigation delivers water directly to the plant roots, minimizing water wastage. It's particularly efficient for water-scarce areas, reducing water use and improving crop yields.
- **Precision Agriculture**: Precision agriculture is a modern farming approach that leverages technology, data, and advanced techniques to optimize the use of resources and enhance agricultural productivity while minimizing environmental impact. The essence of precision agriculture lies in its ability to fine-tune agricultural practices, tailoring them to specific conditions within a field, reducing waste, and improving overall efficiency. Precision agriculture uses technology like GPS, sensors, and data analytics to optimize resource use. Farmers can tailor irrigation, fertilization, and pesticide application to specific areas, reducing input waste and improving efficiency. The main principles of precision agriculture are presented in Figure 5.13.

    Site-Specific Management. Precision agriculture recognizes that different parts of a field have varying soil properties, nutrient levels, and moisture content. By using technology like GPS and sensors, farmers can precisely target inputs such as fertilizers, water, and pesticides to specific areas, optimizing resource use.

    Data-Driven Decision Making. Precision agriculture relies on data collection and analysis to make informed decisions. Soil samples, aerial imagery, satellite data, and crop monitoring tools provide insights that guide farmers in adjusting their management practices.

    Variable Rate Application. Instead of applying a uniform amount of inputs across an entire field, precision agriculture enables farmers to apply inputs at varying rates based on real-time data. This prevents overuse of resources in areas that may not require as much and ensures optimal growth conditions.

    Reduced Input Waste. Precision agriculture aims to minimize waste by precisely applying inputs where they are needed. This reduces the risk of nutrient runoff, decreases chemical usage, and conserves resources like water and energy.

    Improved Crop Health. By identifying areas of a field that need attention, such as disease outbreaks or nutrient deficiencies, precision agriculture allows for prompt interventions, improving overall crop health and yield potential.

    Economic Efficiency. Targeted resource application and reduced waste lead to economic benefits for farmers. Lower input costs, increased yields, and improved product quality contribute to enhanced profitability.

THE MAIN PRINCIPLES OF PRECISION AGRICULTURE

- Site-Specific Management
- Data-Driven Decision-Making
- Natural Pest Management
- Variable Rate Application
- Reduced Input Waste
- Improved Crop Health
- Economic Efficiency
- Environmental Stewardship
- Adaptation to Climate Variability
- Integration of Technology
- Scalability and Flexibility
- Continuous Learning and Improvement
- Global Food Security

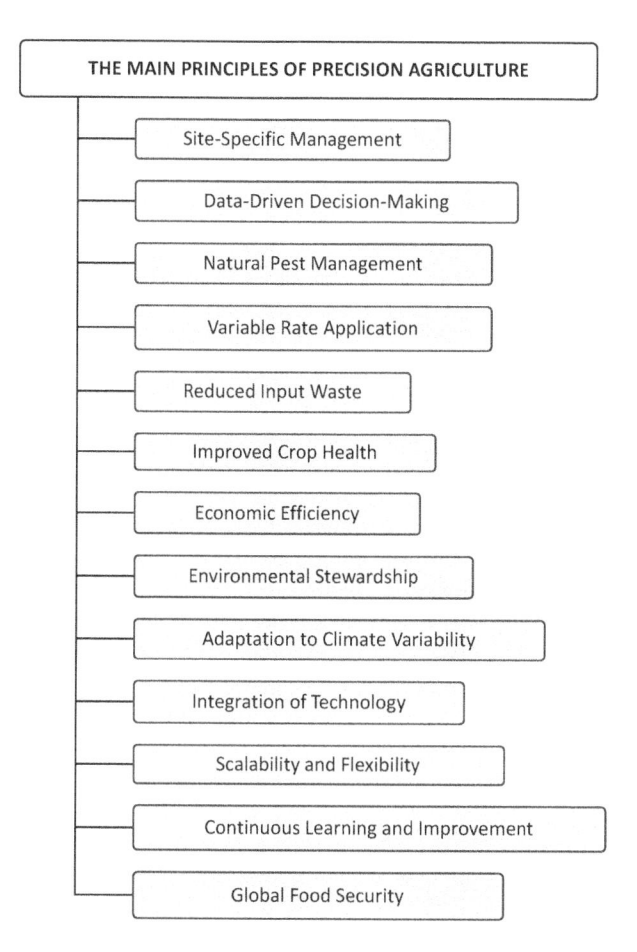

FIGURE 5.13 The main principles of precision agriculture. (Created by authors.)

Environmental Stewardship. Precision agriculture helps reduce the environmental impact of farming. It minimizes pollution from excess fertilizers and pesticides, conserves water resources, and contributes to sustainable land management.

Adaptation to Climate Variability. Precision agriculture helps farmers adapt to changing climatic conditions by adjusting their practices based on real-time data. This resilience is crucial in mitigating the impacts of climate change.

Integration of Technology. Precision agriculture relies on technology like GPS, GIS (Geographic Information Systems), remote sensing, drones, sensors, and data analytics. These tools provide accurate and up-to-date information for decision making.

Scalability and Flexibility. Precision agriculture can be applied on various scales, from small plots to large commercial farms. Its flexibility allows farmers to tailor their approach to their specific needs and available resources.

Continuous Learning and Improvement. Precision agriculture involves a cycle of data collection, analysis, implementation, and assessment. Farmers continually learn from their experiences and adjust their practices for ongoing improvement.

Global Food Security. As the world's population grows, precision agriculture can contribute to global food security by maximizing crop yields on existing agricultural land while minimizing negative environmental impacts.

In essence, precision agriculture represents a shift from traditional blanket approaches to farming toward a more customized, data-driven, and efficient approach. By optimizing resource use, minimizing waste, and adapting to changing conditions, precision agriculture embodies the future of sustainable and productive farming.

- **Sustainable Livestock Management**: This initiative promotes ethical and efficient livestock-rearing practices. It includes rotational grazing, proper waste management, and use of organic feed to minimize environmental impact and enhance animal welfare.
- **Community-Supported Agriculture (CSA)**: CSA connects consumers directly with local farmers, who provide regular shares of their produce. This initiative supports local economies, reduces food miles, and strengthens the relationship between farmers and consumers.
- **Permaculture**: Permaculture is a design approach that mimics natural ecosystems to create sustainable and self-sufficient agricultural systems. It emphasizes diversity, integration, and regenerative practices.
- **No-Till Farming**: No-till farming involves planting crops without plowing or tilling the soil. This minimizes soil disruption, preserves soil structure, and reduces erosion.
- **Seed Banks and Plant Genetic Resources**: Initiatives like seed banks preserve and share diverse crop varieties to ensure genetic diversity for breeding new resilient varieties.
- **Agroecology**: Agroecology is a holistic approach that combines ecological principles with agricultural practices. It focuses on creating resilient, self-sustaining farming systems that work in harmony with nature.

These initiatives collectively contribute to sustainable agriculture by promoting practices that conserve resources, enhance biodiversity, reduce environmental impact, and support rural communities. While the specific initiatives adopted may vary based on local conditions and needs, the overarching goal remains the same: to create a more resilient and ecologically balanced food system.

Governments play a crucial role in promoting and incentivizing the implementation of GAIs. By providing support and creating favorable conditions for farmers to adopt sustainable practices, governments can accelerate the transition toward more environmentally friendly and resilient agricultural systems. Here are some ways governments can offer support and incentives for the implementation of GAIs (Huang et al., 2019; Whitfield, 2006; Luo et al., 2014; Bachynskyi,

2021; Liberalesso et al., 2020; Shazmin et al., 2016; Wang et al., 2022; Migliorelli, 2019; Wasilewski & Wasilewska, 2019; Shen et al., 2020; : Gernego et al., 2022; Khatri-Chhetri et al., 2021; Ulrich, 2021; Borrego et al., 2023; United Nations Environment Programme, 2023):

1. **Financial Support and Subsidies**: Governments can offer direct financial incentives, subsidies, and grants to farmers who adopt sustainable practices. These may include subsidies for organic certification, reduced pesticide use, water conservation, and agroforestry establishment.
2. **Research and Development Funding**: Investing in research and development related to sustainable agricultural practices can lead to the development of new technologies, improved crop varieties, and innovative farming methods that align with green initiatives.
3. **Training and Capacity Building**: Providing training programs, workshops, and extension services to farmers helps educate them about sustainable practices, how to implement them effectively, and how to adapt to changing conditions.
4. **Market Access and Premiums**: Governments can facilitate market access for farmers who adopt green practices by connecting them to eco-friendly certification programs and premium markets that value sustainable products.
5. **Tax Incentives**: Offering tax breaks or deductions for farmers who invest in sustainable infrastructure, equipment, and practices encourages adoption by reducing the financial burden of transitioning to green initiatives.
6. **Low-Interest Loans**: Providing farmers with access to low-interest loans for adopting sustainable practices can help overcome financial barriers to implementation.
7. **Insurance and Risk Management**: Governments can work with insurance companies to offer specialized insurance products that protect farmers from risks associated with climate change and extreme weather events.
8. **Land Tenure and Property Rights**: Clear land tenure and property rights policies encourage farmers to make long-term investments in sustainable practices on their land.
9. **Research and Extension Services**: Supporting agricultural research institutions and extension services ensures that farmers have access to the latest information, technologies, and best practices for sustainable agriculture.
10. **Demonstration Farms**: Establishing demonstration farms or pilot projects showcases the benefits of green initiatives and provides practical examples for farmers to learn from.
11. **Certification and Labeling Programs**: Governments can create or support certification and labeling programs that distinguish sustainably produced agricultural products, making it easier for consumers to make environmentally conscious choices.
12. **Conservation Easements**: Governments can provide financial incentives to farmers who voluntarily enter into conservation easements that preserve natural habitats and encourage sustainable land management.

13. **Eco-Friendly Infrastructure Development**: Governments can invest in infrastructure that supports sustainable agriculture, such as irrigation systems, renewable energy sources, and composting facilities.
14. **Research and Extension Services**: Supporting agricultural research institutions and extension services ensures that farmers have access to the latest information, technologies, and best practices for sustainable agriculture.
15. **Public Awareness Campaigns**: Governments can launch public awareness campaigns to educate consumers about the benefits of supporting farmers who adopt green practices, creating demand for sustainable products.
16. **Carbon Credit Programs**: Farmers who implement practices that sequester carbon dioxide in the soil can participate in carbon credit programs, receiving financial incentives for their contributions to climate change mitigation.
17. **Partnerships with NGOs and Private Sector**: Governments can collaborate with non-governmental organizations (NGOs) and the private sector to pool resources, expertise, and funding for GAIs.

By offering a combination of these support mechanisms, governments can create an enabling environment that encourages farmers to adopt GAIs and contribute to a more sustainable and resilient food system.

5.4 CORPORATE SOCIAL RESPONSIBILITY IN AGRICULTURE

CSR has become an integral aspect of modern businesses, transcending profit-driven motives and emphasizing the role of companies in addressing social and environmental challenges. In the context of agriculture, CSR plays a crucial role in promoting sustainable practices, supporting local communities, and safeguarding the well-being of all stakeholders involved in the food production process. In this teaching material, we explore the significance of CSR in agriculture and how businesses can integrate ethical principles into their operations for the greater good.

5.4.1 Definition of CSR in Agriculture

CSR in agriculture refers to the voluntary actions and initiatives that agricultural businesses undertake to contribute positively to society and the environment. It involves considering the interests and needs of various stakeholders, such as farmers, workers, consumers, local communities, and the environment, beyond the financial bottom line.

CSR in agriculture is a complex and evolving concept, and different scholars may define it in various ways based on their perspectives and research focus. Here are some common definitions and interpretations of CSR in agriculture (Luhmann & Theuvsen, 2016; Mazur-Wierzbicka, 2015; Heyder & Theuvsen, 2009; Vidales et al., 2012; Kloppers & Fourie, 2014; Luhmann & Theuvsen, 2017; Poetz et al., 2013; Soloviova et al., 2022; Coppola et al., 2020; Hartmann, 2011; Nogueira et al., 2023; Tseng et al., 2020; Bunga, 2023; EC, 2022; de Olde &

Valentinov, 2019; Araújo et al., 2023; Vrabcová & Urbancová, 2023; Mijatović et al., 2023; Bavorova et al., 2021; Biró & Szalmáné Csete, 2021):

- **Triple Bottom Line Approach**: CSR in agriculture is often defined as the practice of balancing economic profitability with social and environmental responsibilities. It emphasizes the importance of considering the "triple bottom line" of people (social), planet (environmental), and profit (economic) in agricultural operations.

The TBL approach is especially relevant and impactful in the context of agriculture. Agriculture plays a critical role in providing food, fiber, and raw materials for various industries while simultaneously impacting ecosystems, communities, and economic well-being. In this context, the TBL Approach can be expanded to emphasize the interplay between the three dimensions: economic, social, and environmental, within agriculture. A TBL approach in the context of agriculture is presented in Figure 5.14.

The TBL approach in agriculture underscores the need for a balanced and sustainable approach to farming. It recognizes that economic profitability is interconnected with social well-being and environmental stewardship. By adopting this approach, agricultural stakeholders can work toward achieving a more sustainable and responsible agriculture sector that benefits not only farmers but also communities and the planet.

- **Ethical Farming**: This interpretation focuses on ethical considerations in agriculture, including fair treatment of farm laborers, animal welfare, and responsible land management. CSR in agriculture involves promoting ethical farming practices and treating all stakeholders involved with respect and fairness.

The connection between Ethical Farming and CSR in agriculture is shown in Figure 5.15:

1. **Ethical Farming as a CSR Pillar**.
    - *Respect for Stakeholders*. Ethical Farming, within the context of CSR, involves treating all stakeholders involved in agricultural operations with respect and fairness. This includes not only the farmers but also farm workers, local communities, consumers, and the environment.
    - *Labor Practices*. CSR in agriculture includes commitments to fair labor practices. Ethical Farming ensures that farm workers are paid fair wages, provided safe working conditions, and have access to essential benefits like healthcare and education for their families.
    - *Environmental Responsibility*. Ethical Farming aligns with the environmental dimension of CSR. It promotes practices that reduce the negative impact of farming on the environment, such as minimizing chemical pesticide and fertilizer use, reducing soil erosion, and conserving water resources.

Triple Bottom Line approach in the context of agriculture

PROFIT (Economic Bottom Line)

Financial Viability of Farms. Economic sustainability in agriculture is about ensuring the profitability and long-term viability of farms. This involves efficient resource management, cost-effective production, and fair pricing for agricultural products.
Market Access. Farmers should have equitable access to markets and receive fair compensation for their products. This economic aspect also extends to supply chain actors, ensuring that everyone involved, from seed producers to retailers, receives a fair share of the value created.
Innovation in Agriculture. Promoting innovation within the agricultural sector can enhance productivity, reduce production costs, and open up new markets. This includes adopting modern technologies and sustainable farming practices that can improve economic outcomes for farmers.

PEOPLE (Social Bottom Line)

Farmworker Welfare. The social dimension in agriculture involves ensuring the well-being and fair treatment of farmworkers. This includes fair wages, safe working conditions, and access to social services.
Rural Community Development. Agriculture often forms the backbone of rural communities. A strong social bottom line includes initiatives that support rural development through job creation, infrastructure improvements, and access to education and healthcare.
Food Security and Accessibility. Ensuring that agricultural practices provide access to safe and nutritious food for all segments of society is a key social responsibility. This extends to reducing food deserts and improving food distribution systems.

PLANET (Environmental Bottom Line)

Sustainable Farming Practices. The environmental aspect of the TBL in agriculture emphasizes the adoption of sustainable farming practices that reduce soil erosion, minimize chemical inputs, and conserve water resources. This includes practices like crop rotation, organic farming, and precision agriculture.
Biodiversity Conservation. Agriculture can have a significant impact on biodiversity. The environmental bottom line involves protecting and enhancing biodiversity by preserving natural habitats, promoting agroforestry, and reducing monoculture farming.
Climate Change Mitigation and Adaptation. Given the vulnerability of agriculture to climate change, this dimension includes efforts to reduce greenhouse gas emissions from farming activities and adapt to changing climate conditions. Practices like carbon sequestration in soils and the use of climate-resilient crop varieties are relevant here.

FIGURE 5.14 Triple bottom line approach in the context of agriculture. (Created by authors.)

2. **Transparency and Accountability.**
    - *Transparency.* Ethical Farming practices often require transparency in operations. CSR initiatives in agriculture emphasize disclosing information about farming methods, sourcing, and production processes. This transparency builds trust with consumers and other stakeholders.
    - *Accountability.* CSR encourages agricultural businesses to be accountable for their actions. Ethical Farming practices hold farmers and agribusinesses accountable for adhering to ethical standards, whether in terms of fair labor practices, animal welfare, or environmental stewardship.

The Role of Agricultural Business in Low-Carbon Transition 239

```
┌─────────────────────────────────────────────────────┐
│   The connection between Ethical Farming and Corporate Social │
│              Responsibility (CSR) in agriculture             │
└─────────────────────────────────────────────────────┘
                          ↓
┌─────────────────────────────────────────────────────┐
│            Ethical Farming as a CSR Pillar          │
├──────────────────────────┬──────────────────────────┤
│  Respect for Stakeholders │ Labor Practices and Environmental │
│                          │         Responsibility         │
└──────────────────────────┴──────────────────────────┘
                          ↓
┌─────────────────────────────────────────────────────┐
│            Transparency and Accountability          │
├──────────────────────────┬──────────────────────────┤
│       Transparency       │       Accountability     │
└──────────────────────────┴──────────────────────────┘
                          ↓
┌─────────────────────────────────────────────────────┐
│         Consumer Expectations and Reputation        │
├──────────────────────────┬──────────────────────────┤
│    Consumer Awareness    │   Reputation Management  │
└──────────────────────────┴──────────────────────────┘
                          ↓
┌─────────────────────────────────────────────────────┐
│               Sustainable Agriculture               │
├──────────────────────────┬──────────────────────────┤
│  Long-Term Sustainability│     Economic Viability   │
└──────────────────────────┴──────────────────────────┘
                          ↓
┌─────────────────────────────────────────────────────┐
│         Community and Social Development            │
├──────────────────────────┬──────────────────────────┤
│   Community Engagement   │      Inclusive Growth    │
└──────────────────────────┴──────────────────────────┘
```

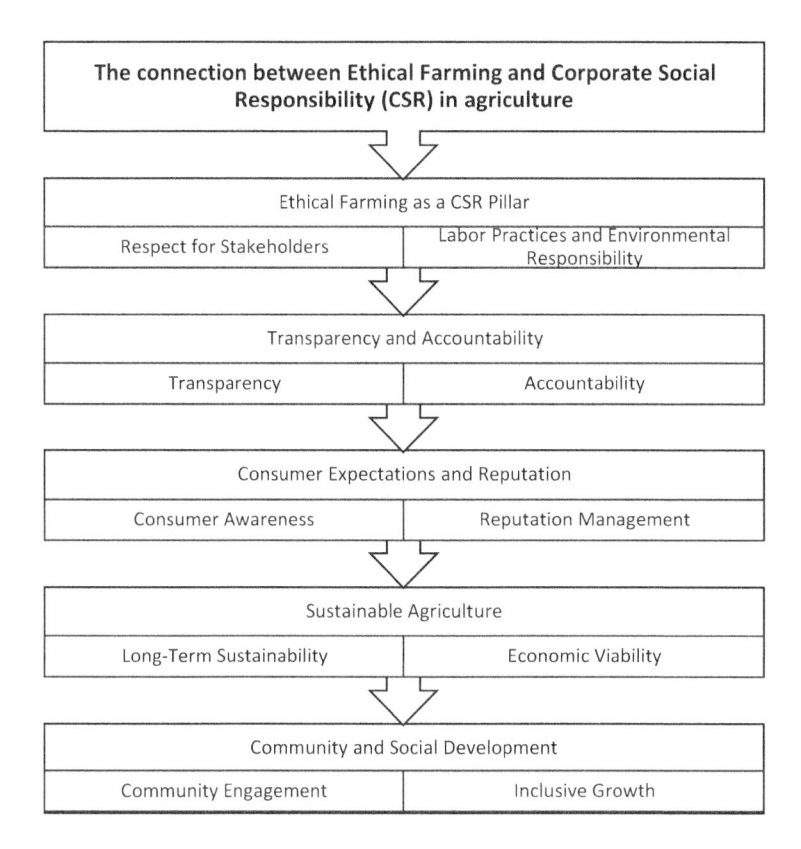

FIGURE 5.15 The connection between ethical farming and corporate social responsibility (CSR) in agriculture. (Created by authors.)

3. **Consumer Expectations and Reputation**.
   - *Consumer Awareness.* Today's consumers are increasingly conscious of where their food comes from and how it's produced. Ethical Farming practices aligned with CSR meet the expectations of consumers who want ethically produced, environmentally friendly, and socially responsible agricultural products.
   - *Reputation Management.* For agricultural businesses, their reputation is closely tied to their CSR efforts, including Ethical Farming. Demonstrating a commitment to ethical practices not only attracts socially conscious consumers but also helps protect the company's reputation in the face of potential controversies.
4. **Sustainable Agriculture**.
   - *Long-Term Sustainability.* Ethical Farming practices are often synonymous with sustainable agriculture. CSR in agriculture seeks to ensure the long-term viability of farming operations, which is inherently linked to sustainable practices that don't deplete resources or harm ecosystems.

- *Economic Viability.* Sustainable farming practices, a key aspect of Ethical Farming, contribute to the economic sustainability of farms. By reducing input costs, conserving soil fertility, and promoting crop diversity, ethical and sustainable practices can enhance farm profitability over the long term.
5. **Community and Social Development**.
    - *Community Engagement.* Ethical Farming practices often involve engaging with and supporting local communities. CSR in agriculture includes initiatives such as education, healthcare, and infrastructure development that directly benefit these communities.
    - *Inclusive Growth.* CSR fosters inclusive growth by ensuring that the benefits of agricultural activities are distributed equitably among stakeholders, including small-scale farmers and marginalized communities.

Ethical Farming is an integral part of CSR in agriculture because it embodies the ethical, social, and environmental responsibilities that agricultural businesses should uphold. By practicing Ethical Farming, agricultural companies not only meet consumer expectations but also contribute to the long-term sustainability and well-being of the farming community and the environment, aligning with the broader principles of CSR.

- **Sustainable Agriculture**: CSR in agriculture can be viewed as a commitment to sustainable farming practices. This includes minimizing the environmental impact of farming activities, conserving natural resources, and promoting long-term agricultural productivity while safeguarding ecosystems.

Sustainable agriculture and CSR in agriculture are closely interconnected and often overlap in their goals and principles. Sustainable agriculture is a fundamental component of CSR in agriculture, and they share common objectives and areas of focus. The links between sustainable agriculture and CSR in agriculture are presented in Figure 5.16.

In essence, sustainable agriculture is a fundamental component of CSR in agriculture because it encompasses responsible environmental stewardship, economic viability, ethical treatment of stakeholders, and a long-term perspective – all of which are central principles of CSR. By adopting sustainable farming practices, agricultural businesses contribute to the broader goals of CSR by promoting a more responsible, ethical, and sustainable agricultural sector that benefits the environment, society, and the economy.

- **Community Engagement**: Some definitions emphasize the engagement of agricultural businesses with local communities. CSR in agriculture involves supporting the communities where farming operations take place, often through initiatives like education, healthcare, and infrastructure development.
- **Supply Chain Responsibility**: Sustainable supply chain management (SSCM) holds significant importance within the agricultural sector.

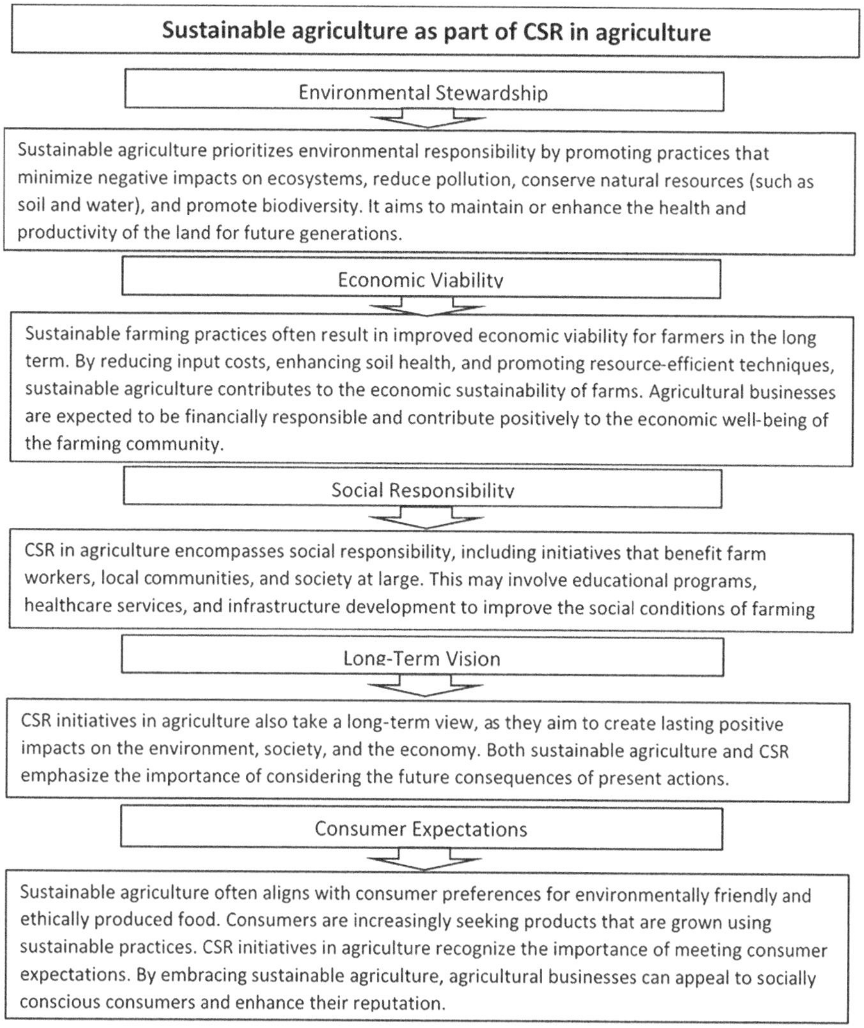

Sustainable agriculture as part of CSR in agriculture

Environmental Stewardship

Sustainable agriculture prioritizes environmental responsibility by promoting practices that minimize negative impacts on ecosystems, reduce pollution, conserve natural resources (such as soil and water), and promote biodiversity. It aims to maintain or enhance the health and productivity of the land for future generations.

Economic Viability

Sustainable farming practices often result in improved economic viability for farmers in the long term. By reducing input costs, enhancing soil health, and promoting resource-efficient techniques, sustainable agriculture contributes to the economic sustainability of farms. Agricultural businesses are expected to be financially responsible and contribute positively to the economic well-being of the farming community.

Social Responsibility

CSR in agriculture encompasses social responsibility, including initiatives that benefit farm workers, local communities, and society at large. This may involve educational programs, healthcare services, and infrastructure development to improve the social conditions of farming

Long-Term Vision

CSR initiatives in agriculture also take a long-term view, as they aim to create lasting positive impacts on the environment, society, and the economy. Both sustainable agriculture and CSR emphasize the importance of considering the future consequences of present actions.

Consumer Expectations

Sustainable agriculture often aligns with consumer preferences for environmentally friendly and ethically produced food. Consumers are increasingly seeking products that are grown using sustainable practices. CSR initiatives in agriculture recognize the importance of meeting consumer expectations. By embracing sustainable agriculture, agricultural businesses can appeal to socially conscious consumers and enhance their reputation.

FIGURE 5.16 The links between sustainable agriculture and CSR in agriculture. (Created by authors.)

According to Wieck (2020), the adoption of supply chain-based approaches must be complemented by enhancements in trade, investment, and agricultural policies to exert influence on agricultural supply chains and address sustainability and human rights issues effectively. Literature reviews by Kalchschmidt and Syahruddin (2011), Syahruddin and Kalchschmidt (2012) emphasize the need for further research in the realm of supply chain management and sustainability within the agricultural sector. They underscore the gaps that exist in the current body of knowledge. Agyemang et al., (2020) introduce socially sustainable supply chain criteria tailored for the cashew industry in West Africa and assess them based on empirical data. The study identifies food safety, labor, and work conditions, traceability, and child and forced/prison labor as crucial criteria for the successful implementation of socially sustainable supply chain initiatives.

The perspective of Supply Chain Responsibility underscores the paramount importance of accountability across the entire agricultural supply chain, spanning from seed suppliers to food retailers. This approach revolves around ensuring that all participants within the supply chain uphold ethical and sustainable practices.

The implementation of agricultural supply chain responsibility standards encounters numerous obstacles. Naik & Suresh (2018) underscore the necessity for the participation of small farmers in sourcing networks and institutional endeavors aimed at meeting food safety and quality regulations. Boström et al. (2015) identify six significant challenges in the governance of sustainability within global supply chains, including gaps in geographical coverage, information flow, communication, compliance, power dynamics, and legitimacy. Wardell et al. (2021) discuss the intricate and varied circumstances within which agri-food and timber supply chains operate, leading to uncertainties and trade-offs between gains and losses. Kamble et al. (2019) suggest the utilization of emerging technologies like the Internet of Things, blockchain, and big data to enable sustainable agriculture supply chains, with a particular emphasis on achieving supply chain visibility and efficiently managing resources as the primary drivers for developing data analytics capabilities and attaining sustainable performance. The successful implementation of agricultural supply chain responsibility standards necessitates addressing challenges related to the involvement of small-scale farmers, the governance of sustainability, the multifaceted and diverse operating conditions, and the integration of emerging technologies.

- **Food Safety and Quality**: CSR in agriculture can be seen as a commitment to producing safe and high-quality food products. This includes measures to ensure food safety, traceability, and adherence to quality standards.

CSR efforts within the agricultural sector can have a beneficial impact on consumer confidence in food products. Mueller and Theuvsen (2014) revealed that consumers make distinctions between a company's economic, internal, and external responsibilities, and the public endorsement of a company's CSR activities positively influences consumers' attitudes toward that company's commitment. Boysselle (2015) demonstrated that effectively conveying a company's environmental message is more successful in bolstering consumers' perceptions of ethical and social value, as well as trust. Civero et al. (2018) noted that the level of awareness among respondents regarding the communication of CSR strategies by food companies is suboptimal, underscoring the necessity for a well-devised communication strategy to heighten awareness on this subject. Zhang (2022) discovered that during the COVID-19 pandemic, a company's image and customer satisfaction play significant mediating roles, and the pandemic amplifies the favorable impact of CSR on customer satisfaction. In summary, CSR initiatives in agriculture can indeed enhance consumer trust in food products, but to do so effectively, well-planned communication strategies are essential to raise awareness and enhance perception.

- **Transparency and Accountability**: Many interpretations of CSR in agriculture stress transparency in operations and accountability for social and environmental impacts. This involves openly disclosing information about sourcing, production methods, and progress toward sustainability goals.
- **Biodiversity Conservation**: CSR in agriculture may encompass efforts to protect and enhance biodiversity on farms. This can involve practices like crop rotation, habitat preservation, and the promotion of pollinator-friendly agriculture.
- **Climate Change Mitigation and Adaptation**: Given the climate challenges in agriculture, CSR often includes strategies to mitigate greenhouse gas emissions, adopt climate-resilient farming practices, and support farmers in adapting to changing climate conditions.
- **Inclusivity and Fair Trade**: Some definitions highlight the importance of inclusivity and fair-trade principles in agriculture. CSR involves ensuring that small-scale and marginalized farmers have access to markets and receive fair compensation for their products.
- **Rural Development**: CSR in agriculture can be interpreted as a commitment to rural development. This includes initiatives to improve the socio-economic conditions of rural communities by providing employment opportunities, skills development, and access to resources.
- **Regulatory Compliance**: Finally, CSR in agriculture may involve strict adherence to local and international regulations related to farming, labor, and environmental protection.

These common interpretations of CSR in agriculture demonstrate the multi-dimensional nature of corporate responsibility in the agricultural sector. It encompasses economic, social, and environmental aspects, with a focus on ethical, sustainable, and responsible practices that benefit both the farming community and the wider society.

Agricultural businesses hold a unique position of influence within the food supply chain, making their engagement in CSR pivotal. By embracing ethical practices, promoting social equity, and fostering environmental stewardship, these businesses can catalyze positive change across the agricultural sector.

Ethical labor practices in agriculture are an important component of CSR in the agricultural sector. Ensuring fair and ethical treatment of farm workers is not only a moral imperative but also contributes to the sustainability and reputation of agricultural businesses. Here's how ethical labor practices in agriculture align with CSR (Maseko, 2023; Hurst, 2023; World Employment and Social Outlook, 2023; Muhie, 2023; Horizon Europe, 2023; de Olde & Valentinov, 2019; Timpanaro et al., 2023; European Union, 2023; Vrabcová & Urbancová, 2023; Mijatović et al., 2023):

Fair Wages and Compensation:

- **CSR Commitment**: Agricultural businesses committed to CSR should provide fair wages and compensation to farm workers, ensuring that they receive a livable income for their labor.

Safe and Healthy Working Conditions:

- **CSR Commitment**: CSR in agriculture involves maintaining safe and healthy working conditions. Agricultural employers must adhere to occupational health and safety standards to protect workers from accidents and health risks associated with agricultural work.

Workplace Diversity and Inclusion:

- **CSR Commitment**: Promoting workplace diversity and inclusion is essential for CSR. Agricultural businesses should provide equal opportunities and treat all workers with respect, regardless of their background or identity.

Child Labor and Forced Labor Prevention:

- **CSR Commitment**: CSR initiatives in agriculture actively work to prevent child labor and forced labor in farming operations. This includes rigorous monitoring of supply chains to ensure compliance with international labor standards.

Access to Education and Training:

- **CSR Commitment**: Providing access to education and training opportunities for farm workers is an integral part of CSR in agriculture. This empowers workers to develop new skills and improve their socio-economic well-being.

Social Benefits and Services:

- **CSR Commitment**: CSR involves offering social benefits and services to farm workers, such as access to healthcare, housing, and childcare services to support their overall well-being.

Freedom of Association and Collective Bargaining:

- **CSR Commitment**: Ethical labor practices in agriculture respect workers' rights to freedom of association and collective bargaining. Agricultural businesses should not interfere with workers' rights to organize or join labor unions.

Non-Discrimination:

- **CSR Commitment**: Agricultural businesses committed to CSR must not engage in discriminatory practices based on factors like race, gender, age, or ethnicity. All workers should be treated equally and fairly.

Working Hour Limits and Rest Periods:

- **CSR Commitment**: CSR in agriculture includes adhering to working hour limits and providing appropriate rest periods to prevent overwork and exhaustion among farm workers.

Regular Auditing and Reporting:

- **CSR Commitment**: Agricultural businesses should conduct regular audits and assessments of labor practices to identify and address any violations. They should also transparently report on their efforts to improve labor conditions.

Training and Awareness:

- **CSR Commitment**: CSR initiatives in agriculture often include training and awareness programs to educate farm managers and workers about ethical labor practices and their importance.

Conflict Resolution Mechanisms:

- **CSR Commitment**: Establishing conflict resolution mechanisms is essential for addressing disputes and grievances among farm workers. Ethical labor practices include providing channels for workers to voice concerns and seek resolutions.

Supply Chain Responsibility:

- **CSR Commitment**: Agricultural businesses should extend their commitment to ethical labor practices throughout their supply chains, ensuring that all suppliers and subcontractors uphold similar standards.

By embracing ethical labor practices in agriculture, agricultural businesses demonstrate their commitment to the well-being and rights of farm workers, contribute to the sustainability of rural communities, and enhance their reputation as socially responsible organizations. These practices align with the broader principles of CSR, which prioritize ethical, social, and environmental responsibilities alongside economic goals.

5.5 CONCLUSIONS

Green growth in agricultural business represents a transformative journey toward sustainable and environmentally conscious farming practices. By adopting innovative strategies, embracing circular business models, and collaborating with stakeholders, agricultural businesses can lead the charge in creating a more resilient and greener food system. As future agricultural leaders and entrepreneurs, understanding and promoting green growth is paramount to ensuring a sustainable and prosperous future for both the agricultural sector and the planet.

The transition to low-carbon energy is a critical component of the broader efforts to combat climate change and foster sustainable development. As key players in the global economy, businesses have a unique opportunity and responsibility to lead the charge in this transformation. By embracing low-carbon energy sources, optimizing energy usage, and investing in renewable solutions, businesses can contribute significantly to building a cleaner, more resilient, and sustainable energy future. The low-carbon energy transition is not just a corporate responsibility but a pathway toward a more prosperous and environmentally conscious world for generations to come.

GAIs represent a transformative force that can revolutionize the agricultural sector and promote sustainable development. By embracing organic farming, agroforestry, climate-smart practices, and green technologies, farmers can enhance productivity, protect natural resources, and contribute to global efforts in mitigating climate change. As stakeholders in the food system, agricultural communities and businesses play a vital role in fostering a greener and more resilient future. Through education, awareness, and collaboration, GAIs can pave the way for a sustainable, healthy, and abundant future for generations to come.

CSR in agriculture is not merely an ethical obligation but a strategic approach that yields long-term benefits for businesses, communities, and the environment. By embracing ethical labor practices, promoting sustainable sourcing, and being environmentally responsible, agricultural businesses can foster a more equitable, resilient, and sustainable food system. Through collaboration, transparency, and stakeholder engagement, CSR initiatives can build bridges between businesses and society, creating a positive impact that extends far beyond the agricultural fields. As responsible stewards of the land and its resources, agricultural businesses have the power to nurture sustainable partnerships that pave the way for a more prosperous and harmonious future.

REFERENCES

Abad-Segura, E., Fuente, A. B., González-Zamar, M., & Belmonte-Ureña, L. J. (2020). Effects of circular economy policies on the environment and sustainable growth: Worldwide Research. *Sustainability*, 12(14), 5792. https://doi.org/10.3390/su12145792

Abman, R., Garg, T., Pan, Y., & Singhal, S. (2020). Agriculture and deforestation. *Agricultural & Natural Resource Economics eJournal*. https://doi.org/10.2139/ssrn.3692682

Adetunji, A. T., Ncube, B., Mulidzi, R. A., & Lewu, F. B. (2020). Management impact and benefit of cover crops on soil quality: A review. *Soil & Tillage Research*, 204, 104717. https://doi.org/10.1016/j.still.2020.104717

Agrawal, R., De Tommasi, L., Lyons, P., Zanoni, S., Papagiannis, G. K., Karakosta, C., Papapostolou, A., Durand, A., Martinez, L., Fragidis, G., & Corbella, M. (2023). Challenges and opportunities for improving energy efficiency in SMEs: Learnings from seven European projects. *Energy Efficiency*, 16, 17. https://doi.org/10.1007/s12053-023-10090-z

Agyemang, M., Kusi-Sarpong, S., Agyemang, J. K., Jia, F., & Adzanyo, M. (2020). Determining and evaluating socially sustainable supply chain criteria in agri-sector of developing countries: Insights from West Africa cashew industry. *Production Planning & Control*, 33, 1115–1133. https://doi.org/10.1080/09537287.2020.1852479

Akkaya, D., Bimpikis, K., & Lee, H. L. (2019, Forthcoming). Government interventions to promote agricultural innovation. *Manufacturing & Service Operations Management.* https://doi.org/10.2139/ssrn.3001342

Alam, A. (2014). Soil degradation: A challenge to sustainable agriculture. *International Journal of Scientific Research, 1,* 50–55. https://doi.org/10.12983/ijsras-2014-p0050-0055

Allan, G. J., Hanley, N., McGregor, P. G., Swales, J. K., & Turner, K. (2007). The impact of increased efficiency in the industrial use of energy: A computable general equilibrium analysis for the United Kingdom. *Energy Economics, 29,* 779–798. https://doi.org/10.1016/j.eneco.2006.12.006

Amede, T., Konde, A. A., Muhinda, J. J., & Bigirwa, G. (2023). Sustainable farming in practice: Building resilient and profitable smallholder agricultural systems in Sub-Saharan Africa. *Sustainability, 15*(7), 5731. https://doi.org/10.3390/su15075731

Araújo, J., Pereira, I. V., & Santos, J. D. (2023). The effect of corporate social responsibility on brand image and brand equity and its impact on consumer satisfaction. *Administrative Sciences, 13,* 118. https://doi.org/10.3390/admsci13050118

Araújo, O. D., & Medeiros, J. L. (2017). Carbon capture and storage technologies: Present scenario and drivers of innovation. *Current Opinion in Chemical Engineering, 17,* 22–34. https://doi.org/10.1016/j.coche.2017.05.004

Arora, P., Peterson, N. D., Bert, F. E., & Podestá, G. P. (2016). Managing the triple bottom line for sustainability: a case study of Argentine agribusinesses. *Sustainability: Science, Practice and Policy, 12,* 60–75. https://doi.org/10.1080/15487733.2016.11908154

Aulakh, C.S., Sharma, S., Thakur, M., & Kaur, P. (2022). A review of the influences of organic farming on soil quality, crop productivity and produce quality. *Journal of Plant Nutrition, 45,* 1884–1905. https://doi.org/10.1080/01904167.2022.2027976

Azar, C., Lindgren, K., Obersteiner, M., Riahi, K., Vuuren, D. V., Elzen, K. M., Möllersten, K., & Larson, E. D. (2010). The feasibility of low CO2 concentration targets and the role of bio-energy with carbon capture and storage (BECCS). *Climatic Change, 100,* 195–202. https://doi.org/10.1007/S10584-010-9832-7

Bachynskyi, R. (2021). International experience in stimulation of green initiatives in agriculture and directions of its implementation in the national practice. https://doi.org/10.33245/2310-9262-2021-162-1-41-49

Badgley, C., Moghtader, J., Quintero, E., Zakem, E. J., Chappell, M. J., Avilés-Vázquez, K., Samulon, A., & Perfecto, I. (2007). Organic agriculture and the global food supply. *Renewable Agriculture and Food Systems, 22,* 86–108. https://doi.org/10.1017/S1742170507001640

Barth, H., & Melin, M. (2018). A green lean approach to global competition and climate change in the agricultural sector – A Swedish case study. *Journal of Cleaner Production, 204,* 183–192. https://doi.org/10.1016/j.jclepro.2018.09.021

Basok, B., & Baseyev, Y. (2022). Low carbon energy (review). 1. Problems and forecasts. *Thermophysics and Thermal Power Engineering, 44*(1), 27–36. https://doi.org/10.31472/ttpe.1.2022.4

Battaglia, M. L., Thomason, W., Ozlu, E., Rezaei-Chiyaneh, E., Fike, J. H., Diatta, A. A., Uslu, O. S., Babur, E., & Schillaci, C. (2023). Short-term crop residue management in no-tillage cultivation effects on soil quality indicators in Virginia. *Agronomy, 13,* 838. https://doi.org/10.3390/agronomy13030838

Bavorova, M., Bednarikova, Z., Ponkina, E. V., & Visser, O. (2021). Agribusiness social responsibility in emerging economies: Effects of legal structure, economic performance and managers' motivations. *Journal of Cleaner Production, 289,* 125157. https://doi.org/10.1016/j.jclepro.2020.125157

Biró, K., & Szalmáné Csete, M. (2021). Corporate social responsibility in agribusiness: Climate-related empirical findings from Hungary. *Environment, Development and Sustainability, 23,* 5674–5694. https://doi.org/10.1007/s10668-020-00838-3

Blake, R. J. (2020). Will the European green deal make agriculture more sustainable? *Outlooks on Pest Management, 31*, 198–200. https://doi.org/10.1564/v31_oct_01

Bobicki, E. R., Liu, Q., Xu, Z., & Zeng, H. (2012). Carbon capture and storage using alkaline industrial wastes. *Progress in Energy and Combustion Science, 38*, 302–320. https://doi.org/10.1016/j.pecs.2011.11.002

Boot-Handford, M. E., Abanades, J. C., Anthony, E. J., Blunt, M. J., Brandani, S., Dowell, N. M., Fernández, J. R., Ferrari, M., Gross, R. J., Hallett, J. P., Haszeldine, R. S., Heptonstall, P. J., Lyngfelt, A., Makuch, Z., Mangano, E., Porter, R. T., Pourkashanian, M., Rochelle, G. T., Shah, N., Yao, J. G., & Fennell, P. S. (2014). Carbon capture and storage update. *Energy and Environmental Science, 7*, 130–189. https://doi.org/10.1039/c3ee42350f

Borrego, A. C., Abreu, R., Carreira, F. A., Caetano, F., & Vasconcelos, A. L. (2023). Environmental taxation on the agri-food sector and the farm to fork strategy: The Portuguese case. *Sustainability, 15*, 12124. https://doi.org/10.3390/su151612124

Boström, M., Jönsson, A. M., Lockie, S., Mol, A. P., & Oosterveer, P. (2015). Sustainable and responsible supply chain governance: Challenges and opportunities. *Journal of Cleaner Production, 107*, 1–7. https://doi.org/10.1016/j.jclepro.2014.11.050

Boysselle, J. M. (2015). The influence of CSR (Corporate Social Responsibility) communication on brand perceived value and trust: The case of SME in the food industry. *Economics and Finance.* Université Montpellier (English). https://theses.hal.science/tel-01360174/file/2015_martinezboysselle_arch.pdf

Braden, J. B., & Shortle, J. S. (2013). Agricultural sources of water pollution. *Encyclopedia of Energy, Natural Resource, and Environmental Economics, 3*, 81–85 https://doi.org/10.1016/B978-0-12-375067-9.00111-X

Bunga, Y. (2023). The importance of triple bottom line based corporate social responsibility and sustainability in the modern era (July 21, 2023). https://doi.org/10.2139/ssrn.4517342

Burch, D., Lyons, K., & Lawrence, G. (2001). What do we mean by green? Consumers, agriculture and the food industry. Corpus ID: 53486375.

Cao, J., & Chen, S. (2022). *Perspectives on the use of advanced nuclear energy systems for new energy vehicles.* 2022 IEEE 5th International Electrical and Energy Conference (CIEEC), pp. 2343–2348. https://doi.org/10.1109/cieec54735.2022.9846127

Capasso, M., Hansen, T., Heiberg, J., Klitkou, A., & Steen, M. (2019). Green growth – A synthesis of scientific findings. *Technological Forecasting and Social Change. 146*, 390–402. https://doi.org/10.1016/j.techfore.2019.06.013

Cardinael, R., Mao, Z., Chenu, C., & Hinsinger, P. (2020). Belowground functioning of agroforestry systems: Recent advances and perspectives. *Plant and Soil, 453*, 1–13. https://doi.org/10.1007/s11104-020-04633-x

Chaudhary, A., Pfister, S., & Hellweg, S. (2016). Spatially explicit analysis of biodiversity loss due to global agriculture, pasture and forest land use from a producer and consumer perspective. *Environmental Science & Technology, 50*(7), 3928–3936. https://doi.org/10.1021/acs.est.5b06153

Chaudhary, A., Timsina, P., Suri, B., Karki, E., Sharma, A., Sharma, R., & Brown, B. (2022). Experiences with conservation agriculture in the Eastern Gangetic Plains: Farmer benefits, challenges, and strategies that frame the next steps for wider adoption. *Frontiers in Agronomy, 3*, 103. https://doi.org/10.3389/fagro.2021.787896

Civero, G., Rusciano, V., Scarpato, D. (2018). Orientation of agri-food companies to CSR and consumer perception: A survey on two Italian companies. *Recent Patents on Food, Nutrition & Agriculture, 9*(2). https://dx.doi.org/10.2174/2212798410666180508103619

Coppola, A., Ianuario, S., Romano, S., & Viccaro, M. (2020). Corporate social responsibility in agri-food firms: The relationship between CSR actions and firm's performance. *AIMS Environmental Science, 7*(6), 542–558. https://doi.org/10.3934/environsci.2020034

Corsten, M., Ramírez, A. R., Shen, L., Koornneef, J., & Faaij, A. P. (2013). Environmental impact assessment of CCS chains – Lessons learned and limitations from LCA literature. *International Journal of Greenhouse Gas Control*, *13*, 59–71. https://doi.org/10.1016/j.ijggc.2012.12.003

Dang, X., Gao, S., Tao, R., Liu, G., Xia, Z., Fan, L., & Bi, W. (2020). Do environmental conservation programs contribute to sustainable livelihoods? Evidence from China's grain-for-green program in northern Shaanxi province. *The Science of the Total Environment*, *719*, 137436. https://doi.org/10.1016/j.scitotenv.2020.137436

Dang, Y. P., Page, K. L., Dalal, R. C., & Menzies, N. W. (2020). No-till farming systems for sustainable agriculture: An overview. https://doi.org/10.1007/978-3-030-46409-7_1

de Olde, E. M., & Valentinov, V. (2019). The moral complexity of agriculture: A challenge for corporate social responsibility. *Journal of Agricultural and Environmental Ethics*, *32*, 413–430. https://doi.org/10.1007/s10806-019-09782-3

Devi, O. R., Laishram, B., Singh, S., Paul, A., Sarma, H. H., Bora, S. S., & Devi, S. B. (2023). A review on mitigation of greenhouse gases by agronomic practices towards sustainable agriculture. *International Journal of Environment and Climate Change*, *13*(8), 278–287. https://doi.org/10.9734/ijecc/2023/v13i81952

Donner, M., Verniquet, A., Broeze, J., Kayser, K., & de Vries, H. (2021). Critical success and risk factors for circular business models valorising agricultural waste and by-products. *Resources, Conservation and Recycling*. *165*, 105236 https://doi.org/10.1016/j.resconrec.2020.105236

EC. (2022). EU agricultural outlook for markets, income and environment, 2022-2032. European Commission, DG Agriculture and Rural Development, Brussels. https://agriculture.ec.europa.eu/system/files/2023-04/agricultural-outlook-2022-report_en_0.pdf

EFI-KAPSARC Workshop Report. (2023, March). A global hydrogen future. https://energyfuturesinitiative.org/wp-content/uploads/sites/2/2023/03/A-Global-Hydrogen-Future-Workshop-Report_FINAL_3-6-2023-2.pdf

Eriksson, O. (2021). The importance of traditional agricultural landscapes for preventing species extinctions. *Biodiversity and Conservation*, *30*, 1341–1357. https://doi.org/10.1007/s10531-021-02145-3

Ertop, H., Kocięcka, J., Atilgan, A., Liberacki, D., Niemiec, M., & Rolbiecki, R. (2023). The importance of rainwater harvesting and its usage possibilities: Antalya example (Turkey). *Water*, *15*, 2194. https://doi.org/10.3390/w15122194

European Commission. (2023). Organic farming in the EU – A decade of organic growth, DG Agriculture and Rural Development, Brussels. https://agriculture.ec.europa.eu/system/files/2023-04/agri-market-brief-20-organic-farming-eu_en.pdf

European Union. (2023). The new common agricultural policy: 2023–27. The new common agricultural policy will be key to securing the future of agriculture and forestry, as well as achieving the objectives of the European Green Deal. 2022 The new common agricultural policy: 2023–27. European Commission (europa.eu).

Evans, J. R., & Lawson, T. (2020). From green to gold: Agricultural revolution for food security. *Journal of Experimental Botany*, *71*(7), 2211–2215. https://doi.org/10.1093/jxb/eraa110

Eyhorn, F., Muller, A., Reganold, J. P., Frison, E., Herren, H. R., Luttikholt, L., Mueller, A., Sanders, J., Scialabba, N. E., Seufert, V., & Smith, P. (2019). Sustainability in global agriculture driven by organic farming. *Nature Sustainability*, *2*, 253–255. https://doi.org/10.1038/s41893-019-0266-6

Fan, L., Pan, S., Liu, G., & Zhou, P. (2017). Does energy efficiency affect financial performance? Evidence from Chinese energy-intensive firms. *Journal of Cleaner Production*, *151*, 53–59. https://doi.org/10.1016/j.jclepro.2017.03.044

Feng, D., Bao, W., Yang, Y., & Fu, M. (2021). How do government policies promote greening? Evidence from China. *Land Use Policy*, *104*, 105389. https://doi.org/10.1016/j.landusepol.2021.105389

Figus, G., Turner, K., McGregor, P. G., & Katris, A. (2016). Making the case for supporting broad energy efficiency programmes: Impacts on household incomes and other economic benefits. *Energy Policy*, *111*, 157–165 https://doi.org/10.1016/j.enpol.2017.09.028

Finco, A., Bentivoglio, D., Belletti, M., Chiaraluce, G., Fiorentini, M., Ledda, L., & Orsini, R. D. (2023). Precision technologies adoption contribute to the economic and agri-environmental sustainability of Mediterranean wheat production? An Italian case study. *Agronomy*, *13*, 1818. https://doi.org/10.3390/agronomy13071818

Findlay, A. J. (2020). Dark side of low carbon. *Nature Climate Change*, *10*, 184. https://doi.org/10.1038/s41558-020-0724-1

Food and Agriculture Organization (FAO). (2017). *The future of food and agriculture –Trends and challenges*. Rome: FAO. https://www.fao.org/3/i6583e/i6583e.pdf

Foran, B. (2011). Low carbon transition options for Australia. *Ecological Modelling*, *223*, 72–80. https://doi.org/10.1016/j.ecolmodel.2011.05.008

Fragkos, P., Laura van Soest, H., Schaeffer, R., Reedman, L., Köberle, A. C., Macaluso, N., Evangelopoulou, S., De Vita, A., Sha, F., Qimin, C., Kejun, J., Mathur, R., Shekhar, S., Dewi, R. G., Diego, S. H., Oshiro, K., Fujimori, S., Park, C., Safonov, G., & Iyer, G. C. (2021). Energy system transitions and low-carbon pathways in Australia, Brazil, Canada, China, EU-28, India, Indonesia, Japan, Republic of Korea, Russia and the United States. *Energy*, *216*, 119385. https://doi.org/10.1016/j.energy.2020.119385

Freire-González, J., Vivanco, D. F., & Puig-Ventosa, I. (2017). Economic structure and energy savings from energy efficiency in households. *Ecological Economics*, *131*, 12–20. https://doi.org/10.1016/j.ecolecon.2016.08.023

Ge, X., Hou, H., Hou, T., Fang, R., Chen, F., & Tang, J. (2022). *Review of key technologies of low-carbon transition on the power supply side*. 2022 IEEE 5th International Conference on Electronics Technology (ICET), pp. 275–279. https://doi.org/10.1109/icet55676.2022.9825357

Gennitsaris, S., Oliveira, M. C., Vris, G., Bofilios, A., Ntinou, T., Frutuoso, A. R., Queiroga, C., Giannatsis, J., Sofianopoulou, S., & Dedoussis, V. (2023). Energy efficiency management in small and medium-sized enterprises: Current situation, case studies and best practices. *Sustainability*, *15*, 3727. https://doi.org/10.3390/su15043727

Geoghegan, C., O'Donoghue, C., & Loughrey, J. (2023). The local economic impact of climate change mitigation in agriculture. *Bio-Based and Applied Economics*, *11*(4), 323–337. https://doi.org/10.36253/bae-13289

Gernego, I., Urvantseva, S., & Sandulskyi, R. (2022). Green investment opportunities for sustainable agriculture. *Management Theory and Studies for Rural Business and Infrastructure Development*, *44*, 185–194. https://doi.org/10.15544/mts.2022.19

Ghosh, S., Das, T. K., Sharma, D. K., & Gupta, K. (2019). Potential of conservation agriculture for ecosystem services: A review. *The Indian Journal of Agricultural Sciences*, *89*(10), 1572–1579. https://doi.org/10.56093/ijas.v89i10.94578

Gibon, T., Hertwich, E. G., Arvesen, A., Singh, B., & Verones, F. (2017). Health benefits, ecological threats of low-carbon electricity. *Environmental Research Letters*, *12*, 034023. https://doi.org/10.1088/1748-9326/aa6047

Gołasa, P., Wysokiński, M., Bieńkowska-Gołasa, W., Gradziuk, P., Golonko, M., Gradziuk, B., Siedlecka, A., & Gromada, A. (2021). Sources of greenhouse gas emissions in agriculture, with particular emphasis on emissions from energy used. *Energies*, *14*(13), 3784. https://doi.org/10.3390/en14133784

Gorji, S. A. (2023). Challenges and opportunities in green hydrogen supply chain through metaheuristic optimization, *Journal of Computational Design and Engineering*, *10*(3), 1143–1157. https://doi.org/10.1093/jcde/qwad043

Görlach, B., Michael, J., & de la Vega, R. (2023). *Advancing a green hydrogen agenda in the G20. Challenges and opportunities of the G20 as a green hydrogen forum*. Berlin: Ecologic Institute.

Grove, B., & Peridas, G. (2023). Sharing the benefits: How the economics of carbon capture and storage projects in California can serve communities, the economy, and the climate, May, 2023. LLNL-TR-848983. https://gs.llnl.gov/sites/gs/files/2023-05/ca-ccs-economic-study-report.pdf

Gul, M. S., & Chaudhry, H. N. (2022). Energy efficiency, low carbon resources and renewable technology. *Energies, 15*(13), 4553. https://doi.org/10.3390/en15134553

Guo, H., Gu, F., Peng, Y., Deng, X., & Guo, L. (2022). Does digital inclusive finance effectively promote agricultural green development? – A case study of China. *International Journal of Environmental Research and Public Health, 19*, 6982. https://doi.org/10.3390/ijerph19126982

Guyomard, H., Détang-Dessendre, C., Dupraz, P., Delaby, L., Huyghe, C., Peyraud, J. L., Reboud, X., & Sirami, C. (2023). How the green architecture of the 2023–2027 common agricultural policy could have been greener. *Ambio, 52*, 1327–1338. https://doi.org/10.1007/s13280-023-01861-0

Halberg, N., & Müller, A.W. (2013). *Organic agriculture for sustainable livelihoods.* London: Routledge. https://doi.org/10.4324/9780203128480

Hallegatte, S., Heal, G., Fay, M., & Treguer, D. O. (2011). From growth to green growth – A framework. *Economics of Innovation eJournal.* https://doi.org/10.1596/1813-9450-5872

Hao, L., Umar, M., Khan, Z., & Ali, W. (2020). Green growth and low carbon emission in G7 countries: How critical the network of environmental taxes, renewable energy and human capital is? *The Science of the Total Environment, 752*, 141853. https://doi.org/10.1016/j.scitotenv.2020.141853

Hartmann, M. (2011, August). Corporate social responsibility in the food sector. *European Review of Agricultural Economics, 38*(3), 297–324. https://doi.org/10.1093/erae/jbr031

Haszeldine, R. S. (2009). Carbon capture and storage: How green can black be? *Science, 325*, 1647–1652. https://doi.org/10.1126/science.1172246

Heal, G. (2020). Economic aspects of the energy transition. *Environmental and Resource Economics, 83*, 5–21. https://doi.org/10.1007/s10640-022-00647-4

Helga, W., Schlatter, B., & Trávníček, J. (Eds.). (2023). *The world of organic agriculture. Statistics and emerging trends 2023.* Bonn: Research Institute of Organic Agriculture FiBL, Frick, and IFOAM - Organics International. Online Version 2 of February 23, 2023. https://www.fibl.org/fileadmin/documents/shop/1254-organic-world-2023.pdf

Hermelin, B., & Andersson, I. (2018). How green growth is adopted by local policy – A comparative study of ten second-rank cities in Sweden. *Scottish Geographical Journal, 134*, 184–202. https://doi.org/10.1080/14702541.2018.1541474

Heyder, M., & Theuvsen, L. (2009). Corporate social responsibility in agribusiness: Empirical Findings from Germany. *Research Papers in Economics.* https://doi.org/10.22004/ag.econ.58152

Heyl, K., Ekardt, F., Sund, L., & Roos, P. (2022). Potentials and limitations of subsidies in sustainability governance: The example of agriculture. *Sustainability. 14*(23), 15859. https://doi.org/10.3390/su142315859

Hirschfeld, S., & van Acker, R. C. (2021). Review: Ecosystem services in permaculture systems. *Agroecology and Sustainable Food Systems, 45*, 794–816. https://doi.org/10.1080/21683565.2021.1881862

Horizon Europe. (2023). Work Programme 2023–2024 food, bioeconomy, natural resources, agriculture and environment, 9. Food, Bioeconomy, Natural Resources, Agriculture and Environment. https://sciencebusiness.net/sites/default/files/inline-files/First%20draft%20horizon-cl6-wp-2023-2024%20for%20PC%20final.pdf

Hristov, G. V., Kinaneva, D., Georgiev, G., Zahariev, P. Z., & Kyuchukov, P. (2020). An overview of the use of unmanned aerial vehicles for precision agriculture. *2020 International Conference on Biomedical Innovations and Applications (BIA)*, pp. 137–140. https://doi.org/10.1109/bia50171.2020.9244519

Hrubovcak, J., Vasavada, U., & Aldy, J. E. (1999). Green technologies for a more sustainable agriculture. *Agricultural Information Bulletins*. https://doi.org/10.22004/ag.econ.33721

Hsu, K., & Liao, S. Z. (2018). *The application of economic value added on green facilities of urban agriculture*. 3rd International Conference on Sustainable and Renewable Energy Engineering (ICSREE 2018). https://doi.org/10.1051/e3sconf/20185705001

Huang, Z., Liao, G., & Li, Z. (2019). Loaning scale and government subsidy for promoting green innovation. *Technological Forecasting and Social Change*, *144*, 148–156 https://doi.org/10.1016/j.techfore.2019.04.023

Hurst, P. (2023). *Agricultural workers and their contribution to sustainable agriculture and rural development*. Food and Agriculture Organization International Labour Organization International Union of Food, Agricultural, Hotel, Restaurant, Catering, Tobacco and Allied Workers' Associations. https://www.ilo.org/wcmsp5/groups/public/@ed_dialogue/@actrav/documents/publication/wcms_113732.pdf

Hysa, E., Kruja, A., Rehman, N. U., & Laurenti, R. (2020). Circular Economy Innovation and Environmental Sustainability Impact on Economic Growth: An Integrated Model for Sustainable Development. *Sustainability*. *12*(12), 4831 https://doi.org/10.3390/su12124831

International Atomic Energy Agency (IAEA). (2021). Transitions to low carbon electricity systems: Key economic and investments trends changing course in a post-pandemic world. https://www.iaea.org/sites/default/files/21/06/transitions-to-low-carbon-electricity-systems-changing-course-in-a-post-pandemic-world.pdf

International Energy Agency (IEA). (2023). *Energy technology perspectives*, 464 p. https://iea.blob.core.windows.net/assets/a86b480e-2b03-4e25-bae1-da1395e0b620/energy-technologyperspectives2023.pdf

Jacobsen, G. D. (2015). Do energy prices influence investment in energy efficiency? Evidence from energy star appliances. *Journal of Environmental Economics and Management*, *74*, 94–106. https://doi.org/10.1016/j.jeem.2015.09.004

Ji, H., & Hoti, A. (2021). Green economy based perspective of low-carbon agriculture growth for total factor energy efficiency improvement. *International Journal of System Assurance Engineering and Management*, *13*, 353–363. https://doi.org/10.1007/s13198-021-01421-3

Jianqiang, B., Qifang, S., Feng, C., & Liang, F. (2011). *The development strategies and path selections of low carbon energy technologies*. 2011 International Conference on Materials for Renewable Energy & Environment, vol. 1, pp. 34–38. https://doi.org/10.1109/icmree.2011.5930759

Kalchschmidt, M., & Syahruddin, N. (2011). Towards sustainable supply chain management in agricultural sector. https://aisberg.unibg.it/retrieve/handle/10446/25452/4331/020-0072%20towards%20sustainable%20supply%20chain%20management.pdf

Kamble, S. S., Gunasekaran, A., & Gawankar, S. (2019). Achieving sustainable performance in a data-driven agriculture supply chain: A review for research and applications. *International Journal of Production Economics*, *219*, 179–194 https://doi.org/10.1016/j.ijpe.2019.05.022

Khatri-Chhetri, A., Sapkota, T. B., Sander, B. O., Arango, J., Nelson, K. S., & Wilkes, A. (2021). Financing climate change mitigation in agriculture: Assessment of investment cases. *Environmental Research Letters*, *16*, 124044. https://doi.org/10.1088/1748-9326/ac3605

King, L. C., & van den Bergh, J. C. J. M. (2018). Implications of net energy-return-on-investment for a low-carbon energy transition. *Nat Energy*, *3*, 347. https://doi.org/10.1038/s41560-018-0144-x

Kloppers, E., & Fourie, L. M. (2014). Defining corporate social responsibility in the South African agricultural sector. *African Journal of Agricultural Research*, *9*, 3418–3426. https://doi.org/10.5897/ajar12.806

Koohafkan, P., Altieri, M. A., & Gimenez, E. H. (2012). Green agriculture: Foundations for biodiverse, resilient and productive agricultural systems. *International Journal of Agricultural Sustainability, 10*, 61–75. https://doi.org/10.1080/14735903.2011.610206

Koutsos, T. M., & Menexes, G. C. (2019). Economic, agronomic, and environmental benefits from the adoption of precision agriculture technologies: A systematic review. *International Journal of Agricultural and Environmental Information Systems, 10*, 40–56. https://doi.org/10.4018/ijaeis.2019010103

Kowalska, A., & Bieniek, M. (2022). Meeting the European green deal objective of expanding organic farming. *Equilibrium, 17*(3), 607–633. https://doi.org/10.24136/eq.2022.021

Kumar, A. R., Yadav, L., & Sudha, P. N. (2020). Precision agriculture: A review on its techniques and technologies. https://www.irjmets.com/uploadedfiles/paper/volume2/issue_9_september_2020/3926/1628083155.pdf

Kumar, S., Meena, R. S., & Jhariya, M. K. (2020). Resources use efficiency in agriculture. https://doi.org/10.1007/978-981-15-6953-1

Kyriakopoulos, G.L. (2021). Should low carbon energy technologies be envisaged in the context of sustainable energy systems? In: *Low carbon energy technologies in sustainable energy systems* (pp. 357–389). https://doi.org/10.1016/B978-0-12-822897-5.00015-8

Lempert, R. J., & Trujillo, H. R. (2018). Deep decarbonization as a risk management challenge. https://doi.org/10.7249/PE303

Li, Y., Zhu, Z., & Xu, P. (2022). Research on the green production motivation of new agricultural business entities: Benefit perception and environmental regulation. *Journal of Environmental and Public Health*. https://doi.org/10.1155/2022/9182725

Li, Z. W., Jin, M., & Cheng, J. (2021). Economic growth of green agriculture and its influencing factors in china: Based on emergy theory and spatial econometric model. *Environment, Development and Sustainability, 23*, 15494–15512. https://doi.org/10.1007/s10668-021-01307-1

Liberalesso, T., Cruz, C. O., Silva, C. M., & Manso, M. (2020). Green infrastructure and public policies: An international review of green roofs and green walls incentives. *Land Use Policy, 96*, 104693. https://doi.org/10.1016/j.landusepol.2020.104693

Liu, Y., Sun, D., Wang, H., Wang, X., Yu, G., & Zhao, X. (2020). An evaluation of China's agricultural green production: 1978–2017. *Journal of Cleaner Production, 243*, 118483. https://doi.org/10.1016/j.jclepro.2019.118483

Liu, Z., Li, R., Zhang, X., Shen, Y., Yang, L., & Zhang, X. (2021). Inclusive green growth and regional disparities: Evidence from China. *Sustainability. 13*(21), 11651. https://doi.org/10.3390/su132111651

Loviscek, V. (2021). Triple bottom line toward a holistic framework for sustainability: A systematic review. *Revista de Administração Contemporânea*. https://doi.org/10.1590/1982-7849rac2021200017.en

Lu, Y., Norse, D., & Powlson, D. S. (2020). Agriculture green development in China and the UK: Common objectives and converging policy pathways. *Frontiers of Agricultural Science and Engineering*. https://doi.org/10.15302/j-fase-2019298

Luhmann, H., & Theuvsen, L. (2016). Corporate social responsibility in agribusiness: Literature review and future research directions. *Journal of Agricultural and Environmental Ethics, 29*, 673–696. https://doi.org/10.1007/S10806-016-9620-0

Luhmann, H., & Theuvsen, L. (2017). Corporate Social Responsibility: Exploring a Framework for the Agribusiness Sector. *Journal of Agricultural and Environmental Ethics, 30*, 241–253. https://doi.org/10.1007/S10806-017-9665-8

Luo, L., Wang, Y., & Qin, L. (2014). Incentives for promoting agricultural clean production technologies in China. *Journal of Cleaner Production, 74*, 54–61. https://doi.org/10.1016/j.jclepro.2014.03.045

Magaudda, S., D'Ascanio, R., Muccitelli, S., & Palazzo, A. L. (2020). 'Greening' green infrastructure. Good Italian practices for enhancing green infrastructure through the common agricultural policy. *Sustainability, 12*(6), 2301. https://doi.org/10.3390/su12062301

Malecha, Z. M. (2022). Risks for a successful transition to a net-zero emissions energy system. *Energies. 15*(11), 4071. https://doi.org/10.3390/en15114071

Maseko, S. (2023). 2023 Sustainable agriculture market intelligence report. Green Cape. https://greencape.co.za/wp-content/uploads/2023/04/agri_mir_2023_digital_singles.pdf

Mauleón, I. (2020). Economic issues in deep low-carbon energy systems. *Energies, 13*(16), 4151. https://doi.org/10.3390/en13164151

Mazur-Wierzbicka, E. (2015). The application of corporate social responsibility in European agriculture. *Miscellanea Geographica, 19*, 19–23. https://doi.org/10.1515/mgrsd-2015-0001

Medina, G. D., Potter, C. A., & Pokorny, B. (2015). Farm business pathways under agri-environmental policies: Lessons for policy design. *Estudos Sociedade E Agricultura, 23*. Corpus ID: 154175407.

Meemken, E., & Qaim, M. (2018). Organic agriculture, food security, and the environment. *Annual Review of Resource Economics, 10*, 39–63. https://doi.org/10.1146/annurev-resource-100517-023252

Migliorelli, M. (2019). The development of green finance in EU agriculture: Main obstacles and possible ways forward. In: M. Migliorelli & P. Dessertine (Eds.), *The rise of green finance in Europe. Palgrave studies in impact finance*. Cham: Palgrave Macmillan. https://doi.org/10.1007/978-3-030-22510-0_8

Mijatović, Marijana Dukić, Uzelac, Ozren, Stoiljković, Aleksandra (2023). Agricultural sustainability and social responsibility. *Economics of Agriculture, Year 68*, No. 4, 2021, (pp. 1109–1119), Belgrade https://doi.org/ 10.5937/ekoPolj2104109D

Mili, S. & Loukil, T. (2023). Enhancing Sustainability with the Triple-Layered Business Model Canvas: Insights from the Fruit and Vegetable Industry in Spain. *Sustainability* 2023, *15*, 6501. https:// doi.org/10.3390/su15086501

Moranta, J., Torres, C., Murray, I., Hidalgo, M., Hinz, H., & Gouraguine, A. (2021). Transcending capitalism growth strategies for biodiversity conservation. *Conservation Biology, 36*, e13821. https://doi.org/10.1111/cobi.13821

Morgan, J. A., Follett, R., Allen, L. H., Del Grosso, S. J., Derner, J. D., Dijkstra, F. A., Franzluebbers, A. J., Fry, R. C., Paustian, K., & Schoeneberger, M. M. (2010). Carbon sequestration in agricultural lands of the United States. *Journal of Soil and Water Conservation, 65*, 6A–13A. https://doi.org/10.2489/jswc.65.1.6A

Mouysset, L. (2014). Agricultural public policy: Green or sustainable? *Ecological Economics, 102*, 15–23. https://doi.org/10.1016/j.ecolecon.2014.03.004

Mueller, H., & Theuvsen, L. (2014). Influences on consumer attitudes towards CSR in agribusiness. https://doi.org/10.22004/ag.econ.166108

Muhie, S. H. (2023). Novel approaches and practices to sustainable agriculture. *Journal of Agriculture and Food Research, 10*, 100446. https://doi.org/10.1016/j.jafr.2022.100446

Murthy, I., Dutta, S., Varghese, V., Joshi, P. P., & Kumar, P. (2017). Impact of agroforestry systems on ecological and socio-economic systems: A review. *Global Journal of Science Frontier Research.* https://globaljournals.org/GJSFR_Volume16/3-Impact-of-Agroforestry-Systems.pdf

Naik, G., & Suresh, D. (2018). Challenges of creating sustainable agri-retail supply chains. *IIMB Management Review, 30*(3), 270–282. https://doi.org/10.1016/j.iimb.2018.04.001

Nair, P. K., Nair, V. D., Kumar, B., & Showalter, J. M. (2010). Chapter five. Carbon sequestration in agroforestry systems. *Advances in Agronomy, 108*, 237–307. https://doi.org/10.1016/S0065-2113(10)08005-3

Nazir, M. S., Ali, Z. M., Bilal, M., Sohail, H. M., & Iqbal, H. (2020). Environmental impacts and risk factors of renewable energy paradigm – A review. *Environmental Science and Pollution Research*, 1–11. https://doi.org/10.1007/s11356-020-09751-8

Nistor, C. (2015). Green agriculture – Features and agricultural policy measures for the transition to a sustainable agriculture. *Manager Journal, 22*, 112–127. Corpus ID: 155764202.

Nogueira, E., Gomes, S., & Lopes, J. M. (2020). The key to sustainable economic development: A triple bottom line approach. *Resources, 11*, 46. https://doi.org/10.3390/resources11050046

Nogueira, E., Gomes, S., & Lopes, J. M. (2023). Triple -bottom line, sustainability, and economic development: What binds them together? A bibliometric approach. *Sustainability, 15*, 6706.. https://doi.org/10.3390/su15086706

Norton, G. W., Alwang, J., Kassie, M., & Muniappan, R. (2019). Economic impacts of integrated pest management practices in developing countries. The economics of integrated pest management of insects. https://doi.org/10.1079/9781786393678.0140

Nurdiawati, A., & Urban, F. (2021). Towards deep decarbonisation of energy-intensive industries: A review of current status, technologies and policies. *Energies, 14*(9), 2408 https://doi.org/10.3390/en14092408

Nwachukwu, C. S. (2023). *Green agriculture and food security, a review*. IOP Conference Series: Earth and Environmental Science, p. 1178. https://doi.org/10.1088/1755-1315/1178/1/012005

O'Neill, D.W. (2020). Beyond green growth. *Nature Sustainability, 3*, 260–261. https://doi.org/10.1038/s41893-020-0499-4

Pan, C., Jiang, Y., Wang, M., Xu, S., Xu, M., & Dong, Y. (2021). How can agricultural corporate build sustainable competitive advantage through green intellectual capital? A new environmental management approach to green agriculture. *International Journal of Environmental Research and Public Health, 18*(15), 7900. https://doi.org/10.3390/ijerph18157900

Pan, S., Di, C., Chandio, A. A., Sargani, G. R., & Zhang, H. (2022). Investigating the impact of grain subsidy policy on farmers' green production behavior: Recent evidence from China. *Agriculture. 12*(8), 1191 https://doi.org/10.3390/agriculture12081191

Papadis, E., & Tsatsaronis, G. (2020). Challenges in the decarbonization of the energy sector. *Energy, 205*, 118025. https://doi.org/10.1016/j.energy.2020.118025

Patel, S.K., Sharma, A.K., & Singh, G. (2020). Traditional agricultural practices in India: an approach for environmental sustainability and food security. *Energy, Ecology and Environment, 5*, 253–271. https://doi.org/10.1007/s40974-020-00158-2

Paul, M. (2018). Community-supported agriculture in the United States: Social, ecological, and economic benefits to farming. *Journal of Agrarian Change. 19*(1), 162–180 https://doi.org/10.1111/joac.12280

Pe'er, G., Bonn, A., Bruelheide, H., Dieker, P., Eisenhauer, N., Feindt, P. H., Hagedorn, G., Hansjürgens, B., Herzon, I., Lomba, Â., Marquard, E., Moreira, F., Nitsch, H., Oppermann, R., Perino, A., Röder, N., Schleyer, C., Schindler, S., Wolf, C., Zinngrebe, Y., & Lakner, S. (2020). Action needed for the EU Common Agricultural Policy to address sustainability challenges. *People and Nature, 2*, 305–316. https://doi.org/10.1002/pan3.10080

Poetz, K., Haas, R., & Balzarova, M. A. (2013). CSR schemes in agribusiness: Opening the black box. *British Food Journal, 115*, 47–74. https://doi.org/10.1108/00070701311289876

Pogonyshev, V. A., Torikov, V. E., Mokshin, I., & Pogonysheva, D. A. (2021). *Resource economy in agriculture*. IOP Conference Series: Earth and Environmental Science, p. 723. https://doi.org/10.1088/1755-1315/723/3/032035

Prasad, R. (2006). Towards sustainable agriculture in India. *National Academy Science Letters – India, 29*, 41–44. Corpus ID: 88822634.

Price, R. J., & Hacker, R. B. (2009). Grain & Graze: An innovative triple bottom line approach to collaborative and multidisciplinary mixed-farming systems research, development and extension. *Animal Production Science, 49*, 729–735. https://doi.org/10.1071/EA08306

Purohit, C. K., Singh, L., & Kothari, M. (2017). Performance evaluation of drip irrigation systems. *International Journal of Current Microbiology and Applied Sciences, 6*, 2287–2292. https://doi.org/10.20546/ijcmas.2017.604.266

Puttaswamaiah, S., Manns, I. D., & Shah, A. (2006). Promoting sustainable agriculture: Experiences from India and Canada. *Journal of Social and Economic Development, 8*, 147–176. Corpus ID: 153672653. https://gidr.ac.in/pdf/WP-162.pdf

Qadrdan, M., Abeysekera, M., Wu, J., Jenkins, N., & Winter, B. (2019). Overview of the transition to a low carbon energy system. The future of gas networks. https://doi.org/10.1007/978-3-319-66784-3_1

Qiu, H., van Wesenbeeck, C. F., & van Veen, W. (2020). Greening Chinese agriculture: Can China use the EU experience? *China Agricultural Economic Review, 13*, 63–90. https://doi.org/10.1108/caer-10-2019-0186

Qureshi, K. (2015). *Role of advanced nuclear reactor technologies in meeting the growing energy demands*. 2015 Power Generation System and Renewable Energy Technologies (PGSRET), pp. 1–5. https://doi.org/10.1109/pgsret.2015.7312206

Pan, Chulin, Yufeng Jiang, Mingliang Wang, Shuang Xu, Ming Xu, and Yixin Dong. (2021). How can agricultural corporate build sustainable competitive advantage through green intellectual capital? A new environmental management approach to green agriculture. *International Journal of Environmental Research and Public Health, 18*(15), 7900. https://doi.org/10.3390/ijerph18157900

Reganold, J., & Wachter, J. (2016). Organic agriculture in the twenty-first century. *Nature Plants, 2*, 15221. https://doi.org/10.1038/nplants.2015.221

Research Institute of Organic Agriculture. (2023). The world of organic agriculture 2023. https://www.organic-world.net/yearbook/yearbook-2023/contents/download.html

Reis, A. C. (2016). Island studies and tourism: Diversifying perspectives. *Tourism and Hospitality Research, 16*(1), 3–5. https://doi.org/10.1177/1467358415615244

Resque, L. A., Coudel, E., Piketty, M. G., Cialdella, N., Sá, T., Piraux, M., Assis, W. S., & Le Page, C. (2019). Agrobiodiversity and public food procurement programs in Brazil: Influence of local stakeholders in configuring green mediated markets. *Sustainability, 11*(5), 1425. https://doi.org/10.3390/SU11051425

Rezaee, B., Naderi, N., & Rostami, S. (2020). Developing a green entrepreneurship model in agricultural sector (Case Study: Kermanshah County). https://doi.org/10.30473/EE.2020.7248

Rosegrant, M. W., Ringler, C., & Zhu, T. (2009). Water for agriculture: Maintaining food security under growing scarcity. *Annual Review of Environment and Resources, 34*, 205–222. https://doi.org/10.1146/annurev.environ.030308.090351

Rotz, A., (2020). Environmental sustainability of livestock production. *Meat and Muscle Biology, 4*(2). https://doi.org/ https://doi.org/10.22175/mmb.11103

Rubin, E. S., Davison, J., & Herzog, H. (2015). The cost of CO2 capture and storage. *International Journal of Greenhouse Gas Control, 40*, 378–400. https://doi.org/10.1016/j.ijggc.2015.05.018

Saenchaiyathon, K., & Wongthongchai, J. (2021). Green operation strategy for SMEs: Agri-food business Thailand case study. *TEM Journal, 10*(4), 1803–1812. https://doi.org/10.18421/tem104-43

Sahu, B., Chatterjee, S., Mukherjee, S., & Sharma, C. (2019). Tools of precision agriculture: A review. *International Journal of Chemical Studies, 7*, 2692–2697. Corpus ID: 229957195.

Samoggia, A., Perazzolo, C., Kocsis, P., & Del Prete, M. (2019). Community supported agriculture farmers' perceptions of management benefits and drawbacks. *Sustainability, 11*(12), 3262 https://doi.org/10.3390/SU11123262

Scheffran, J. & Froese, R. (2016). Enabling environments for sustainable energy transitions: The diffusion of technology, innovation and investment in low-carbon societies. In: H. Brauch, Ú. Oswald Spring, J. Grin, & J. Scheffran (Eds.), *Handbook on sustainability transition and sustainable peace. Hexagon series on human and environmental security and peace*, vol 10. Cham: Springer. https://doi.org/10.1007/978-3-319-43884-9_34

Schmidt, C., & Correa, P.G. (2014). Public research organizations and agricultural development in Brazil: How did Embrapa get it right? World Bank Publications, 1–10. https://documents1.worldbank.org/curated/en/156191468236982040/pdf/884900BRI0EP1450Box385225B000PUBLIC0.pdf

Schmidt, T. S. (2014). Low-carbon investment risks and de-risking. *Nature Climate Change, 4*, 237–239. https://doi.org/10.1038/NCLIMATE2112

Scialabba, N. E. (2013). Organic agriculture's contribution to sustainability. *Crop Management, 12*, 1–3. https://doi.org/10.1094/CM-2013-0429-09-PS

Sdrolia, E., & Zarotiadis, G. (2018). A comprehensive review for green product term: From definition to evaluation. *PSN: Sustainable Development (Topic), 33*(1), 150–178. https://doi.org/10.1111/joes.12268

Serafeim, G. (2022). Risks and opportunities from the transition to a low carbon economy: A business analysis framework. Harvard Business School Technical Note 123-014.

Shazmin, S., Sipan, I. A., & Sapri, M. (2016). Property tax assessment incentives for green building: A review. *Renewable & Sustainable Energy Reviews, 60*, 536–548. https://doi.org/10.1016/j.rser.2016.01.081

Shen, J., Zhu, Q., Jiao, X., Ying, H., Wang, H., Wen, X., Xu, W., Li, T., Cong, W., Liu, X., Hou, Y., Cui, Z., Oenema, O., Davies, W. J., & Zhang, F. (2020). Agriculture green development: A model for China and the world. *Frontiers of Agricultural Science and Engineering, 7*, 5–13. https://doi.org/10.15302/j-fase-2019300

Shen, S., Li, J., & Xu, R. (2021). Agricultural ecological environment protection based on the concept of sustainable development. *Acta Agriculturae Scandinavica, Section B – Soil & Plant Science, 71*, 920–930. https://doi.org/10.1080/09064710.2021.1961852

Sher, A., Mazhar, S., Zulfiqar, F., Wang, D., & Li, X. (2019). Green entrepreneurial farming: A dream or reality? *Journal of Cleaner Production, 220*, 1131–1142 https://doi.org/10.1016/j.jclepro.2019.02.198

Sills, E. O., & Caviglia-Harris, J. L. (2015). Evaluating the long-term impacts of promoting "green" agriculture in the Amazon. *Agricultural Economics, 46*, 83–102. https://doi.org/10.1111/agec.12200

Singh, R., Singh, H., & Raghubanshi, A.S. (2019). Challenges and opportunities for agricultural sustainability in changing climate scenarios: A perspective on Indian agriculture. *Tropical Ecology, 60*, 167–185. https://doi.org/10.1007/s42965-019-00029-w

Singh, S., & Srivastava, S. K. (2022), Decision support framework for integrating triple bottom line (TBL) sustainability in agriculture supply chain, *Sustainability Accounting, Management and Policy Journal, 13*(2), 387–413. https://doi.org/10.1108/SAMPJ-07-2021-0264

Smidt, S. J., Haacker, E. M., Kendall, A. D., Deines, J. M., Pei, L., Cotterman, K. A., Li, H., Liu, X., Basso, B., & Hyndman, D. W. (2016). Complex water management in modern agriculture: Trends in the water-energy-food nexus over the High Plains Aquifer. *The Science of the Total Environment, 566–567*, 988–1001. https://doi.org/10.1016/j.scitotenv.2016.05.127

Soini Coe, E, & Coe, R. (2023). Agroecological transitions in the mind. *Elem Sci Anth, 11*, 1. https://doi.org/10.1525/elementa.2022.00026

Solazzo, R., Donati, M., Tomasi, L., & Arfini, F. (2016). How effective is greening policy in reducing GHG emissions from agriculture? Evidence from Italy. *The Science of the total environment, 573*, 1115–1124. https://doi.org/10.1016/j.scitotenv.2016.08.066

Soloviova, O., Krasnyak, O., Cherkaska, V., & Revkova, A. (2022). Strategic development of international corporate social responsibility in agribusiness. *Economics Ecology Socium, 6*, 51–64. https://doi.org/10.31520/2616-7107/2022.6.4-5

Sovacool, B. K., Martiskainen, M., Hook, A., & Baker, L. (2019). Beyond cost and carbon: The multidimensional co-benefits of low carbon transitions in Europe. *European Economics: Agriculture*. https://doi.org/10.1016/j.ecolecon.2019.106529

Stagnari, F., Ramazzotti, S., & Pisante, M. (2009). Conservation agriculture: A different approach for crop production through sustainable soil and water management: A review. In: Lichtfouse, E. (Eds.), *Organic farming, pest control and remediation of soil pollutants* (Sustainable Agriculture Reviews, vol. 1). Dordrecht: Springer. https://doi.org/10.1007/978-1-4020-9654-9_5

Stein, S., Hartung, J., Perkons, U., Möller, K., & Zikeli, S. (2023). Plant and soil N of different winter cover crops as green manure for subsequent organic white cabbage. *Nutrient Cycling in Agroecosystems, 127*(2), 285–298. https://doi.org/10.1007/s10705-023-10306-9

Stoicea, P., Basa, A. G., Stoian, E., Toma, E., Micu, M. M., Gidea, M., Dobre, C. A., Iorga, A. M., & Chiurciu, I. A. (2023). Crop rotation practiced by romanian crop farms before the introduction of the "environmentally beneficial practices applicable to arable land" eco-scheme. *Agronomy, 13*, 2086. https://doi.org/10.3390/ agronomy13082086

Stoyanova, Z., & Harizanova, H. (2019). Impact of agriculture on water pollution. *Agrofor – International Journal, 4*(1). https://doi.org/10.7251/agreng1901111s

Svensson, G., Ferro, C., Høgevold, N. M., Padín, C., Varela, J. C., & Sarstedt, M. (2018). Framing the triple bottom line approach: Direct and mediation effects between economic, social and environmental elements. *Journal of Cleaner Production, 197*(1), 972–991. https://doi.org/10.1016/j.jclepro.2018.06.226

Syahruddin, N., & Kalchschmidt, M. (2012). Sustainable supply chain management in the agricultural sector: A literature review. *International Journal of Engineering Management and Economics, 3*, 237–258. https://doi.org/10.1504/ijeme.2012.049894

Szabo, L., Madai, H., & Nábrádi, A. (2022). Potential impact of the European Green Agreement on EU and Hungarian crop production. *Applied Studies in Agribusiness and Commerce. 16*(2). https://doi.org/10.19041/apstract/2022/2/9

Taghizadeh-Hesary, F., & Yoshino, N. (2020). Sustainable solutions for green financing and investment in renewable energy projects. *Energies, 13*(4), 788. https://doi.org/10.3390/en13040788

Tavares, P. D., Uzêda, M. C., & Pires, A. D. (2019). Biodiversity conservation in agricultural landscapes: The importance of the matrix. *Floresta e Ambiente, 26*(4), e20170664. https://doi.org/10.1590/2179-8087.066417

Theuer, S. L. H., & Olarte, A.. (2023). *Emissions trading systems and carbon capture and storage: Mapping possible interactions, technical considerations, and existing provisions.* Berlin: International Carbon Action Partnership. https://icapcarbonaction.com/system/files/document/La%20Hoz%20Theuer%20%26%20Olarte%20%282023%29.%20ETSs%20and%20CCS_ICAP.pdf

Thiesmeier, A. & Zander, P. (2023). Can agroforestry compete? A scoping review of the economic performance of agroforestry practices in Europe and North America. *Forest Policy and Economics, 150*, 102939. https://doi.org/ 10.1016/j.forpol.2023.102939

Timpanaro, G., Scuderi, A., Guarnaccia, P., & Foti, V. T. (2023). Will recent world events shift policy-makers' focus from sustainable agriculture to intensive and competitive agriculture? *Heliyon, 9*(7), e17991. https://doi.org/ 10.1016/j.heliyon.2023.e17991

Tinsley, E. A., & Agapitova, N. (2018). Private sector solutions to helping smallholders succeed: Social enterprise business models in the agriculture sector. Social Enterprise Business Models in the Agriculture Sector. The World Bank group. https://doi.org/10.1596/29543

Toman, M.A. (2012). Green growth: An exploratory review. *Economic Growth eJournal.* https://doi.org/10.1596/1813-9450-6067

Tseng, M. L., Chang, C. H., Lin, C. W. R., Wu, K. J., Chen, Q., Xia, L., & Xue, B. (2020). Future trends and guidance for the triple bottom line and sustainability: A data driven bibliometric analysis. *Environmental Science and Pollution Research, 27*, 33543–33567. https://doi.org/10.1007/s11356-020-09284-0

Uchida, E., Rozelle, S., & Xu, J. (2008). Conservation payments, liquidity constraints, and off-farm labor: Impact of the grain-for-green program on rural households in China. *Development Economics: Women*, *91*(1), 70–86. https://doi.org/10.1111/j.1467-8276.2008.01184.x

Ulrich, R. (2021). Funding agricultural emission mitigation. *Nature Reviews Earth & Environment*, *3*, 2–2. https://doi.org/10.1038/s43017-021-00260-x

Unay-Gailhard, İ., & Bojnec, Š. (2019). The impact of green economy measures on rural employment: Green jobs in farms. *Journal of Cleaner Production*, *208*, 541–551. https://doi.org/10.1016/j.jclepro.2018.10.160

United Nations Environment Programme. (2023). Driving finance for sustainable food systems. A roadmap to implementation for financial institutions and policy makers. Geneva. https://www.unepfi.org/wordpress/wp-content/uploads/2023/04/Driving-Finance-for-Sustainable-Food-Systems.pdf

Velasco Muñoz, J. F., López Felices, B., Balacco, G., and Aznar Sánchez, J. A. (2023). Adoption of rainwater harvesting systems for agricultural irrigation to improve water management, EGU General Assembly 2023, Vienna, Austria, 24-28 Apr 2023, EGU23-133, https://doi.org/10.5194/egusphere-egu23-133

Velasco-Muñoz, J. F., Aznar-Sánchez, J. A., López-Felices, B., & Balacco, G. (2022). Adopting sustainable water management practices in agriculture based on stakeholder preferences. *Agricultural Economics (Zemědělská ekonomika)*, *68*(9), 317–326. https://doi.org/10.17221/203/2022-agricecon

Vidales, B. K. V., Jorge, L., Alcaraz, V., & García, J. O. (2012). Exploratory analysis of corporate social responsibility practices in Mexican agricultural companies. *China-USA Business Review*, *11*(9), 1277–1285. ISSN 1537-1514. https://doi.org/10.17265/1537-1514/2012.09.010

Vrabcová, P., & Urbancová, H. (2023): Sustainable innovation in agriculture: Building a strategic management system to ensure competitiveness and business sustainability. *Agricultural Economics*, *69*, 1–12. https://agricecon.agriculturejournals.cz/pdfs/age/2023/01/01.pdf

Wambugu, P. W., Nyamongo, D. O., & Kirwa, E. C. (2023). Role of seed banks in supporting ecosystem and biodiversity conservation and restoration. *Diversity*, *15*, 896. https://doi.org/10.3390/d15080896

Wang, C., Chen, P., Hao, Y. X., & Dagestani, A. (2022). Tax incentives and green innovation-The mediating role of financing constraints and the moderating role of subsidies. *Frontiers in Environmental Science*, *10*, 2353. https://doi.org/10.3389/fenvs.2022.1067534

Wang, S. (2022). The positive effect of green agriculture development on environmental optimization: Measurement and impact mechanism. *Frontiers in Environmental Science*. *10*. https://doi.org/10.3389/fenvs.2022.1035867

Wang, X., Liu, W., & Wu, W. (2009). A holistic approach to the development of sustainable agriculture: Application of the ecosystem health model. *International Journal of Sustainable Development & World Ecology*, *16*, 339–345. https://doi.org/10.1080/13504500903106675

Ward, C., & Reynolds, L. (2013). Organic agriculture contributes to sustainable food security. In: *Vital Signs* (vol. 20). Washington, DC: Island Press. https://doi.org/10.5822/978-1-61091-457-4_16

Wardell, D. A., Piketty, M. G., Lescuyer, G., & Pacheco, P. (2021). Reviewing initiatives to promote sustainable supply chains: The case of forest-risk commodities. https://doi.org/10.17528/CIFOR/007944

Wasilewski, A., & Wasilewska, A. (2019). Financing the research and development activity for the agri-food sector and rural areas. *Western Balkan Journal of Agricultural Economics and Rural Development*, *1*, 29–39. https://doi.org/10.5937/wbjae1901029w

Wezel, A., Casagrande, M., Celette, F., Vian, J., Ferrer, A., & Peigné, J. (2013). Agroecological practices for sustainable agriculture. A review. *Agronomy for Sustainable Development, 34*, 1–20. https://doi.org/10.1007/s13593-013-0180-7

Whitfield, J. A. (2006). Agriculture and environment: How green was my subsidy? *Nature, 439*, 908–909. https://doi.org/10.1038/439908A

Wieck, C. (2020). Sustainable supply chains in the agricultural sector: Adding value instead of just exporting raw materials. https://www.swp-berlin.org/publications/products/comments/2020C43_SustainableSupplyChains.pdf

World Economic Forum. (2023). Fostering effective energy transition 2023 Edition Insight Report. https://www3.weforum.org/docs/WEF_Fostering_Effective_Energy_Transition_2023.pdf?_gl=1*4gwmah*_up*MQ..&gclid=CjwKCAjw8symBhAqEiwAaTA_FtFf3El9Sl-Qxexs8Ztwa_g3eOXZlcy5quYwNni0yl2oyqzcCSz_RoCIUoQAvD_BwE

World Employment and Social Outlook. (2023). *The value of essential work*. Geneva: International Labour Office. https://www.ilo.org/wcmsp5/groups/public/---dgreports/---dcomm/---publ/documents/publication/wcms_871016.pdf

Wu, X., Xu, Y., Lou, Y., & Chen, Y. (2018). Low carbon transition in a distributed energy system regulated by localized energy markets. *Energy Policy. 122*: 474–485 https://doi.org/10.1016/j.enpol.2018.08.008

Yang, P., Wu, L., Cheng, M., Fan, J., Li, S., Wang, H., & Qian, L. (2023). Review on drip irrigation: Impact on crop yield, quality, and water productivity in China. *Water, 15*, 1733. https://doi.org/10.3390/w15091733

Yıldırım, H. G., & Yılmaz, N. (2023). *Field crops and sustainable agriculture practices*. International Research in Agriculture, Forestry and Aquaculture Sciences. International Research in Agriculture, Forestry and Aquaculture Sciences. ISBN: 978-625-6971-39-4 Available from: https://www.researchgate.net/publication/369356800_Field_Crops_and_Sustainable_Agriculture_Practices (Accessed 05 August 2023].

Yu, T., Mahe, L., Li, Y., Wei, X., Deng, X., & Zhang, D. (2022). Benefits of crop rotation on climate resilience and its prospects in China. *Agronomy, 12*(2), 436. https://doi.org/10.3390/agronomy12020436

Zanin, A., Dal Magro, C. B., Kleinibing Bugalho, D., Morlin, F., Afonso, P., & Sztando, A. (2020). Driving sustainability in dairy farming from a TBL perspective: Insights from a case study in the West Region of Santa Catarina, Brazil. *Sustainability. 12*(15), 6038. https://doi.org/10.3390/su12156038

Zegada-Lizarazu, W., & Monti, A. (2011). Energy crops in rotation. A review. *Biomass & Bioenergy, 35*, 12–25. https://doi.org/10.1016/j.biombioe.2010.08.001

Zegar, J. S., & Wrzaszcz, W. (2017). The holism principle in agriculture sustainable development. *Environmental Engineering Science, 17*, 1051–1069. https://doi.org/10.25167/ees.2017.44.26

Zhang, N. (2022). How does CSR of food company affect customer loyalty in the context of COVID-19: a moderated mediation model. *International Journal of Corporate Social Responsibility, 7*, 1. https://doi.org/10.1186/s40991-021-00068-4

Zhao, N., & You, F. (2020). Can renewable generation, energy storage and energy efficient technologies enable carbon neutral energy transition? *Applied Energy, 279*, 115889. https://doi.org/10.1016/j.apenergy.2020.115889

Zhou, P., Lv, Y., & Wen, W. (2023). The low-carbon transition of energy systems: A bibliometric review from an engineering management perspective. *Engineering*. https://doi.org/10.1016/j.eng.2022.11.010

Zhu, B., & Wei, C. (2022). A green hydrogen era: Hope or hype? *Environmental Science & Technology, 56*(16), 11107–11110. https://doi.org/10.1021/acs.est.2c04149

Zhu, Y., & Chen, J. (2022). Small-Scale farmers' preference heterogeneity for green agriculture policy incentives identified by choice experiment. *Sustainability, 14*(10), 5770. https://doi.org/10.3390/su14105770

6 Assessment of Low-Carbon Transition Policy in Agriculture

Indre Siksnelyte-Butkiene

6.1 LOW-CARBON ENERGY TRANSITION CHALLENGES TO AGRICULTURE

According to the projections of the United Nations Food and Agriculture Organization (2019), 70% increase in global food production will be by 2050 compared to 2005–2007 levels to meet the food demand of a growing world population, which is expected will reach 9.6 billion people. Accordingly, the increased need of food production will result an increase in energy demand in agricultural sector. Energy is critical to almost every process in the food production value chain.

Today agricultural sector requires much higher input of energy than traditional agriculture, which is strongly dependent on fossil fuels to drain grain, produce fertilizers, drive or run various techniques and equipment, generate electricity for heating and lighting, etc. Such energy-intensive activities in farming are major contributors for climate-related emissions. Various clean energy technologies and services provide sustainable alternatives to meet agricultural demand. By employing clean energy technologies, farmers can carry out energy-intensive operations with less impact to the environment.

The agricultural sector is being increasingly recognized as one of the main drivers of climate change and one of the most important parts in mitigating of climate change. Currently, there is growing awareness that the way food is produced and consumed must be transformed ensuring the low possible impact to the environment and whole society. But just as stakeholders of fossil-fuel industry are threatened by the transition from fossil fuels, many farmers' ordinary lives are also threatened by the proposed changes in agriculture. Despite of that, significant changes are necessary for the reduction of the impact to the environment, because today's agricultural sector is highly industrialized. Many different governmental and non-governmental organizations, scientists, and stakeholders in agriculture sector are actively discussing a "just transition" in agriculture and its possible challenges.

A "just transition" should reduce negative impact from agriculture to the environment while ensuring needs of farmers and various other stakeholders related to the sector. The discussions of shifts away from fossil fuels have been prevalent for more than half a century. However, the debates regarding the changes in agriculture and food systems emerged relatively recently.

DOI: 10.1201/9781003460589-7

The need for changes in agriculture sector is paramount and urgent. According to the findings of the study by Xu et al. (2021), where the authors estimated greenhouse gas (GHG) emissions from production and consumption of human food worldwide, 57% of GHG emissions correspond to the production of animal-based food (e.g., as meat and dairy), 29% of GHG emissions come from plant-based food production, and 14% of GHG emissions correspond to the other utilizations. Among the largest contributors from plant-based food is rice (12%), while from animal based the first position takes beef (25%). The findings of the assessment showed that the highest production-based GHG emitters from agricultural sector are South Asia, Southeast Asia, and South America.

Also, it is necessary to highlight that a third of the food produced is wasted after harvest. Paradoxically, but in 2021, around 2.3 billion people in the world were moderately or severely food insecure, 11.7% of the global population were food insecure at severe levels and between 702 and 828 million people of the world faced hunger (FAO et al., 2022). The projections made by Nelson et al. (2018) showed an increase of 65 million people in food insecurity due to consequences of climate change to mid-century.

Agricultural sector is recognized as one of the most affected to climate change. Such natural phenomena as droughts, floods, wildfires, and heat waves have significant negative impact on harvest and on the stability and profitability of farmers' activities. Thus, the consequences of climate change put farmers and other people working in agricultural sector at risk. The undoubted consequences of climate change are also for the entire population through the increase in prices of food products and their affordability issues. Therefore, the transition to low-carbon energy in agriculture will directly affect the agriculture sector through lower consequences of climate change.

As agriculture sector is both a victim and a driver of climate change, there is an urgent need to invest in GHG emissions mitigation options and to play a crucial role in the climate change mitigation. Prevailing practices in agriculture sector have a significant impact on the environment and contribute to climate change to the extent that they are systemic – the negative effects that amount to structural injustices are the result of the system-wide processes, where agriculture products are produced, placed in the market, consumed, and regulated (Lehtonen et al., 2022).

With the new technologies currently under development and its rapidly falling costs low-carbon energy technologies have been increasingly applied in agriculture. However, the transition from traditional fossil fuel to clean energy raises some challenges. Three main categories of challenges facing the agricultural sector today can be distinguished are economic, technical, and information and knowledge (Figure 6.1).

Economic challenges can be divided into three main groups, which are related to:

- **Investments and Costs of Clean Energy Transition**: A fundamental economical challenge for the implementation of a clean energy-based agriculture system is the high costs of installation and quite a long payback period. Huge investments and high costs can have a disproportionate effect, especially on small farmers, because many of them can be unable to raise the finance required to invest in low-carbon energy sources, as well as

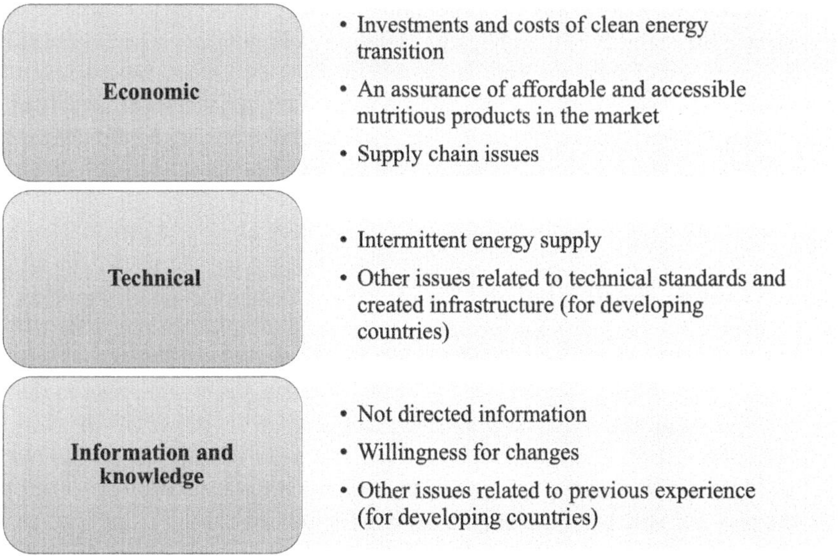

Economic
- Investments and costs of clean energy transition
- An assurance of affordable and accessible nutritious products in the market
- Supply chain issues

Technical
- Intermittent energy supply
- Other issues related to technical standards and created infrastructure (for developing countries)

Information and knowledge
- Not directed information
- Willingness for changes
- Other issues related to previous experience (for developing countries)

FIGURE 6.1 The low-carbon energy transition challenges to agriculture. (Created by authors.)

sustainable farming practices. Typically, small farmers cannot afford new technologies without the financial support from the government or other direct investments.

- **An Assurance of Affordable and Accessible Nutritious Products in the Market**: The other economical challenge is a tension between the investments and cost of transitioning to clean energy and production of affordable and accessible nutritious products. Usually, the increased costs are being passed onto consumers and this is also a challenge for food producers because demand may decrease.
- **Supply Chain Issues**: The global COVID-19 pandemic exacerbated supply chain issues across many sectors; however, logistics continues to be a challenge for agriculture sector. The geopolitical instability because of the war in Ukraine has caused problems for many food producers reliant on production grown in Ukraine and for farmers using various agrochemicals. In many cases, the logistics chain has grown significantly along with the carbon footprint generated. The pandemic has also affected the energy sector, disrupting the supply of renewable energy technologies and thus slowing their penetration in the market. Geopolitical aspects have a significant impact on many energy-related issues such as energy supply, energy prices, the development of energy infrastructure, and the assurance of energy infrastructure maintenance. The example of European countries dependence on energy sources from Russia revealed the importance to cooperate with reliable partners and allocate the risk. The COVID-19 pandemic revealed the fragility of the global supply chains (Emenike & Falcone, 2020; Goldthau & Hughes, 2020). Low-carbon energy transition requires a rapid development

of clean energy infrastructure. Therefore, it is very important to ensure supply chains stability. This obligation falls on policymakers when creating agriculture and energy policy that would guarantee both the smooth development of the new energy infrastructure and its maintenance. Scientific studies showed that supply difficulties in disruptive cases can be managed with the effective centralized decision making (Ekinci et al., 2022).

Technical challenges are related to several aspects, such as:

- **Intermittent Energy Supply**: An intermittent energy supply is another challenge for smooth low-carbon energy transition in agriculture (Gorjian et al., 2021). Here, energy-storage devices can supplement existing grid capacity by storing surplus energy during off-peak hours and given to the network when energy is not produced. Such storage devices allow to leverage intermittent resources, such as wind and solar generators, efficiently and in ways that impact the grid system more consistently. However, such devices are still very expensive and their wide application is still in the future. From the perspective of a country scale, minimizing the costs linked with intermittent energy variability will be one of the major challenges for countries seeking low-carbon energy transition in the coming decades.
- **Other Issues Related to Technical Standards and Created Infrastructure (for Developing Countries)**: The transition from a fossil fuels-based energy system to a clean energy-based agriculture system rises more challenges for some countries. Rahman et al. (2022) analyzed challenges to develop renewable energy sources in agriculture in developing countries. According to the study, in developing countries, there are much more challenges than in developed countries. For example, as technical challenge, the authors singled out:
  - the difficulties to obtain a proper system design with appropriate site planning;
  - the absence of an effective quality control process for energy infrastructure;
  - the absence of energy-related infrastructure technical standards;
  - difficulties with grid integration of off-grid networks;
  - a lack of skilled specialists and a lack of high-level training programs to prepare them.

Information and knowledge challenges are related to:

- **Not Directed Information**: Many countries still do not have active directed and appropriate promotion campaign on low-carbon energy technologies that would increase public awareness of the technology's capabilities and long-term benefits from various perspectives. Often, farmers may be unaware of the variety and benefits of new technologies that may be suitable for them. Also, farmers have limited access to distributors for their installation and service providers. Therefore, it is important, that tools and

information basis directed to agriculture sector will be created to help farmers make financially and environmentally based decisions regarding energy development in their farms.
- **Willingness for Changes**: Although there are active discussions about the use of clean energy in agriculture and there is a growing concern about the environment and the impact of agricultural activities on the environment, the economic aspect of agricultural activities often remains in the first place. Therefore, the spread of clean technologies and various other initiatives reducing GHG emissions and mitigating against climate change in agriculture should be supported by various effective policy measures. For example, the study performed by Pascaris (2021) showed that integration of multi-level and multi-sector policy is important for the development of solar energy in United States agriculture.
- **Other Issues Related to Previous Experience (for Developing Countries)**: Most of the low-carbon energy technologies are imported from other developing countries with no standards regulating this equipment. Therefore, a lot of low-quality equipment is imported, which leads consumers not to trust the suppliers and not to feel the benefits of the technologies.

It can be stated that the increase in the capacity and potential of low-carbon energy technologies in agriculture depends on such elements as initial cost; direct investments; social, environmental, and agriculture policy; fulfilment of technical conditions; and support policy. While the acceptance of various clean technologies and penetration in the market are mostly influenced by the issues of technology awareness, assessment of profit and loss, and local-level conditions (Gorjian et al., 2022).

It is also necessary to highlight that there are clear linkages between the development of low-carbon energy technologies in agriculture and positive impact to the development of rural areas. The usage of clean and more affordable energy in farms can have significant positive impact for the community through the creation of new jobs places because of increased and diversified production in farms. The increase in people income can positively affect many other economical and social aspects in the whole community, such as people health, wellbeing of people, and poverty level.

6.2 POLICIES TO PROMOTE LOW-CARBON ENERGY TRANSITION IN AGRICULTURE

The projections made by Millar et al. (2017) showed that limiting global warming to 1.5°C and achieving the targets of the Paris Agreement is possible, but it requires to implement strengthened commitments. The agriculture sector has a huge potential to contribute for the implementation of the climate change goals. GHG reduction strategies such as changes in dietary, food loss and waste reduction, carbon sequestration, installation of renewable energy systems in agriculture, and others can play an essential role in contribution to the reduction of climate emissions in the agriculture sector. Some of the strategies can act as mitigation instruments or as adaptation strategies, which contribute to cope with risk and reduce climate emissions.

The strategies for GHG reduction in agriculture should focus on both demand and supply. Strategies on the demand side are related to the changes in dietary and reduction of food wasting. The supply side includes strategies focusing on agricultural practices which allow to reduce GHG emissions, such as low-carbon energy transition or manure management.

However, the various climate change mitigation strategies come with direct and external costs. First, significant attention should be made and investments in the design, scaling, and implementation of agricultural strategies to mitigate climate change are required. Various scientific studies show that the major barriers for scaling various climate change mitigation technologies among smallholder farmers are the direct costs and the lack of capacity, knowledge, and skills. Second, the created and implemented strategies could have an external effect on different aspects. There are considerations that some of GHG reduction technologies can negatively affect types and quantities of production in the market and have impact on food security issues, income and livelihood of people in rural areas, price of agricultural products, the related industries, and the earnings from export (on agriculture-dependent economies) (Rahut et al., 2023). Whereas the effect of GHG reduction on agricultural production and food security varies among countries and regions. Therefore, it is important to select such strategies which have as high as possible impact on emissions reduction and less as possible negative effect on agricultural production.

But speaking exclusively only about low-carbon energy development in agriculture, negative effects are harder to find here, except for the challenges faced at the farm and government levels, which are discussed in the previous section. It can be said that now policies to promote low-carbon energy transition in agriculture are mainly based on the common climate change mitigation policies and strategies. The main measures which are used are presented in Figure 6.2.

- *Renewable energy subsidies* are measures offered by the government that encourage renewable energy technologies installation serving agricultural operations to reduce GHG emissions in agriculture and energy sectors as well. The measure is really widely implemented in different regions of the world under different support programs. Renewable energy subsidies are favorably regarded both in the business and household sectors, as well as in the farms. The results of the study by Majewski et al. (2023), where 94 middle-income countries were analyzed, showed that such policy tools as subsidies and low-interest loans are effective measures to support renewable energy development in middle-income countries. Also, the authors found that 1% increase in renewable electricity output decreases CO_2 emissions for about 0.18%. The effectiveness of renewable energy subsidies in the EU was analyzed in the study by Saman (2021). The author examined the effect of agricultural subsidies to the achievements in environmental outcomes. According to the analysis, subsidies increased the use of renewable energy sources in agriculture in all EU countries with the exception of Sweden and the United Kingdom.
- The *Net Metering* and Feed-in Tariffs are models for energy prosumers. Energy prosumers generate, use, and sell the energy they produce. The Net

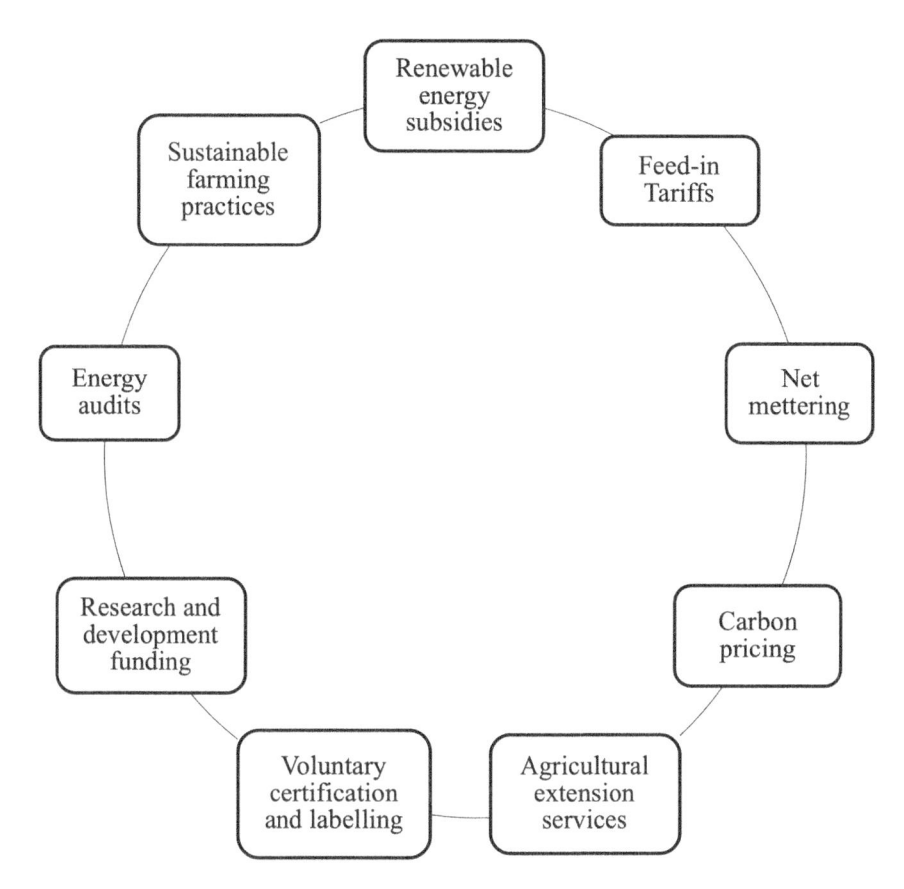

Assessment of Low-Carbon Transition Policy in Agriculture

FIGURE 6.2 The main measures to promote low-carbon energy transition in agriculture. (Created by authors.)

Metering allows energy prosumer to supply the surplus energy to the grid and return the energy supplied to cover a shortfall of energy demand when it needed. The efficiency of renewable energy technologies is very dependent on climatic conditions; therefore, the model is useful for energy prosumer to meet their energy needs throughout entire year. The prosumer is interested to generate as much energy as possible to cover the shortfall in the months they cannot generate as much energy as required. Usually, the limit for the energy supplied under the credit to the grid is one year (Roux & Shanker, 2018). Also, the prosumer does not pay some energy charges paid by all other consumers buying energy from the grid. Necessary to mention that often the prosumers use other support measures proposed by the government to buy the technology itself (Londo et al., 2020). According to Aleksiejuk-Gawron et al. (2020), however, despite the financial benefits and attractiveness net-metering model faces economical and technological challenges. For example, this model does not affect energy prosumers behavior as in motivation to decrease the self-consumption of energy. Also, the

future technological achievements in energy storage systems and affordable battery prices are essential for the successful implementation of the model in the future and low-carbon energy transition itself (Shaw-Williams and Susilawati, 2020).
- The *Feed-in Tariffs* model offers above-market prices for certain kinds of energy generated. The model was very popular and used as the main support measure in the beginning of the deployment of renewable technologies to stimulate energy consumers to start to be involved in renewable energy generation quickly. However, feed-in tariffs are very expensive for the state as energy bought at above-market rates is finally sold at the market price. The model is really an effective instrument to promote renewable energy generation. Despite of that it has received a lot of criticism. For example, Winter and Schlesewsky (2019) found that feed-in tariffs are unfair for every household because they finally pay bigger price for energy, while the energy prosumers acting under this model get return. Hitaj and Loschel (2019) found that feed-in tariffs positively affected investments in wind energy in Germany. These investments have hypothetical positive effect on GHG emissions reduction in the country. However, looking at it from another perspective, emissions can be reduced by using the Emissions Trading Scheme.
- *Carbon pricing* is implemented through a tax on the carbon content of fossil fuels or on their CO_2 emissions. The measure is simple to administer as an extension of existing fuel taxes. This measure provides certainty about future prices of emissions and encourages to mobilize investments in clean technologies. Carbon pricing can be implemented through emissions trading systems, where emitters must acquire allowances for each ton of GHG emitted. The supply of permits is limited by the government. These permits can be bayed and sold by establishing a certain price. Although this measure has been implemented for quite a long time in different countries around the world and is popular for taxing carbon emissions in other economical sectors, the carbon pricing is rarely applied to emissions from the agricultural sector. Only in recent years have some countries begun to bring the agricultural sector alongside the other sectors into carbon taxation (e.g., Canada and New Zealand). Meanwhile, in many countries, the agricultural sector exclusively avoids this measure, but both at the political level and in the scientific literature, the discussions on the application of carbon pricing to the agricultural sector are gaining more and more attention. For example, Alig et al. (2010) analyzed the possible impact of carbon pricing policy to the urban development rates, and land transfers between forestry and agriculture in order to find out the potential efficiency of carbon pricing policy to mitigate against climate change in the United States. The results of the projections showed that carbon pricing has a significant impact on forest and agriculture sectors and can significantly affect future land use patterns, the level of terrestrial carbon sequestration, condition of forest resources, trends of agricultural production, and generation of bioenergy. Jansson et al. (2023) created five policy scenarios for the analysis of possible impact of

carbon tax in agriculture on GHG emissions reduction. The projections made by authors showed that a global tax of 120 EUR per ton CO_2-eq could contribute to 19% decrease of global emissions from agricultural sector. However, it could also increase the risk to face food security issues in some countries of the world. A global tax of 12 EUR could contribute to 3.2% cut of GHG emissions. The opposite conclusions are provided in the study by Mardones and Lipski (2020), where the authors assessed the implementation of carbon tax in Chile agricultural sector. The results of the assessment showed that the application of carbon tax to the agricultural sector does not significantly reduce GHG emissions and more efficient way is to apply a tax to all economic sectors. Alkaabneh et al. (2021) stressed that carbon pricing can affect the final production price for consumers.

- *Agricultural extension services* are relevant tools to spread the information and affect farmers' behavior. Usually, agricultural extension services are provided by formal institutions that play an important role in supporting small-scale farms and in meeting national policy goals (Rickards et al., 2018; Indraningsih et al., 2023). Agricultural extension services contribute for the improvements of farmers' agricultural knowledge and skills, spread information regarding new technologies and their practical application and contribute for changes in farmers' attitudes and behaviour. Also, the measure accelerates the development of the whole community through human and social capital development, facilitates access to markets, works with farmers toward sustainable management (Antwi-Agyei et al., 2018) of natural resources, and manages climate risks (Afsar & Idrees, 2019). The effectiveness of the measure is mainly dependent on the knowledge of extension specialists on the various agricultural innovations they disseminate to farmers. Therefore, the competencies of such specialists are critical for the high innovation adoption rates among farmers (Babalola & Oladele, 2011; Antwi-Agyei & Stringer, 2021). This measure is widely applied globally. Various agencies, organizations cooperating with farmers, and rural development organizations are supported under different policies and programs in different countries. It is important to mention that agricultural extension services are combined with other measures supporting the development of low-carbon energy technologies in the agricultural sector, such as subsidies or business models for energy prosumers.
- Various environment-related certification schemes and carbon labeling schemes were developed to address growing concerns regarding the issues of climate change. *Voluntary certification and labeling* provide reliable information for consumers about the impact to the environment of agricultural-related products and services. Basically, the certification and labeling schemes can be categorized into five different groups, which are: carbon footprint certification, carbon-neutral product certification, carbon reporting schemes, low-carbon farming certification, and other certification schemes (Tuomisto et al., 2013). The wide spread of certification can help to create a lucrative and environmentally sustainable market for climate-friendly products. Along with the other government support, changes

in consumer behavior, and broad farmer adoption the agricultural sector could effectively fight against climate change (Thurman, 2022).
- *Research and development funding* is an important measure in seeking to reduce GHG emissions in agriculture. It is calculated that annual rate of returns for renewable energy, energy efficiency, and agriculture research and development programs are more than 20% for research expenditures (Technology Executive Committee, 2017). Also, it should be noted that these calculations of the economic returns may not take into account all the benefits of research and development funding. These external benefits can be increase of agricultural production, decrease of GHG emissions, reduction of water use, decrease in poverty indicators, economic development, growth of productivity, acceleration of learning, improvements in knowledge, new patents, etc. Therefore, investments in research and development in agriculture are critical in seeking to minimize negative impact of climate change on the global food system in upcoming decades. The importance of research and development funding in agriculture is justified in the scientific literature as well. For example, Rosegrant et al. (2022) estimated the investment gap in research and innovation to sustainable agriculture to meet Sustainable Development Goals in the Global South countries. According to the projections, $4 billion per year investments in research and development can end hunger by 2030. The additional $6.5 billion per year directed to the technical climate-smart options can implement targets of GHG emission reductions for 2030 consistent with the Paris Agreement 2°C and 1.5°C pathways. Baldos et al. (2020) estimated the research costs required to adapt agriculture to climate change damages globally by 2050. According to the authors, investments in research and development are critical to ensure food and environmental security. Alkaabneh et al. (2021) analyzed carbon policy for fruit supply chains in the United States and found that the biggest potential has research and development funding, which leads to developing new technologies with the greatest emission reduction potential.
- An *energy audit* is the assessment of energy consumption in a whole farm or of an equipment used. Farm energy audits allow to reconsider energy savings options, energy usage patterns and reduce the amount of energy demand in the farm. In the process of assessment, all the forms of energy are quantified and measured in order to determine energy use seasonal trends, inefficient energy consumption, and target areas to improve energy efficiency. Depending on the country, the energy audit can be either fully or partially supported by the government, or initiated by the farm owners themselves at their own expense. Scientific studies show that adoption of energy efficiency solutions is impacted by economical factors such as initial cost, income, ability to obtain profits, existence of financial incentives and support instruments, and cognitive factors such as knowledge, education attainment, and age (Snow et al., 2021).
- *Sustainable farming practices* are farming using methods that are economically viable, environmentally friendly and are safe for public health. These practices do not only focus on the economic aspects but also responsible

use of non-renewable resources thoughtfully and effectively. Such farming contributes to the growth of nutritious and healthy food and brings up the standards of farming. There are many different sustainable agricultural practices. Basically, they can be classified into five categories (Coulibaly et al., 2021), which are: (i) pest control; (ii) agriculture mechanization; (iii) integrated nutrient management; (iv) systems for better nutrient flows, energy cycles, and carbon footprints; and (v) soil and water.

The factors affecting adoption of sustainable agriculture practices are the subject of the scientific research. For example, Waseem et al. (2020) examined the significance of socioeconomic and psychosocial factors in the adoption of sustainable agricultural practices. It was found that such aspects as economic status, information about agricultural training programs, participation in various extension programs, personal perception of sustainable agriculture, and the feasibility of sustainable agricultural practices were significant factors affecting farmers' decision to adapt sustainable agriculture practices. The study performed by Hyland et al. (2018) showed that the perceptions of farmers are an important factor in determining the adoption of new technologies and practices. Therefore, both linear and social approaches to transfer knowledge and information should be enabled. The farming sector is heterogeneity; therefore, the communication should be well planned and targeted at the different types of farmers. Only in such a way positive changes in farm management practices can be achieved. Liu et al. (2019) analyzed the adoption of technical training for low-carbon management practices in China. It was found that technology adoption depends on the age and gender of the farmers. Young farmers are more prepared to change farming practices and adopt new technologies. As for low-carbon technologies, their adaptation is more likely among women farmers than among men farmers. Also, the importance of sharing good examples and experience of sustainable farming practices among farmers are stressed in the study. Jiang et al. (2022) analyzed the motivations of farmers to adopt low-carbon technologies in agriculture. According to the findings of the research, farmers' attitude on low-carbon production and behavioral perception has direct positive impact on the adoption of low-carbon technologies in agriculture. However, the implementation cost and socio-environmental factors can contribute as negative factors reducing readiness to implement these technologies.

Although it is possible to assign a fairly wide range of various measures aimed at promoting low-carbon energy technologies in agriculture, they are still not enough used to achieve a successful and fast enough energy transition in the sector. In summary, four main categories of influencing factors can be distinguished as having an impact on the adoption of sustainable farming, these are characteristics of farmers, characteristics of farms, behavioral factors of farm owners, and external factors (Figure 6.3).

The increase in people concerns regarding environmental problems, the knowledge about benefits of energy independence and possibility to control its own energy generated, the spread of affordable new and user-friendly technologies, the possible

Farmers characteristics	Farms characteristics	Behavioural factors	External factors
• Age
• Gender
• Level of education
• Farm experience
• Level of economic freedom (income)
• Others	Size
Geographical features
Type of farming
Level of farming diversification
Others	• Perception
• Norms
• Environmental concerns
• Knowledge
• Others	• Access to resources
• Access to markets
• Training
• Agricultural services
• Memberships
• Existance and suitability of support measures
• Acquisition of information
• Others

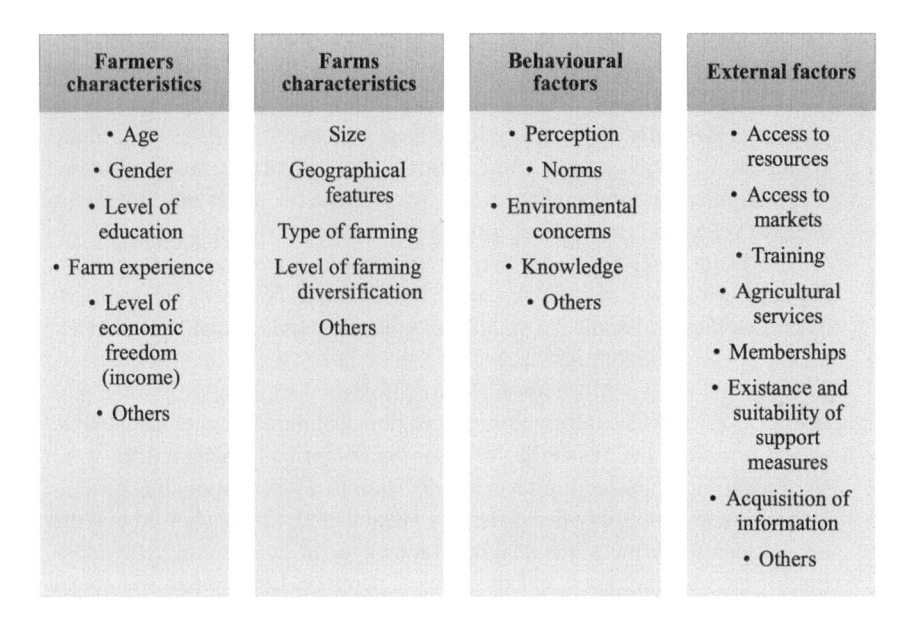

FIGURE 6.3 Factors affecting adoption of sustainable farming. (Created by authors.)

energy savings and monetization of surplus energy, and stable and clear regulatory framework in a country are the main influencing factors which could help to encourage empower low-carbon energy technologies and participation in the energy market of households and farms.

The profitability of the technology used depends on the investments, operation, and maintenance costs in comparison to the amount of the energy generated. Renewable energy has a great potential to contribute as a huge part into energy balance in the upcoming decades as the equipment and installation costs are decreasing significantly in recent years. For example, production of solar PV energy decreased by 56% over the period 2010–2015 (Balakrishnan et al., 2020). Solar energy is the most popular choice among energy prosumers, because of the quite fast photovoltaic technologies cost decline (Lang et al., 2016). The profitability aspects of solar energy are examined widely at country level (e.g., Prol & Steininger, 2020) or across several countries (e.g., Rodrigues et al., 2016; De Boeck et al., 2016) in the scientific literature. However, financial aspects are the main factors for the decision to invest in low-carbon energy technologies (Scarpellini et al., 2021). Therefore, the development of alternative and effective financial measures is critical for the increase of farm owners' willingness to invest in these technologies.

The research performed by Ge et al. (2017) showed that diversified farms are more inclined to adopt low-carbon energy technologies, especially solar and biomass energy. Also, it was found that the adoption of new technologies is impacted by local energy demand and suitable conditions for renewable energy generation. Therefore, it is important that support policy would be adapted to each renewable energy option, because possible energy prosumers differ in farm characteristics and geographic location.

The results of the study performed by Siksnelyte-Butkiene et al. (2023) showed that such factors as market barriers, scarcity of effective support measures, slow penetration of new effective technologies in the market, quite low public awareness, and acceptability of new technologies should be resolved in order to achieve the rapid growth of energy prosumers and their participation in the market. The authors stressed, despite the fact, that the number of energy prosumers are growing each year, the achievements made are not enough in order to reach ambitious climate change and energy goals. Therefore, much more political effort should be done in all sectors, which are dependent on energy. The agricultural sector is undoubtedly dependent on energy resources and is one of the main drivers of climate change.

6.3 MULTI-CRITERIA DECISION MAKING (MCDM) INSTRUMENTS FOR SUSTAINABLE AGRICULTURE DECISION MAKING

This section discusses the variety of MCDM methods for sustainable agriculture decision making and its application in the scientific literature as well as provides comparative assessment of the most popular multi-criteria techniques to solve sustainable agriculture issues. Finally, the guidelines for criteria selection and for the whole process of the assessment are proposed in order to fully follow the methodical recommendations for sustainability assessment.

6.3.1 APPLICATION OF MCDM METHODS FOR SUSTAINABLE AGRICULTURE DECISION MAKING

The issues of agricultural sector sustainability started to become highly important in decision making since the early 1980s with the aim to sustain the economic viability, solve food security problems, improve the quality of life of people and farmers, use non-renewable resources efficiently, reduce impact on the environment, satisfy nutrition needs, etc. The term "sustainable agriculture" was defined in 1977 by the United States Department of Agriculture. The huge and still growing negative impact on the environment and concerns due to climate change has changed aspects and priorities of decision makers, adding various environmental issues to the agricultural sector planning programs and decision making. The growing concerns due to wellbeing of people started to pay attention to such questions as people health, nutrition needs, and others in shaping future agricultural sector. However, it must be recognized that there are many contradictions between different goals when solving sustainable agriculture or sustainable energy development issues. For example, the most favorable option from an environmental perspective may be very cost-intensive; the most cost-effective alternative may be unacceptable from social point of view, etc.

Recognized and reasonably selected mathematical models can help to combine conflicting aspects and different needs of different groups of stakeholders according to the criteria selected. MCDM or multiple-criteria decision analysis (MCDA) is an operations research sub-discipline, which uses mathematical models to assess multiple conflicting criteria in making decision. Various MCDM methods allow to tackle compromises, using decision makers' criteria selected and preferences between possible alternatives.

The relevance of idea of sustainable development has been increasing in society, business, political documents, and strategies. However, despite the significance of this concept, the understanding and requirements for sustainable development continue to be undetermined. This is not only because of the potential technical difficulties in measuring sustainability but also because the goal of sustainability is unattainable, and when the selected target is achieved, others must be set. It can be said that sustainability is like a horizon or a perfection that cannot be reached but should always be tried to be as close as possible.

Today, sustainability assessment is actively used as the analysis instrument and preventative measure to assess the possible impact on the environment. Many different frameworks, methodologies, and assessment systems have been developed in the scientific literature and many of them are based on different MCDM methods. The usage of MCDM methods to solve questions in different policy areas started to be popular in the last decade. Nowadays, MCDM assessment is widely applied in dealing with sustainability measurement in various areas such as energy policy (Siksnelyte et al., 2018), evaluation of technologies (Siksnelyte-Butkiene et al., 2020; Bohra & Anvari-Moghaddam, 2022), transport development (Keshavarz-Ghorabaee et al., 2022), civil engineering (Wen et al., 2021), construction sector (Santoso & Darsono, 2019), and agriculture (Cicciu et al., 2022).

The way the decision-making process is performed has a significant impact on the implementation of policy goals. The usage of multi-criteria instruments in decision-making processes allow to deal with contradictory issues, identify the most preferable alternative among others based on the criteria selected, combine policies from different areas. MCDM techniques can be used not only in the process of decision making but also to set the objectives or identify the best measures to attain them. Because of the wide applicability various MCDM approaches are increasingly being used to address different issues in various fields, including agriculture.

Various MCDM methods enable to carry out an assessment, which considers indicators chosen in a comprehensive way and measures the interrelations among them. Although many different categorizations can be found in the literature, the most recognized way is to categorize multi-criteria methods into two main groups: multi-objective decision making (MODM) techniques and multi-attribute decision-making (MADM) techniques. MODM and MADM approaches have quite similar characteristics but also some differences among them can be singled out. MODM is intended to handle questions with the continuous space of decision, where the aim is to select the most favorable alternative within predefined boundaries. Whereas MADM techniques are applied for questions with discrete decision spaces among alternatives selected (Baumann et al. 2019).

MCDM process consists of two main steps (Figure 6.4). The first step is Construction, where the focus is on aim, alternatives, criteria, and stakeholders. The second is Exploitation step, which considers possible techniques, weighting schemes, and other procedures of the evaluation. These two basic steps are interlaced, and it is not possible strictly to separate them (Guitouni & Martel, 1998).

The sustainability assessment of agricultural systems is multifunctional in nature and requires multi-criteria approach. Amongst others, it considers various criteria such as economic, environmental, social, and technical criteria. The assessment of agricultural systems can be characterized by the existence of many and often

Assessment of Low-Carbon Transition Policy in Agriculture 275

CONSTRUCTION step	**EXPLOITATION step**
• Goal definition • Scope definition • Definition of possible alternatives • Consideration of stakeholders engagement • Consideration of stakeholders groups • Consideration of of criteria • Criteria selection	• Measurement of performance of the criteria chosen • Selection of MCDM technique • Criteria aggregation • Determination of criteria weights • Determination of weighting schemes • Results comparison

FIGURE 6.4 The main steps of MCDM procedure. (Created by authors.)

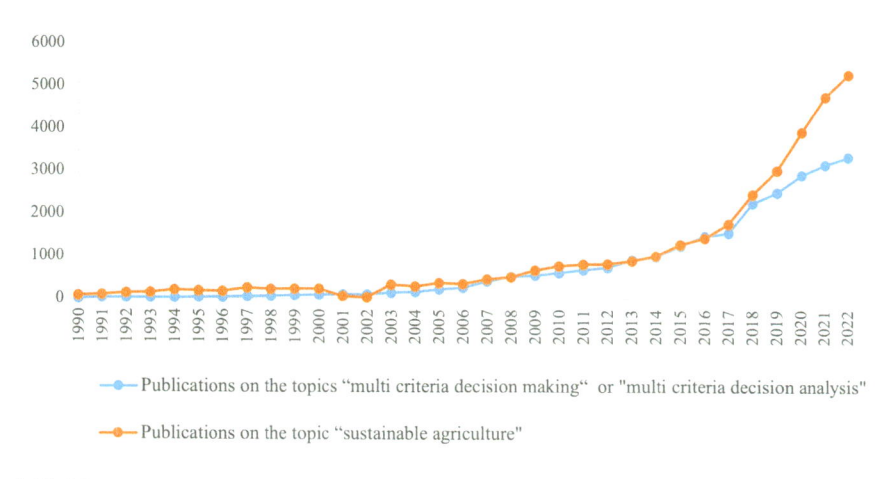

FIGURE 6.5 Number of publications on the different topics in the Web of Science database. (Created by authors.)

conflicting with each other criteria, different groups of stakeholders and decision makers with competing interests, different levels of information and knowledge, etc.

The interest in sustainable agriculture is being seen for the last 20 years. The analysis of scientific studies in the Web of Science database showed that there are more than 33,000 of publications on the topic "sustainable agriculture" in the period 1990–2022 (Figure 6.5). However, the majority of studies (more than 80%) were published during the last ten years (2012–2022). The application of various MCDM techniques started to be very popular in the last ten years. Overall, more than 25,000 of publications on the topics "multi criteria decision making" or "multi criteria decision analysis" have been published in the period of 1990–2022 and 84% of them in the last ten years (2012–2022).

Due to its wide applicability and growing application to solve different decision-making problems, MCDM started to be applied in agriculture. The analysis of the

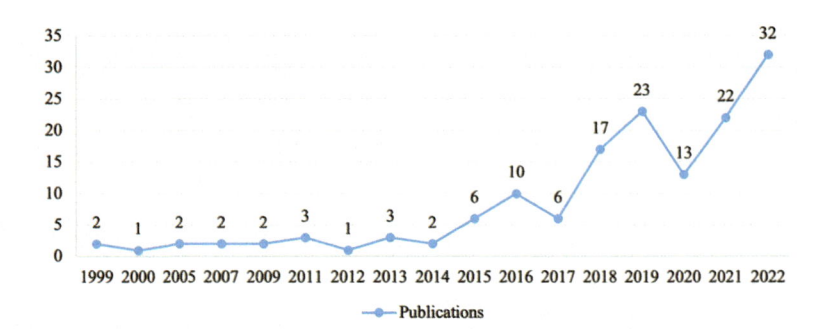

FIGURE 6.6 Number of publications on the topics "multi criteria decision making" or "multi criteria decision analysis" and "sustainable agriculture" in the Web of Science Core Collection database. (Created by authors.)

Web of Science database (2023) showed that there is notable increase in studies applying multi-criteria analysis for sustainable agriculture decision making in the last five years (Figure 6.6). It should be noted that this number of publications is not exhaustive, since by changing the search terms, more examples of the application of MCDM methods can be found. However, the popularity of the application of multi-criteria techniques can be seen in the last years, which clearly shows the trend to grow.

The diversity of problems and issues in sustainable agriculture decision making can be seen by analyzing the keywords and their frequency in the publications using MCDM for sustainable agriculture decision making. Figure 6.7 shows keywords map

FIGURE 6.7 Keywords map of studies using MCDM techniques for sustainable agriculture decision making. (Created by authors.)

and identifies the main instruments and supporting methods used. As it can be seen, the map also reveals that there is a problem of terms inconsistency in the scientific literature. Often, the same terms are written differently, which complicates the processes of analysis and application of one or another instrument or results of the study for further research.

The main research areas of studies dealing with sustainable agriculture issues that utilize MCDM methods can be arranged into many different categories. Table 6.1 provides one of the possible classifications. These research areas demonstrate the application of MCDM methods in various aspects of sustainable agriculture, enabling

TABLE 6.1
The Main Research Areas of Studies Dealing with Sustainable Agriculture Decision Making

Research Area	MCDM Application
Sustainable crop selection and management	MCDM methods can be employed to evaluate and select the most suitable crop species and varieties based on multiple criteria such as yield, resource use efficiency, environmental impact, and economic viability. These methods help in optimizing crop selection and management practices for sustainable agriculture.
Water resource management	MCDM techniques assist in decision-making related to water allocation, irrigation strategies, and water conservation practices in agricultural systems. They consider various criteria such as water availability, water quality, efficiency, and environmental impacts to enhance water resource management in a sustainable manner.
Soil conservation and management	MCDM methods aid in assessing and ranking different soil conservation and management practices. These practices include cover cropping, reduced tillage, organic amendments, and erosion control measures. MCDM helps in selecting the most effective and sustainable soil management techniques.
Agrochemical use and integrated pest management	MCDM approaches support decision making regarding the use of agrochemicals and the implementation of integrated pest management strategies. They consider criteria such as pest control efficacy, environmental impact, cost-effectiveness, and human health risks to promote sustainable and reduced chemical input practices.
Agroforestry and sustainable land use	MCDM methods can be applied to assess the suitability and benefits of integrating trees and shrubs into agricultural landscapes. They consider multiple criteria such as biodiversity conservation, soil fertility improvement, carbon sequestration, and economic viability to guide sustainable land use decisions.
Livestock production and management	MCDM techniques assist in evaluating different livestock production systems and management practices based on criteria such as animal welfare, resource efficiency, GHG emissions, and economic viability. These methods aid in promoting sustainable and ethical livestock farming.
Sustainable supply chain management	MCDM approaches contribute to sustainable food supply chain management by considering criteria such as environmental impact, social responsibility, economic viability, and food quality. They help in making decisions related to sourcing, processing, distribution, and consumption to achieve sustainability goals.

Source: Created by authors.

decision makers to make informed choices that balance economic, social, and environmental considerations.

Different MCDM methods have been used to address various agriculture sustainability issues. But it is necessary to state, despite the fact, that numerous MCDM techniques have been widely applied to address sustainability issues in many different sectors, MCDM application is a quite new approach in the field of agriculture. However, after analysis of the scientific literature, it can be stated that the most widely applied method surely is the Analytical Hierarchy Process (AHP) presented by Saaty (1980). The MCDM approaches used for sustainable agriculture decision making can be singled out into several categories, these are presented in Figure 6.8.

As it was mentioned before, the most popular multi-criteria technique applied in agriculture studies is AHP. Also, AHP has been employed to solve various questions in different areas such as farming and growing practices, indicators selection, identification of preferences, and development of sustainability frameworks. For example, Rodríguez Sousa et al. (2020) compared a multifunctionality of integrated and ecological olive grove management types considering environmental, economic, and social indicators in Andalusia region of Spain. The multifunctional behavior of these different management schemes was assessed by employing AHP technique. The

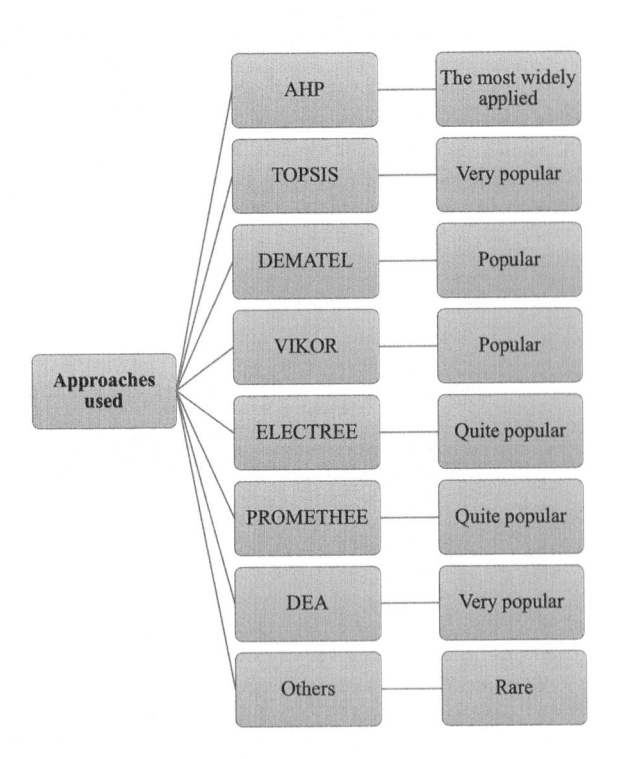

FIGURE 6.8 The main MCDM approaches used for sustainable decision making in agriculture. (Created by authors.)

results showed that ecological farming leads to a higher multifunctionality of olive groves. Abdallah et al. (2018) performed comparative sustainability assessment of conventional and organic systems of olive growing in Tunisia. The AHP method was used considering economic, environmental, and socio-territorial aspects. The findings showed the need to strengthen the economic performance of organic growing through political strategies by increase of farms productivity and demand of organic products. A comparative evaluation of the multifunctional performance of alternative olive growing systems in a region of Spain was also performed in the study by Parra-López et al. (2008). The analysis relies on AHP technique and engagement of different groups of experts.

Srinivasa Rao et al. (2019) sought to identify indicators for the assessment of climate-resilient agriculture in India. The authors overviewed 1209 indicators and employed the experts to develop suitable set of 30 indicators for India. The authors applied AHP and Weighted Sum Method (WSM) introduced by Zadeh (1963) to analyze the results obtained during the experts' survey. The other study related to criteria selection was published by Sajadian et al. (2017), where the authors sought to develop a set of organic farming indicators and quantify them. AHP and experts' interviews were used to reach the goal of the research. It was found that indicators such as: management of pest and disease, yield, management of soil nutrient, rate of water consumption, consumption of chemical fertilizer, and transgenic materials usage are the most significant aspects when assessing organic farming. Veisi et al. (2016) employed AHP technique to develop an ethics-based approach for the analysis of sustainability of Iran agriculture and food systems strategies sustainability. Rezaei-Moghaddam and Karami (2008) assessed sustainable agricultural development models for Iran using AHP method. It was determined that ecological criteria are more important than economic or social for the development of sustainable agriculture in the country. Kazemi and Hosseinpour (2022) used AHP and Geographic Information System (GIS) to identify the most preferable sites for urban agriculture in a city of Iran.

Olveira et al. (2016) used AHP for the prioritization of milk production systems based on their sustainability performance. Tzouramani et al. (2020) evaluated the performance of dominant agricultural systems sustainability at the farm level in Greece. The authors used AHP to range the alternatives and gain the main insights to support strategic farm choices. According to the results, such choices as permanent crops, olive trees, and sheep farms are better choices in terms of sustainability than arable and intensive crop farms. Olguín et al. (2019) proved the sustainability of quinoa grain in rainfed agricultural systems. The authors compared two types of grain cultures quinoa and wheat employing AHP technique and considering economic, technical, political, and environmental criteria. Tran et al. (2018) applied AHP to find out the preferences and views of rice farmers and experts in the field regarding flood-based solutions from a sustainable livelihood perspective in a province of Vietnam. Renwick et al. (2019) applied AHP method and evaluated factors and their importance for the determination of the need to change land use system from the perspective of the land use managers.

Among the other quite widely used MCDM methods for the sustainable decision making in agriculture can be mentioned application of the Technique for Order of Preference by Similarity to the Ideal Solution (TOPSIS) introduced by Hwang and Yoon (1981); Decision-Making Trial and Evaluation Laboratory (DEMATEL) presented by Gabus and Fontela (1972); Vlse Kriterijumska Optimizacija Kompromisno Resenje) (VIKOR) developed by Opricovic (1998); Preference Ranking Organization Method for Enriching Evaluation (PROMETHEE) presented by Mareschal and Brans (1992); and Elimination and Choice Transcribing Reality (ELECTRE) developed by Roy (1978). These MCDM methods together with AHP are widely used in sustainable agriculture studies to provide structured decision-making frameworks and facilitate the evaluation and selection of sustainable agricultural practices, technologies, and policies.

It can be stated that TOPSIS is the second most widely used technique for sustainable decision making in agriculture. For example, Morkunas and Volkov (2023) analyzed the progress of climate-smart agriculture in the EU countries, where the composite indicator was constructed based on SAW, TOPSIS, and VIKOR techniques. Wang et al. (2022) measured the level of agricultural sustainability in a province of China. The authors used TOPSIS method to consider four groups of criteria reflecting economy, environment, society, and resources aspects. Fu et al. (2022) analyzed sustainable development of agriculture regarding the efficiency of resource usage in a province of China. The authors applied TOPSIS for the evaluation of six groups of aspects related to resource endowment, production, science and technology levels of agriculture, economic, ecological, and social benefits. Zhou and Wen (2023) examined the achievements of green development, disparities among regions, and the main patterns in Chinese agriculture. The authors developed a comprehensive evaluation system based on several techniques, while for multi-criteria analysis the authors used TOPSIS. Khan and Ali (2022) analyzed factors having an impact on the implementation of circular bio-economy in Pakistan agriculture. The authors combined Strength, Weakness, Opportunities, and Threats (SWOT) analysis and fuzzy TOPSIS method. It was found that financial support for both the local farmers and investors was the most preferable strategy in order to successful implement circular bio-economy in agricultural sector of the country. Di Bene et al. (2022) sought to find out the perception of various stakeholders regarding the implementation of sustainable farming practices in Italy. The research is based on stakeholders' survey and multi-criteria TOPSIS technique. Mng'ong'o et al. (2022) applied fuzzy TOPSIS and fuzzy AHP to evaluate agro-processed organic crop alternatives for sustainable agriculture in a region of Tanzania. Chang and Liang (2023) presented a model for the ecological agriculture project management from a low-carbon perspective based on artificial intelligence algorithm and hybrid technique for risk assessment, using TOPSIS and AHP.

Also, the DEMATEL method started to be applied in sustainable agriculture decision making. Usually, DEMATED is applied for the analysis of barriers, enablers, and drivers for the sustainable agriculture. For example, Wang (2022) sought to identify drivers having an impact on the sustainable development of agro-industrial parks in a province of China. The author considered four groups of indicators (economic, social, sustainability, and technical) and applied DEMATEL for quantitative analysis. The results showed that economic criteria play the most important role for the sustainable development of agro-industrial parks. Avikal et al. (2022) sought to identify drivers for the sustainable circular economy of agricultural products supply chain. The authors employed DEMATED and ANP methods to evaluate the relationship among the main aspects and their contribution toward implementation of sustainable circular economy. Mangla et al. (2018) sought to identify and analyze the main enablers for the implementation of sustainable agri-food supply chains initiatives. The authors engaged the experts and used Fuzzy DEMATEL technique for the analysis and classification. Barbosa et al. (2022) applied Fuzzy DEMATEL technique to identify barriers having an impact for the adoption of sustainable agriculture. Xia and Ruan (2020) analyzed barriers having an impact to develop sustainable circular economy in China agriculture. The authors identified barriers for farmers, government, and business and their correlation with the application DEMATEL technique. Finally, a support framework for decision-makers was provided. Kumar et al. (2022) identified and analyzed barriers for adoption of Industry 4.0 in sustainable food supply chain from a perspective of circular economy. The authors performed a cause-effect analysis and ranked the barriers identified using Rough-DEMATEL method. The results showed that aspects related to technological maturity, investment cost, public acceptance, and lack level of innovations are the most influencing barriers for adoption of Industry 4.0 in food supply chain. Dixit et al. (2022) evaluated the barriers having an impact on organic farming development in developing countries. Eighteen barriers were identified and DEMATEL technique was applied for ranking and analysis. According to the results such aspects as level of information and knowledge, financial and institutional support have the most significant impact for adoption of organic farming in developing countries.

In recent years, VIKOR method is also more widely applied. For example, Jain et al. (2023) evaluated different models for the adoption of sustainable technologies of Industry 4.0 in agro supply chain. The evaluation framework is based on combination of AHP and fuzzy VIKOR techniques. Yin et al. (2022) applied fuzzy VIKOR technique to deal with digital green innovation issues in implementing Agriculture 5.0 in agricultural manufacturing system. Deepa et al. (2019) proposed an evaluation tool based on VIKOR and TOPSIS techniques for the assessment of sustainable sugarcane farms. Volkov et al. (2020) sought to measure the environmental and economic performance of agriculture in several EU countries. Multi-criteria VIKOR method was applied for calculations. The results showed that new EU members distinguish with higher results than old member states. Wang et al. (2019) proposed an evaluation framework for the assessment and selection of sustainable energy conversion technologies for agricultural residues. The authors prioritized seven bioenergy technologies considering four groups of criteria, which are technological, environmental, economic, and social. The proposed evaluation framework relies on LCA, AHP for determination of criteria weights, and VIKOR for ranking the selected technologies.

PROMETHEE and ELECTRE techniques can be also distinguished as quite popular used to deal with sustainable agriculture issues. However, the application of these methods is often found in previous studies. Talukder and Hipel (2018) used PROMETHEE to compare sustainability of five agricultural systems in Bangladesh. Zekic et al. (2018) analyzed the impact of agricultural sector on the environment in the EU countries and Serbia. PROMETHEE method was used to rank the countries based on their achievements. Jeong (2018) developed an evaluation framework for the analysis of sustainable land-use adaptation in the context of climate change. The proposed framework relies on GIS and PROMETHEE techniques. Palma et al. (2007) employed PROMETHEE to assess the integrated economic and environmental performance of silvoarable agroforestry on hypothetical farms in landscape test sites in three European countries. Cardoso et al. (2018) analyzed different sugarcane production technologies from the economic, social, and environmental perspectives using PROMETHEE for ranking.

Maydana et al. (2020) performed an integrated assessment of the agricultural ecosystem land use scenarios in a region of Argentina applying ELECTRE III for scenario ranking. Aouadi et al. (2020) measured the environmental and socioeconomic performance of wine-growing systems and assessed different agro-ecological transition scenarios using ELECTRE III and ELECTRE Tri-C techniques. The authors analyZed the usage of pesticides in the Bordeaux region (France) and proposed scenarios for reduction the reliance on pesticides. Mendas et al. (2014) and Mendas and Delali (2012) sought to identify the most preferable lands for sustainable agriculture in Algeria. The authors developed land suitability maps for wheat growing using GIS and ELECTRE TRI techniques. Moulogianni and Bournaris (2017) assessed the potential to produce biomass energy in Central region of Macedonia. The authors used ELECTRE III method for seven agro-energy regions ranking. The results showed that regions where cereals and arable crops are cultivated distinguished with better results than regions with fruit trees or other crops. Arondel and Girardin (2000) proposed an evaluation system based on ELECTRE TRI technique in order to sort cropping systems based on their impact on groundwater quality.

Among the quite rarely used methods can be mentioned: Multiple Attribute Value Theory (MAVT) presented by Fishburn (1967); Complex Proportional Assessment (COPRAS) proposed by Zavadskas et al. (1994); Simple Additive Weighting (SAW) presented by MacCrimon (1968); Multicriteria Optimization and Compromise Solution (original language: Step-Wise Weight Assessment Ratio Analysis (SWARA) introduced by Kersuliene et al., 2010); the Evaluation Based on Distance from Average Solution (EDAS) presented by Keshavarz-Ghorabaee et al. (2015).

Several studies using MAVT method can be found in the literature. For example, Talukder et al. (2016) sought to assess sustainability of agricultural systems in the five coastal areas of Bangladesh. The authors developed set of indicators, which consider productivity, efficiency, stability, durability, compatibility, and equity aspects of agriculture systems and applied MAVT technique for the assessment. Troiano et al. (2019) performed a quantitative sustainability assessment of three different farming systems (conventional, organic, and biodynamic). Talukder et al. (2018) applied MAVT and six categories of composite indicators to evaluate and compare sustainability of different agricultural systems in Bangladesh. Streimikiene and Mikalauskiene (2023) measured the sustainability of the Baltic States agricultural sector and applied COPRAS method for countries ranking. Yucenur et al. (2022) proposed a framework for the analysis of Agriculture 4.0 applications of Turkey's agricultural sector based on SWARA and EDAS methods.

However, although these methods are rarely used, in recent years there has been a significant increase in the application of MCDM methods and an increase in the variety of methods used. Therefore, it is likely that in the next few years' agricultural studies will apply not only currently used, but also rarely used and new methods.

As a separate group can be singled out studies employing Data Envelopment Analysis (DEA) (Charnes et al., 1978), which aims to assess the performance of set of decision-making units. DEA is primarily a method for evaluating the relative efficiency of decision-making units based on their input and output data. DEA can be extended to incorporate multiple criteria by including them as additional inputs or outputs in the analysis. This allows for a multi-criteria perspective, where the efficiency scores are based on multiple criteria or dimensions. In such extended form, the method considers the efficiency of decision-making units in relation to multiple criteria simultaneously. It allows to identify decision making units that achieve the best performance based on all criteria, rather than just efficiency in terms of inputs and outputs. Therefore, while DEA itself is not inherently a multi-criteria approach, it can be adapted to incorporate multiple criteria.

An example can be mentioned in the study performed by Mu et al. (2018), where the authors assessed the effect of uncertainty on the eco-efficiency of 55 dairy farms in Western Europe. Fuzzy DEA was applied to measure the eco-efficiency accounting for multiple indicators simultaneously. The results showed that the uncertainty due to the environmental and economic indicators can have an impact on the eco-efficiency of dairy farms. Gerdessen and Pascucci (2013) proposed a methodological approach for the sustainability assessment of agricultural systems, considering three dimensions of sustainability (economic, social, and environmental). The developed framework was applied for the categorization of European agricultural regions using DEA. Picazo-Tadeo et al. (2011) evaluated eco-efficiency of farming with the application of DEA approach in a county of Spain. According to the results, the more eco-efficient are farmers, who have higher level of education and those, who are participating in agro-environmental programs. Zhu et al. (2023) used DEA for the analysis of productivity growth measures for energy-related GHG emissions and their application for the EU agriculture. According to the findings of the study, after accounting for energy-related GHG emissions and structural inefficiencies, 16% decline in agricultural multipurpose productivity in the period 2003–2019 was identified. Streimikis et al. (2023) sought to disentangle the main patterns in energy-related GHG emissions performance in agriculture by assuming different optimization directions. The authors used DEA and provided a case study of selected EU countries. In the other study, Streimikis et al. (2022) sought to measure the achievements of the EU countries toward the development of sustainable agriculture and evaluated the environmental performance of agricultural sector using DEA. Fusco et al. (2023) applied DEA for the assessment of economic and environmental performance of agriculture in terms of eco-efficiency for the case study of Italy. Data-driven assessment of regional agricultural production efficiency toward sustainable agricultural development was performed in the study by Liu et al. (2022). The authors used DEA Malmquist model to construct data modeling for a province of China. The results showed that technical efficiency is the main driver for the improvement in regional agricultural production efficiency.

Many different supporting methods have been combined with MCDM approaches to deal with sustainable agriculture decision making. The most widely used can be singled out such methods as methods for criteria weighting (e.g., Pairwise Comparison, Entropy Weighing); Fuzzy Set Theory; Sensitivity Analysis; LCA; GIS; Strengths, Weaknesses, Opportunities, and Threats (SWOT) analysis; Goal Programming; Neutrosophic Cognitive Maps; and Grey Relational Analysis.

It's important to note that these are just a few examples, and there are many other MCDM methods that can be applied in sustainable agriculture depending on the specific objectives and constraints of decision making. The choice of method should be based on the specific research context and the preferences of the stakeholders involved. Also, the popularity of MCDM methods used in sustainable agriculture decision making may vary depending on the research area and context.

6.3.2 Comparison of the Most Popular MCDM Techniques

Various multi-criteria assessment techniques have been used to support decision making for sustainable agriculture. After the analysis of scientific literature, it can be stated that some of the methods are more commonly used. The most widely applied MCDM techniques are compared in this subsection, distinguishing their main advantages and disadvantages and providing the comparison among them. The most widely used MCDM approaches to solve sustainable agriculture issues can be classified in pairwise comparison-based methods (AHP), distance to an ideal solution-based methods (TOPSIS and VIKOR), outranking methods (PROMETHEE and ELECTRE), cause and effect analysis methods (DEMATEL), efficiency measurement-based methods (DEA). The AHP method is commonly applied for criteria weighting. TOPSIS and VIKOR methods assessed the alternatives based on their distances to the best solution. DEMATEL allows to perform cause and effect analysis. The ELECTRE and PROMETHEE methods evaluate the selected alternatives by means of outranking relations between them. DEA technique is for efficiency measurement. Each technique differs and has its own features as also its own advantages and disadvantages. Therefore, in each case, the most applicable technique should be selected taking into account all the features, advantages and disadvantages, the available data, decision maker experience, the precision of the results, time cost, etc. Comparison of the most widely applied MCDM techniques used for sustainable agriculture decision making is provided in Table 6.2.

AHP method relies on pairwise comparison and is widely used in various fields, including business, engineering, and social sciences. In many fields, this method is the most widely applied. AHP technique distinguishes its structured approach, allowing decision makers to break down complex problems into a hierarchy of criteria and alternatives. The method can handle both qualitative and quantitative data, making

TABLE 6.2
Comparison of the Most Popular MCDM Techniques

	AHP	TOPSIS	VIKOR	DEMATEL	ELECTRE	PROMETHEE	DEA
Simple enough to calculate	x	x	X				
Computation process is fast enough	x	x	X				
Non-compensatory method					x	x	x
Clear calculation logic		x	X		x		
Do not require subjective assumptions							x
The method is more suitable for experts because of complex calculations				x	x	x	x
Suitable for significantly different values between alternatives				x			x

Source: Created by authors.

it suitable for a wide range of decision-making scenarios. AHP enables decision makers to consider multiple criteria simultaneously and assign relative weights to each criterion based on their importance. The subjective judgments in the process of evaluation allow decision makers to express their preferences and opinions. Despite these advantages, some disadvantages of the AHP method also can be singled out. The subjectivity on which AHP relies can introduce bias and inconsistency in the decision-making process, especially if decision makers have different perspectives or interests. AHP involves a series of calculations and transformations, which may make it difficult for decision makers to understand and interpret the results. The method is sensitive to changes in the input data and may produce different results with slight variations in the pairwise comparison judgments. This sensitivity can make the method less reliable and robust. However, despite these disadvantages, the AHP method has been widely adopted and continues to be a valuable tool for decision makers in various domains. Its structured approach and ability to handle multiple criteria make it a useful method for prioritizing alternatives and facilitating decision-making processes.

TOPSIS is commonly used to rank alternatives based on their similarity to an ideal solution and their distance from a negative ideal solution. The method is recognized as simple and intuitive: TOPSIS is relatively easy to understand and apply, and it does not require extensive mathematical knowledge or expertise. Also, TOPSIS takes into account both the positive aspects (similarities to the ideal solution) and the negative aspects (distances from the negative ideal solution) of alternatives, providing a more comprehensive evaluation. The method provides a clear ranking of alternatives, making it easier for decision makers to compare and prioritize the options based on their relative performance. Also, it can handle both quantitative and qualitative data, allowing decision makers to incorporate different types of criteria in the decision-making process. This technique enables decision makers to conduct sensitivity analysis by examining the impact of changes in criteria weights or alternative performance on the rankings. Despite these advantages, some disadvantages also can be singled out. The process of assigning weights to criteria in TOPSIS involves subjective judgment. Different decision makers may assign different weights, leading to potential bias and inconsistency in the results. The performance scores used in TOPSIS are highly dependent on the accuracy and reliability of the input data. Errors or inconsistencies in the data can significantly impact the rankings and final decisions. Also, TOPSIS neglects interdependencies among criteria selected and treats criteria as independent of each other. This can limit its ability to capture the complex relationships among criteria in certain decision-making situations. However, despite these limitations, TOPSIS remains one of the most popular and widely used methods in decision-making processes. Its simplicity, ability to handle both qualitative and quantitative data, and provision of a clear ranking make it a valuable tool for decision makers in various domains.

Like TOPSIS, VIKOR follows the similar logic and measures the distance between the ideal positive and the ideal negative solution. While other methods may focus on identifying the best alternative based on a single criterion or a weighted sum of criteria, VIKOR aims to strike a balance between the best and worst outcomes. It can be useful in situations where stakeholders have conflicting preferences. As TOPSIS, VIKOR method is also relatively easy to understand and implement compared to other MCDM techniques. Also, both qualitative and quantitative data can be used. The method considers the decision-maker's risk attitude by incorporating a weight coefficient that reflects their level of risk aversion. Despite these advantages, the process of assigning weights to criteria is subjective, which can introduce bias and inconsistency. VIKOR assumes that the criteria are independent and that there is no interaction between them and does not account for changes in the relative importance of criteria during the decision-making process. This method is most suitable to solve issues with a moderate number of alternatives and criteria, while to handle large-scale decision problems with VIKOR can be challenging and time-consuming.

DEMATEL is a methodology used for analyzing complex systems and decision-making processes. The method provides a systematic framework for analyzing complex systems and understanding the relationships between various components or factors involved. DEMATEL uses visual representations, such as cause-and-effect diagrams and influence matrices, which help in understanding the interdependencies between different factors in a clear and concise manner. The analysis of the cause-and-effect relationships among criteria allows to identify the interdependencies and influences among criteria, which can be valuable in situations where values between alternatives vary significantly. By considering the relationships and dependencies among criteria, DEMATEL can provide insights into how strongly each criterion affects others and help in understanding the impact of these relationships on the overall decision. This makes DEMATEL particularly useful when there are significant differences in values between alternatives, as it can help uncover the underlying factors contributing to these differences and provide a more comprehensive understanding of the decision problem. However, as previously mentioned methods DEMATEL also is based on expert opinions and subjective judgments. Therefore, the accuracy and reliability of the results heavily depend on the expertise and knowledge of the experts engaged. The technique requires a significant amount of data on the relationships between various factors, which can be a challenging and time-consuming task. Also, the method is a complex methodology and requires expertise in the underlying concepts and techniques. Therefore, the method is basically suitable only for experts.

ELECTRE technique belongs to an outranking family of MCDM methods, allowing the direct alternatives comparison. ELECTRE focuses on the concept of outranking, compares the alternatives, and determines their relative superiority or inferiority. ELECTRE allows to consider multiple criteria simultaneously, which is essential in cases where decisions are influenced by many aspects. The method provides a transparent and systematic approach to decision making, ensuring that the decision process is well-documented and easily understandable. Also, ELECTRE is well-suited to handle imprecise or uncertain data, as it uses fuzzy sets and linguistic variables to represent subjective judgments. However, the method is complex and time-consuming, especially in cases with a large number of criteria and alternatives. Also, as previously mentioned methods, ELECTRE also relies on subjective judgments and preferences. It can also be distinguished that the process of assigning criteria weights can be challenging, as it requires accurately to assess the relative importance of each criterion. ELECTRE does not provide numerical values as output, making it difficult to compare the results with other methods that rely on quantitative analysis. While ELECTRE is useful for decision making under uncertainty, it may not be suitable for all types of decision problems. It is more suitable for qualitative or semi-quantitative analysis rather than purely quantitative analysis.

The other used outranking technique in sustainable agriculture studies is PROMETHEE, which allows to incorporate different criteria and their weights, making it suitable for complex decision-making problems. The method provides graphical representations, such as radar charts and preference graphs, which help decision makers to understand and compare alternatives easily. The method allows to define thresholds to be considered when comparing alternatives, ensuring that only viable options are included in the ranking. However, the method is subjective, because relies on decision makers to assign weights to criteria. The process of evaluation (collecting data for multiple criteria and assigning preference functions for each criterion) can be time-consuming and challenging, especially when dealing with a large number of alternatives. PROMETHEE is suitable to deal with problems where alternatives can be objectively compared using numerical data. It may not be suitable for subjective or qualitative evaluations. Overall, PROMETHEE is a useful method for multi-criteria decision making, offering flexibility and visual representation. However, it also has limitations related to subjectivity in criteria weighting and data collection complexity.

DEA is a non-parametric method used for evaluating the relative efficiency of decision-making units based on their input and output data. DEA does not require any assumptions about the functional form or distribution of the data, making it suitable for situations where traditional parametric methods may not be applicable. The technique provides a relative efficiency score for each decision-making unit, allowing to compare and identify the best performing. DEA can handle multiple inputs and outputs simultaneously, making it useful for evaluating complex systems with multiple dimensions. The technique identifies benchmark units that are the most efficient and can serve as role models for improving the performance of inefficient units. DEA is particularly useful when values between alternatives differ strongly because it does not rely on explicit mathematical functions or assumptions about the relationship between inputs and outputs. Instead, it compares the relative efficiency of each alternative based on their observed values. This allows DEA to handle cases where there are significant differences in values between alternatives, as it focuses on relative performance rather than absolute values. By considering the efficiency scores obtained from DEA, decision-makers can identify the most efficient alternatives and gain insights into improving the performance of less efficient alternatives, even when values between alternatives differ strongly. However, DEA requires precise and accurate data on inputs and outputs for each decision-making unit, which can be challenging to gather, especially when dealing with large-scale or complex systems. Also, the technique does not provide sensitivity analysis to assess the impact of changes in input or output values on efficiency scores. Overall, DEA is recognized as a powerful tool to perform efficiency assessment, offering a non-parametric approach and the ability to handle multiple inputs and outputs.

6.3.3 Guidelines for Criteria Selection and the Whole Process of the Agriculture Sustainability Assessment

The assessment of sustainability seeks to apply the popular and together very ambitious concept of sustainability for the analysis of specific issues and support decision making. However, sustainability assessment is not being strictly defined methodology and has a lot of space for interpretation and practice, covering numerous possible instruments and processes (Halla & Binder, 2020). However, as the analysis of the scientific literature shows, a large part of the developed assessment methodologies may have various methodological gaps. For example, Arulnathan et al. (2020) analyzed decision support tools dedicated to farm-level assessment and their consistency with sustainability principles. The results showed that only 3 decision support tools from 19 partially meet principles of sustainability assessment. Meanwhile, Siksnelyte-Butkiene et al. (2021) evaluated composite indicators and indices for energy poverty measurement. According to the results, also only a small part of developed indicators is in line with sustainability assessment principles. This sub-section proposes general recommendations for the indicators selection and the whole process of assessment in sustainable way to support agricultural decision making. It is shown how the whole stages of the evaluation process are important from the determination of the goal of the assessment to communication of the final results of the assessment.

Many evaluation frameworks have been developed to assess sustainability in different areas. However, a very helpful and recognized (Bartke & Schwarze, 2015) tool can be singled out the Bellagio Sustainability Assessment and Measuring Principles (STAMP). The Bellagio STAMP can be named as guidelines for the whole process of sustainability assessment and a tool to support sustainable decision making. The first version of these guidelines was developed in 1996 with the aim to provide guidance for sustainability assessment. In 2009, the principles were reviewed and updated by the Organisation for Economic Co-operation and Development and the International

Institute for Sustainable Development (Pinter et al., 2012). An evaluation performed following the Bellagio STAMP allows to ensure that the assessment performed reflects sustainability issues and is performed in sustainable way as much as possible. These guidelines provide recommendations for the whole process of the assessment, including definition of the main vision of the assessment, inclusion of all dimensions of sustainable development, determination of the evaluation scope, indicators selection process, transparency assurance of the whole assessment, engagement of stakeholders, results communication, and continuity assurance of the assessment. It is necessary to stress that all eight principles are interrelated and should be used jointly. The Bellagio STAMP principles and their interpretation are presented in Figure 6.9.

Selection of indicators is one of the most important steps in the process of sustainability assessment. The most popular way is to consider three dimensions of

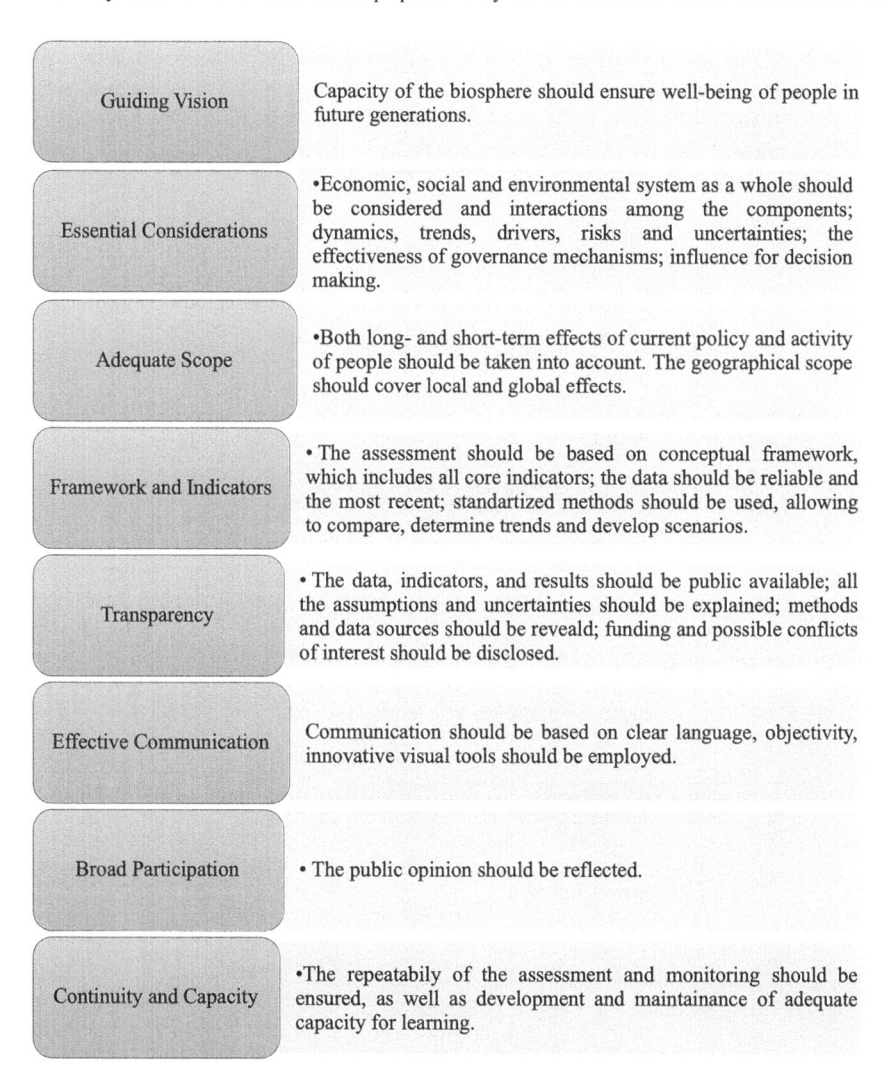

Guiding Vision	Capacity of the biosphere should ensure well-being of people in future generations.
Essential Considerations	•Economic, social and environmental system as a whole should be considered and interactions among the components; dynamics, trends, drivers, risks and uncertainties; the effectiveness of governance mechanisms; influence for decision making.
Adequate Scope	•Both long- and short-term effects of current policy and activity of people should be taken into account. The geographical scope should cover local and global effects.
Framework and Indicators	• The assessment should be based on conceptual framework, which includes all core indicators; the data should be reliable and the most recent; standartized methods should be used, allowing to compare, determine trends and develop scenarios.
Transparency	• The data, indicators, and results should be public available; all the assumptions and uncertainties should be explained; methods and data sources should be reveald; funding and possible conflicts of interest should be disclosed.
Effective Communication	Communication should be based on clear language, objectivity, innovative visual tools should be employed.
Broad Participation	• The public opinion should be reflected.
Continuity and Capacity	•The repeatabily of the assessment and monitoring should be ensured, as well as development and maintainance of adequate capacity for learning.

FIGURE 6.9 The Bellagio STAMP principles. (Created by authors.)

sustainability (economic, social, and environmental) and select indicators based on this traditional sustainability approach. However, depending on specific decision-making problem these dimensions can vary, reflecting different important aspect for each specific problem under analysis. It is necessary to highlight that selected indicators should not only reflect the main aspects of dimensions important and different preferences and needs of various stakeholders but also should be in line with principal requirements for indicators. The principal requirements for indicators set (Hirschberg et al., 2008) are presented in Figure 6.10.

Representativeness requires that indicators set should consider the main characteristics of the problem under analysis and all the indicators should be precise in their content and value. Comprehensibility and specificity principle require that there would be no double counting between indicators and they would reflect the problem as directly as possible. Sensitivity requires that indicators would be comparable and reflective to changes. Verifiability principle requires that indicators would be easily applicable, the data used would be public available, and methods used should be justified and recognized and would ensure repeatability of the study. The development of indicators set consists of four basic steps (Figure 6.11).

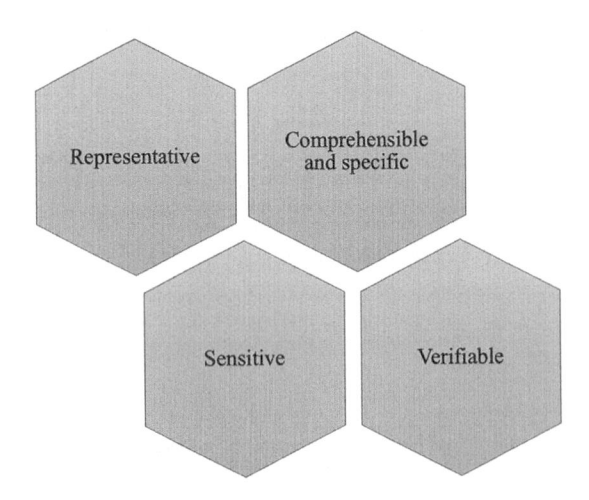

FIGURE 6.10 The principal requirements in selecting the indicators. (Created by authors.)

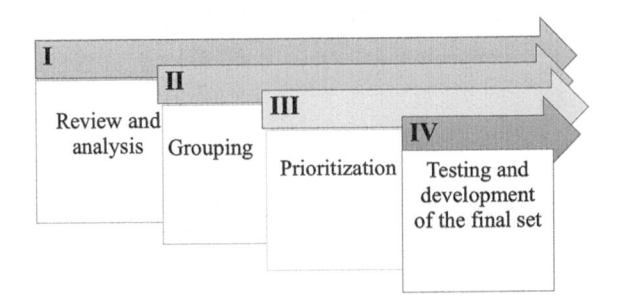

FIGURE 6.11 The main steps of indicators set development. (Created by authors.)

The first step is dedicated to the indicators review and analysis and their reflection of the sustainability dimensions. The second step is intended to divide indicators into categories. The third step aims to prioritize the indicators according the time horizon and the impact areas. Also, it is necessary to pay significant attention to aspects such as indicators comparability, adequacy, and availability. The fourth is dedicated to test, supplement, or correct the indicators set developed and includes tasks such as checking of reliability and double counting analysis.

The sustainability assessment guidelines sustainable agriculture studies in line with the Bellagio STAMP principles are presented in Table 6.3. Assessment in accordance with these guidelines will help ensure that the sustainability assessment reflects the concept of sustainable development and that the assessment process itself is carried out in accordance with recognized sustainable development recommendations.

By applying these Bellagio STAMP principles in the assessment of sustainable agriculture research, a comprehensive, transparent, and participatory assessment process can be established. This approach aligns assessment objectives with sustainable agriculture goals, considers essential aspects of sustainability, defines the appropriate scope, develops frameworks and indicators, promotes transparency and effective communication, engages diverse stakeholders, and ensures continuity and capacity for ongoing assessment activities.

TABLE 6.3
The Methodological Recommendations for Sustainable Agriculture Studies

The Bellagio STAMP	Implementation in Sustainable Agriculture Studies
Guiding vision	A clear vision and goals for the agricultural sustainability assessment must be established. The objectives of the assessment can be aligned with broader sustainable agriculture goals.
	The assessment should take into account the needs of future generations.
	The guiding vision provides a framework for the assessment and helps prioritize research efforts toward achieving sustainable agricultural outcomes.
Essential considerations	Critical aspects of sustainable agriculture in the assessment should be considered, covering fundamental dimensions of sustainability. Such aspects as: soil health, water management, biodiversity conservation, climate change resilience, and socio-economic impacts crucial for sustainable agriculture decision making.
	The dynamics of indicators and interactions between them, the current trends, and drivers should be considered.
	It will allow to ensure that the assessment covers the fundamental elements of sustainability and provides a comprehensive evaluation of agriculture and it practices.
Adequate scope	The appropriate scale and level of detail in the assessment should be determined, whether the assessment focuses on specific farming practices, regional agricultural systems, or global sustainability issues. The scope should be determined based on the research objectives and the relevance of the assessment findings to the target audience.
	Short- and long-term effects of the current agricultural performance should be considered.

(Continued)

TABLE 6.3 (*Continued*)
The Methodological Recommendations for Sustainable Agriculture Studies

The Bellagio STAMP	Implementation in Sustainable Agriculture Studies
Framework and indicators	The framework should outline the key components of the agricultural system under analysis. The fundamental requirements for indicators selection are presented in Figure 6.10. Indicators selected should measure and monitor the performance of these components, allowing for the evaluation of sustainable agriculture practices and the identification of areas for improvement. The agricultural sustainability assessment should be based on new and reliable data sources. Indicators selected should be standardized and comparable, methods should be scientific recognized in order to ensure repeatability and comparability of the research and its results.
Transparency	It is necessary to ensure that the methods, data sources, relevant literature and indicators used in the assessment are transparent and accessible. The weights of indicators should be justified. Transparent assessment processes foster trust, enable peer review, and allow for the replication and validation of findings.
Effective communication	The findings of the assessment should be presented for various stakeholders of agriculture. The results should be presented in a clear, objective, visual, concise, and accessible manner. Effective communication ensures that the findings are understood by policymakers, farmers, researchers, and other stakeholders, facilitating informed decision-making and action in agricultural sector.
Broad participation	Various stakeholders, such as farmers, policymakers, researchers, consumers, and local communities in the design, implementation, and interpretation of the assessment should be engaged. Broad participation ensures that the assessment incorporates different perspectives, local knowledge, and practical experiences, leading to more relevant and inclusive outcomes.
Continuity and capacity	It is necessary to ensure that resources and expertise will be available in the future in order to repeat the assessment and monitor the progress achieved. Capacity building initiatives can also be implemented to enhance the skills, knowledge, and capabilities of researchers, practitioners, and stakeholders involved in sustainable agriculture assessment.

Source: Created by authors.

6.4 THE ASSESSMENT OF SUCCESS OF LOW-CARBON TRANSITION IN AGRICULTURE: A CASE STUDY OF THE EU COUNTRIES

This sub-section presents an illustrative example of the assessment of success of low-carbon transition in agriculture of the EU countries with the application of MCDM technique and the fundamental requirements for sustainability assessment. The purpose of this illustrative case study is quite narrow, as it does not include a full assessment of the sustainability of the agricultural sector, but only the measurement and ranking of the achievements of low-carbon transition in the sector. Although the concept of sustainability varies in different fields based on the context, time, and decision-making problem, the properly selected instruments for the research allow to handle with complicated decision-making problems in logical and scientifically justified way.

6.4.1 SELECTION OF INDICATORS

In order to measure the achievements of each country made during the last years' countries achievements in the period of 2012–2020 were measured. Mostly, the last available data of the selected indicators are presented for 2020. While the first data provided by some countries is only from 2012. These conditions defined the research period.

Five indicators were selected for the comparative assessment of the success of the EU countries for low-carbon transition, which meet the principal requirements for indicators selection (Figure 6.10). The process of indicators selection was also aimed at meeting not only the public availability requirement but also the ease to finding the data, i.e., the all required data should be presented systematically in one database. This allows to ensure a simple process of monitoring and repeatability of the study in the future. The selected indicators provide insights into the efforts and progress made in transitioning the agricultural sector toward low-carbon practices and reducing its contribution to climate change. The indicators framework for comparative assessment of the EU countries is presented in Table 6.4.

Undoubtedly, one of the most important indicators is GHG emissions, which partially reflect the use of renewable energy in the agricultural sector and the application of sustainable farming practices, improvements in supply chain, the penetration of new efficient technologies, etc.

Agricultural activities affect the quality of air mainly via GHG and ammonia (NH3) emissions. Therefore, the second selected indicator shows ammonia emissions

TABLE 6.4
Indicators Framework for Comparative Assessment of the EU Countries

Indicators	Measurement	Target	Data Source
GHG emissions from agriculture	Percentage	Min	Eurostat
Ammonia emissions from agriculture	Kilograms per hectare	Min	Eurostat
Final consumption of energy in agriculture and forestry	Kilograms of oil equivalent per hectare of utilized agricultural area	Min	Eurostat
Organic farming are	Percent of total utilized agricultural area	Max	Eurostat
Support from the government for agricultural research and development	EUR per inhabitant	Max	Eurostat

Source: Created by authors.

from agriculture/kg per a hectare. Agriculture accounts for about 80%–90% of global ammonia emissions. The ammonia emissions from agricultural production include manure management, inorganic N-fertilizers, and animal manure applied to soil as well as urine and dung deposited by grazing animals. These emissions are associated with two types of environmental problems such as acidification and eutrophication.

The third indicator selected indicates the amount of energy used for agricultural purposes. As the majority of energy in all sectors is still produced using fossil fuels, growing energy consumption also has a significant impact on climate change.

Indicator reflecting area under organic farming includes all the elements characteristic of organic farming, including adoption of sustainable farming practices, environmental certification, and other measures more balanced with the environment than traditional farming.

It is scientifically and in real application recognized that support from the government for agricultural research and development is one of the main measures to promote low-carbon transition in agriculture. Therefore, this indicator is selected as reflecting government support to achieve the improvements toward low-carbon transition in the sector.

In this illustrative assessment, all indicators are given the same weight, but depending on the assessment objectives, the dimensions reflected and the specificity of the indicators, the weights of the indicators can be determined by various different methods. One of the most acceptable for such type of evaluation is the involvement of various stakeholders, which can be used both in the process of selecting, evaluating, and assigning weights to indicators.

6.4.2 GENERAL REGION OVERVIEW

The EU launched its common agricultural policy (CAP) as a partnership between players in agricultural sector and European society in 1962. The CAP determines the conditions allowing farmers to fulfil their functions in three main areas: food production; development of rural community; and environmentally sustainable farming. There are around 10 million farms in the EU offering safe, high-quality products at affordable prices. Around 22 million people are permanently employed in the agricultural sector in the EU. With a well-known food and culinary tradition, the EU is one of the largest producers and exporters of agri-food products and plays an important role in ensuring food security for the world's population.

The development of rural communities is very important for ensuring stable agricultural activity. Carrying out various tasks related to farming is integral to the smooth running of the farm. Together with the food sector, agriculture generates almost 40 million jobs in the region. Therefore, it is very important to ensure the efficiency and productivity of farms and to reflect recent challenges. For example, CAP funds for 2014–2020 were focused on improving high-speed technologies, Internet services, and infrastructure for 18 million rural residents, which represent a total of 6.4% of the rural population in the EU.

Environmentally friendly farming is one of the challenges for today's farmers, so regional policy is focused on providing multifaceted support to encourage farming practices friendly with the environment.

Assessment of Low-Carbon Transition Policy in Agriculture

The CAP seeks the several main tasks (European Commission, 2023), which can be grouped into a few categories:

- provide support for farmers and make improvements in agricultural productivity, ensuring a stable food supply and affordable prices;
- protect the EU farmers;
- support to tackle the issues of climate change and assist with the management of natural resources in a sustainable way;
- preserve the landscape and rural areas across the whole region;
- maintain economic vitality in rural areas.

The CAP is the main pillar through which climate change can be systematically tackled on a regional scale. In the 2014–2020 period, 104 billion EUR was allocated under the CAP financing measures, which accounted for 25% of the total funding, to address issues related to climate change such as reducing GHG emissions, increasing carbon sequestration, supporting farmers adapt to the recent environmental challenges. The 2023–2027 CAP further strengthened the focus on climate change mitigation with the ambition to contribute to the Green Deal's climate action goal of 2050 to decarbonize the sector across the region.

The selected indicators for the measurement of achievements regarding the low-carbon transition provide general overview and understanding of success of recent agricultural policy implemented in the whole region. Figure 6.12 shows GHG emissions from agriculture in the EU in the period of 2010–2021. Although the agricultural sector has significantly reduced GHG emissions over the past three decades, compared to other sectors, these achievements are not so great as to reduce the sector's share of total GHG emissions.

Ammonia emissions from agriculture in the period of 2010–2021 are provided in Figure 6.13. As can be seen in the figure, ammonia emissions for one hectare of utilized agricultural area in kilograms of oil equivalent have a trend to decrease. However, the decrease rate is very small and accounts for only 5% in the last ten years.

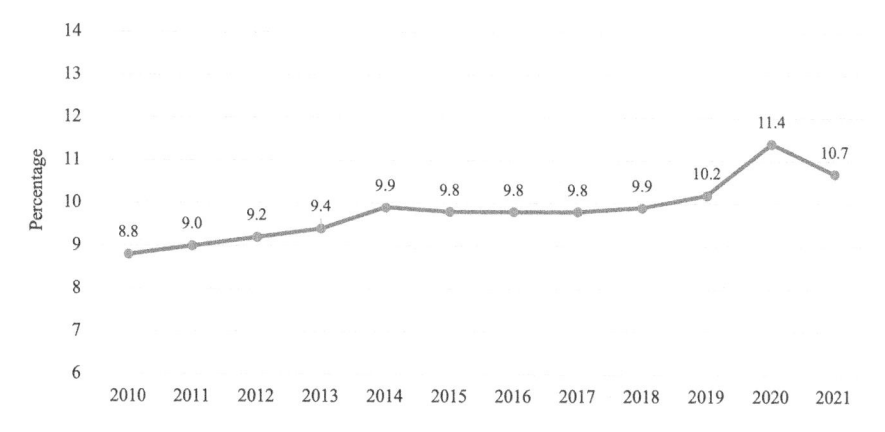

FIGURE 6.12 GHG emissions from agriculture in the EU, 2010–2021. (Prepared based on data from Eurostat, 2023).

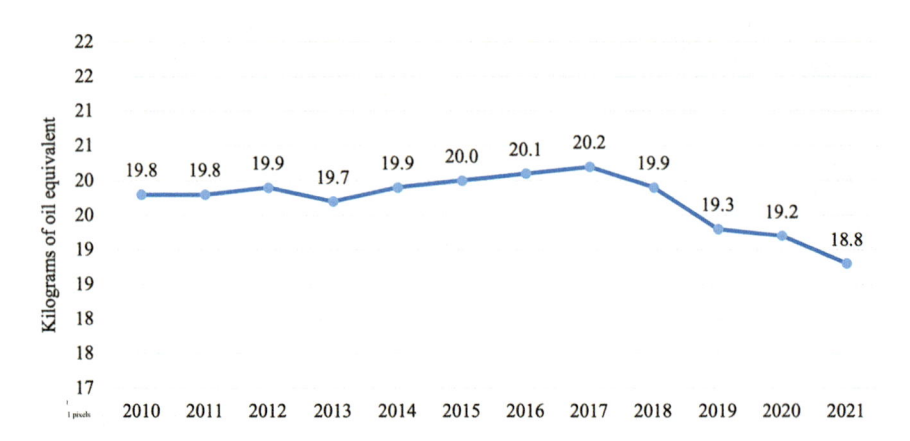

FIGURE 6.13 Ammonia emissions per hectare of utilized agricultural area in the EU, 2010–2021. (Prepared based on data from Eurostat, 2023.)

FIGURE 6.14 Final energy consumption by agriculture/forestry per hectare of utilized agricultural area in the EU, 2010–2021. (Prepared based on data from Eurostat, 2023.)

Figure 6.14 presents final energy consumption by agriculture and forestry per hectare of utilized agricultural area in the period of 2010–2021. Energy consumption in the agriculture and forestry sector has increased by more than 12% over the last decade. Although the statistics do not provide separate data on what share of the total energy consumed in agriculture is renewable energy, nevertheless, today fossil fuels make up the majority of the energy balance, so the growing energy consumption is a negative indicator when measuring the success of low-carbon transition in the region.

The share of organic farming in the period of 2000–2020 is provided in Figure 6.15. As it can be seen from the figure, there is a clear trend of increase during the last two decades. However, the European Green Deal target for organic farming

Assessment of Low-Carbon Transition Policy in Agriculture 297

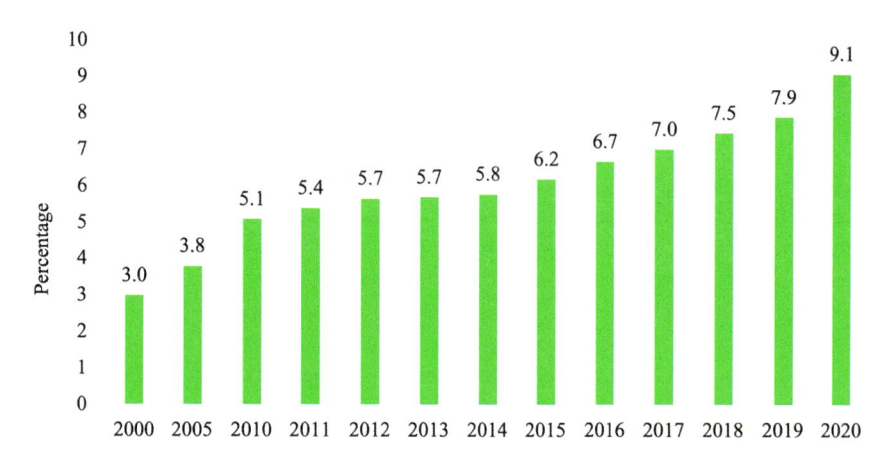

FIGURE 6.15 Share of organic farming of total utilized agricultural area in the EU, 2000–2020. (Prepared based on data from Eurostat, 2023.)

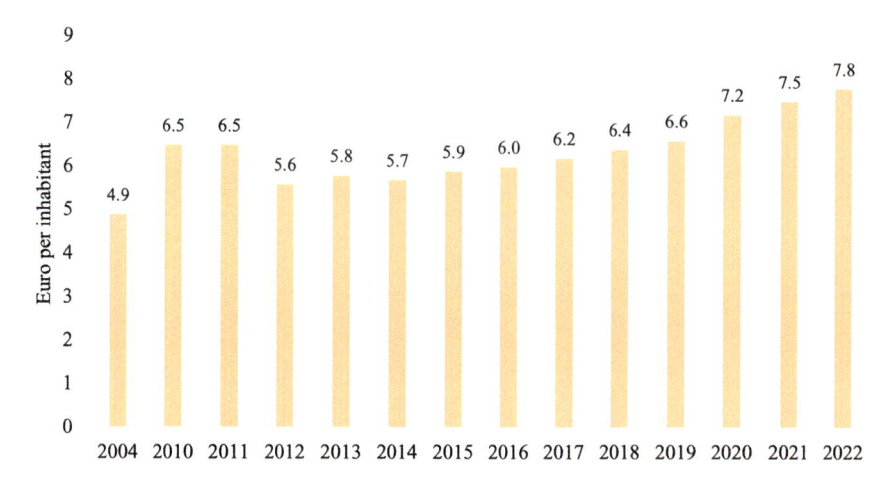

FIGURE 6.16 Support from the government to agricultural research and development in the EU, 2004–2022. (Source: prepared based on data from Eurostat, 2023.)

by 2030 is 25%. Today's achievements and growth rate are very small compared to the goal to be achieved in the next decade. Therefore, significant improvement in policy measures and adaptation of suitable financial mechanisms should be done to encourage organic farming in the region.

Support from the government to agricultural research and development in the EU in the period of 2004–2022 is presented in Figure 6.16. The data shows support in EUR per one inhabitant. Although steady growth is seen each year, it should be kept in mind that this data does not reflect the rate of inflation. Considering this, the support can be considered approximately constant.

6.4.3 EVALUATION TOOL

The measurement of success of low-carbon transition in agriculture and countries comparison is performed using the multi-criteria TOPSIS technique (Hwang & Yoon, 1981). The selected approach is one of the most widely applied in sustainability assessment and agriculture studies.

The TOPSIS technique relies on the ideal solution concept and can be described as having quite simple calculation process and possibility to quickly obtain the results of the assessment compared with other MCDM methods. Therefore, it is simple for the decision maker to understand the results and the significance of the indicators and their weights for the results and final rank. The clear and understandable calculation process is convenient to both scientists and policymakers. These features create favourable conditions for repeatability of the assessment and allow easy to track the progress made during the years. Calculation process by TOPSIS relies on the measurement of the distances to the most ideal and the worst ideal solutions. The following seven steps of calculation can be singled out:

Step I: The creation of a decision matrix with m alternatives and n criteria:

$$D = [x_{ij}] = \begin{array}{c} a_1 \\ a_2 \\ \ldots \\ a_m \end{array} \begin{bmatrix} x_1 & x_2 & \ldots & x_n \\ x_{11} & x_{12} & \ldots & x_{1n} \\ x_{21} & x_{22} & \ldots & x_{2n} \\ \ldots & \ldots & \ldots & \ldots \\ x_{m1} & x_{m2} & \ldots & x_{mn} \end{bmatrix} \quad (6.1)$$

Step II: Obtainment of the normalized matrix:

$$\overline{x}_{ij} = \frac{x_{ij}}{\sqrt{\sum_{i=1}^{m} x_{ij}^2}}; i = \overline{1,m},; j = \overline{1,n}, \quad (6.2)$$

Step III: Computation of the weighted normalized matrix:

$$V = \begin{bmatrix} w_1 r_{11} & w_2 r_{12} & \ldots & w_n r_{1n} \\ w_1 r_{21} & w_2 r_{22} & \ldots & w_n r_{2n} \\ \ldots & \ldots & \ldots & \ldots \\ w_1 r_{m1} & w_2 r_{m2} & \ldots & w_n r_{mn} \end{bmatrix}; \sum_{j=1}^{n} w_j = 1 \quad (6.3)$$

Step IV: Determination of the best ideal solution A⁺ and the worst ideal solution A⁻:

$$A^+ = \{\max_j v_{ij} \mid i \in I), (\min_j v_{ij} \mid i \in I^c), j = \overline{1,n},\} = \{v^+_1, v^+_2 \ldots v^+_n\}; \quad (6.4)$$

$$A^- = \{\min_j v_{ij} \mid i \in I), (\max_j v_{ij} \mid i \in I^c), j = \overline{1,n},\} = \{v^-_1, v^-_2 \ldots v^-_n\} \quad (6.5)$$

Step V: Computation of the relative distance of each solution from the most and the worst ideal:

$$S_i^+ = \sqrt{\sum_{j=1}^{n}\left(v_{ij} - v_j^+\right)^2}, j = \overline{1,m}, \tag{6.6}$$

$$S_i^- = \sqrt{\sum_{j=1}^{n}\left(v_{ij} - v_j^-\right)^2}, i = \overline{1,m}, \tag{6.7}$$

Step VI: **Computation of** the relative distance of each alternative to the ideal solution:

$$C_i = \frac{S_i^+}{S_i^+ + S_i^-}; \tag{6.8}$$

Step VII: The alternatives selected are ranked based on the computed values for each alternative, where the alternative with the best result has the highest C_i value.

6.4.4 Data for Comparative Assessment

Data for the comparative assessment of the EU countries achievements regarding low-carbon transition in the period of 2012–2020 are presented in Tables 6.5 and 6.6.

The analysis of the statistical data selected for evaluation showed that the values of several indicators of several countries differ strongly enough from the other countries under assessment. Malta can be singled out here, where ammonia emissions are significantly higher than those of other EU countries and account to 125.5 kg per hectare of utilized agricultural area in 2020. Meanwhile, the EU average is 19.2 in the same year. The Netherlands also stands out very strongly due to the amount of final energy consumed in the agricultural sector, which, e.g., in 2020 is 2161.9 kg of oil equivalent per hectare of utilized agricultural area. For comparison, the EU average in the same year is 175.7 kg of oil equivalent per hectare of utilized agricultural area. Countries such as Belgium and Malta are also very energy intensive, with 635.25 and 598.22 kg of oil equivalent per hectare of agricultural land, respectively. The share of organic farming also differs significantly among countries from 0.62% of total utilized agricultural area in Malta to 25.69% in Austria in 2020. All these differences have a decisive influence on the assessment of countries' achievements.

6.4.5 Results

The results of comparative assessment of the EU countries success in implementing low-carbon transition in agriculture in the period of 2012–2020 are presented in Table 6.7. In the period under analysis, the best-ranked countries are Estonia and Finland, which occupy the first and second places, respectively, and have not changed their positions. Meanwhile, the last positions are occupied by the same four countries throughout the entire period: Luxembourg, Belgium, Malta, and the Netherlands.

TABLE 6.5
Data for the Comparative Assessment, 2012

Countries	GHG	Ammonia	Final Energy	Organic Farming	Government Support
Belgium	7.7	50.9	570.51	4.48	3.60
Bulgaria	8.3	6.4	38.79	0.76	2.30
Czech Republic	5.6	17.5	159.81	13.29	3.90
Denmark	21.0	25.8	251.54	7.31	12.30
Germany	6.4	32.1	79.04	5.76	8.60
Estonia	7.0	10.3	115.12	14.86	12.10
Ireland	31.9	25.5	55.28	1.16	19.90
Greece	7.7	11.9	55.40	9.01	2.50
Spain	8.8	17.9	114.02	7.49	8.30
France	14.3	19.5	138.42	3.55	4.40
Croatia	11.6	21.9	155.78	2.40	0.50
Italy	6.6	28.4	209.22	9.30	4.50
Cyprus	5.8	48.2	318.48	3.38	10.50
Latvia	17.4	7.0	77.26	10.63	2.40
Lithuania	20.1	12.4	38.18	5.51	2.00
Luxembourg	4.9	41.1	190.10	3.14	0.90
Hungary	9.4	11.6	74.60	2.45	1.90
Malta	2.5	111.7	345.76	0.32	1.00
The Netherlands	8.6	59.6	2,010.60	2.61	8.90
Austria	8.8	20.9	183.02	18.62	4.60
Poland	8.0	19.0	252.35	4.51	1.70
Portugal	9.8	12.6	87.94	5.48	0.90
Romania	13.7	10.2	36.25	2.10	1.50
Slovenia	9.0	35.3	148.05	7.32	3.60
Slovakia	5.7	12.9	74.44	8.53	1.80
Finland	9.7	14.3	338.09	8.65	18.40
Sweden	11.2	14.9	197.68	15.76	6.10

Source: Eurostat (2023).

Compared to other countries, Estonia has relatively low ammonia emissions and a large share of organic farming. Meanwhile, Finland can be characterized as having low ammonia emissions from agriculture and providing a fairly high level of support for agricultural research and development.

The four bottom countries in the ranking (Luxembourg, Belgium, Malta, and the Netherlands) can be characterized by high ammonia emissions, high energy intensity, low share of organic farming, and low public funding for agricultural research and development activities (except the Netherlands). Fundamental changes are necessary for these countries in order to contribute significantly to the decarbonization of the agricultural sector.

There are many different policy measures developed to promote low-carbon transition in agriculture. However, not all of them are successfully implemented in

TABLE 6.6
Data for the Comparative Assessment, 2020

Countries	GHG	Ammonia	Final Energy	Organic Farming	Government Support
Belgium	8.6	45.5	635.25	7.25	5.10
Bulgaria	12.3	7.5	37.27	2.30	3.70
Czech Republic	6.8	17.6	181.01	15.33	5.50
Denmark	28.3	25.9	216.78	11.45	15.50
Germany	7.7	26.3	216.37	9.59	12.40
Estonia	13.7	9.0	110.98	22.41	7.50
Ireland	36.7	27.2	51.16	1.66	18.70
Greece	10.6	11.0	52.64	10.15	4.10
Spain	12.4	19.3	111.12	9.98	10.00
France	16.8	18.2	151.47	8.71	5.50
Croatia	11.2	17.3	155.15	7.21	2.70
Italy	8.6	25.2	210.25	15.96	5.00
Cyprus	6.8	43.3	354.83	4.63	8.00
Latvia	21.1	6.9	104.13	14.79	5.70
Lithuania	22.1	12.8	38.64	8.00	3.60
Luxembourg	6.6	43.1	212.71	4.63	1.10
Hungary	11.3	14.1	140.75	6.03	5.30
Malta	3.9	125.5	598.22	0.62	1.60
The Netherlands	10.7	58.5	2 161.90	3.95	12.60
Austria	9.6	23.1	195.38	25.69	4.40
Poland	9.1	20.3	262.14	3.45	3.80
Portugal	12.2	13.2	103.62	8.05	1.70
Romania	16.9	10.6	40.67	3.59	1.00
Slovenia	11.3	35.4	148.61	10.29	6.00
Slovakia	6.8	12.5	69.00	11.67	2.90
Finland	13.2	12.2	330.15	13.93	10.90
Sweden	14.4	15.0	196.09	20.31	4.80

Source: Eurostat (2023).

the EU region. The EU has set ambitious goals for climate change and agriculture, and in order to implement them, it is necessary to apply the most effective policy measures, as very large improvements in the sector are required in a fairly short period of time.

Recently, there have been frequent discussions among academics and policymakers on whether carbon pricing should be applied to the agricultural sector under different emissions trading systems. Significant increase in interest to examine the potential impact of carbon pricing on EU agriculture can be found in the scientific literature as well. For example, Isbasoiu et al. (2021) examined the impact of carbon pricing as a measure to mitigate agricultural GHG emissions on calorie production in the EU. The results of the simulations show that a moderate increase in the carbon price would lead total areas and outputs of crops increase and decrease in animal

TABLE 6.7
Comparative Assessment of the EU Countries Achievements in the Period of 2012–2020

Countries	2012 Ci	2012 Rank	2020 Ci	2020 Rank	Change ΔCi	Change ΔRank
Estonia	0.835	1	0.786	1	−0.050	0
Finland	0.786	2	0.766	2	−0.021	0
Sweden	0.736	3	0.735	7	−0.001	−4
Austria	0.729	4	0.745	5	0.016	−1
Spain	0.724	5	0.749	4	0.025	1
Czech Republic	0.717	6	0.741	6	0.024	0
Germany	0.705	7	0.759	3	0.054	−4
Greece	0.689	8	0.710	11	0.021	−3
Italy	0.684	9	0.724	8	0.040	1
Slovakia	0.682	10	0.709	12	0.028	−2
Denmark	0.681	11	0.711	10	0.030	1
Latvia	0.670	12	0.717	9	0.047	3
Slovenia	0.656	13	0.694	13	0.039	0
Cyprus	0.653	14	0.656	20	0.003	−6
Portugal	0.648	15	0.672	16	0.023	−1
Bulgaria	0.643	16	0.669	18	0.027	−2
France	0.642	17	0.689	15	0.047	2
Hungary	0.641	18	0.691	14	0.049	4
Ireland	0.636	19	0.648	22	0.013	−3
Poland	0.633	20	0.654	21	0.021	−1
Lithuania	0.631	21	0.668	19	0.037	2
Romania	0.630	22	0.645	23	0.014	−1
Croatia	0.610	23	0.671	17	0.061	6
Luxembourg	0.608	24	0.621	24	0.013	0
Belgium	0.564	25	0.602	25	0.038	0
Malta	0.494	26	0.453	26	−0.040	0
The Netherlands	0.335	27	0.372	27	0.036	0

production. The projections showed that the current net calorie production can be more than twice higher and simultaneously by 10%–15% lower level of GHG emissions. Stepanyan et al. (2023) quantitatively examined the possible impact of carbon pricing on the environment and the economy on agricultural sector of Germany and the whole EU. The results of the projections showed that carbon pricing can contribute to net agricultural emissions reduction significantly in the region. The authors also stressed the significance to stimulate the usage and transferability of the technological achievements not only to mitigate against climate emissions but also to reduce the emissions leakage to third countries. Therefore, the agricultural sector has a huge potential to contribute achieving national, regional, and global targets of climate change.

Every country and region's contribution to reducing emissions is very important. But it is also important to stress that the fight against climate change should take place on a global scale. For example, according to the projections made by Jansson et al. (2023) 12 EUR global carbon tax per ton CO2-eq could have more than 20 times bigger effect on global reduction of GHG emissions in agriculture than 120 EUR carbon tax in the entire EU. Where 12 EUR global carbon tax could contribute to 3.2% of global emissions cut and ten times higher tax in the EU (120 EUR) could reduce global GHG emissions only 0.15%.

6.5 CONCLUSIONS

Today sustainable agriculture development faces many challenges related with low-carbon energy transition. The high initial cost, an intermittent energy supply, supply chain issues, and people willingness for changes play a critical role in low-carbon energy transition in agriculture. Although the cost of low-carbon energy transition is one of a major challenge to the agriculture sector, such investments are also an opportunity to get real benefits in competitiveness because there is an increasingly trend in the market to seek sustainable approaches in food production.

It is necessary to highlight the importance of the role of the government (regional, central, and local). Such measures as investments in clean energy sources or tax breaks for farmers and agriculture sector companies can build up new opportunities for sustainable growth. Therefore, low-carbon energy transition should be done urgent and organized manner with the application of incentive-based policy.

The case study of sustainability assessment in EU agriculture based on in implementing low-carbon transition aims in agriculture revealed that during 2012–2020 period, the best-ranked countries are Estonia and Finland, which occupy the first and second places, respectively, and have not changed their positions. Meanwhile, the last positions are occupied by the same four countries throughout the entire period: Luxembourg, Belgium, Malta, and the Netherlands.

Estonia had relatively low ammonia emissions and a large share of organic farming. Meanwhile, Finland had low ammonia emissions from agriculture and provided a fairly high level of support for agricultural research and development.

The four bottom countries in the ranking based on sustainability of agriculture (Luxembourg, Belgium, Malta, and the Netherlands) had high ammonia emissions, high energy intensity, low share of organic farming, and low public funding for agricultural research and development activities (except the Netherlands). Fundamental changes are necessary for these countries in order to contribute significantly to the decarbonization and sustainable development of the agricultural sector.

REFERENCES

Abdallah, S. B., Elfkih, S., & Parra-López, C. (2018). A sustainability comparative assessment of Tunisian organic and conventional olive growing systems based on the AHP methodology. *New Medit*, *17*, 51–68.

Afsar, N., & Idrees, M. (2019). Farmer's perception of agricultural extension services in disseminating climate change knowledge. *Sarhad Journal of Agriculture*, *35*(3), 942–947.

Aleksiejuk-Gawron, J., Mikiuviene, S., Kirsiene, J., Doheijo, E., Garzon, D., Urbonas, R., & Milcius, D. (2020). Net-metering compared to battery-based electricity storage in a single-case PV application study considering the Lithuanian context. *Energies*, *13*(9), 2286.

Alig, R., Latta, G., Adams, D., & McCarl, B. (2010). Mitigating greenhouse gases: The importance of land base interactions between forests, agriculture, and residential development in the face of changes in bioenergy and carbon prices. *Forest Policy and Economics*, *12*(1), 67–75.

Alkaabneh, F. M., Lee, J., Gomez, M. I., & Gao, H. O. (2021). A systems approach to carbon policy for fruit supply chains: Carbon tax, technology innovation, or land sparing? *Science of the Total Environment*, *767*, 144211.

Antwi-Agyei, P., Dougill, A. J., Stringer, L. C., Codjoe, S. N. A. (2018). Adaptation opportunities and maladaptive outcomes in climate vulnerability hotspots of northern Ghana. *Climate Risk Management*, *19*, 83–93.

Antwi-Agyei, P., & Stringer, L. C. (2021). Improving the effectiveness of agricultural extension services in supporting farmers to adapt to climate change: Insights from northeaster Ghana. *Climate Risk Management*, *32*, 100304.

Aouadi, N., Macary, F., & Ugaglia, A. A. (2020). Multi-criteria assessment of socioeconomic and environmental performance of wine-growing systems and agroecological transition scenarios. *Cahiers Agricultures*, *29*, 19.

Arondel, C., & Girardin, P. (2000). Sorting cropping systems on the basis of their impact on groundwater quality. *European Journal of Operational Research*, *127*(3), 467–482.

Arulnathan, V., Heidari, M. D., Doyon, M., Li, E., & Pelletier, N. (2020). Farm-level decision support tools: A review of methodological choices and their consistency with principles of sustainability assessment. *Journal of Cleaner Production*, *256*, 120410.

Avikal, S., Pant, R., Barthwal, A., Ram, M., & Upadhyay, R. K. (2022). Factors implementing sustainable circular economy in agro-produce supply chain: DEMATEL-DANP-based approach. *Management of Environmental Quality*, *34*(4), 1158–1173.

Babalola, O. O., & Oladele, O. I. (2011). Biotechnology in agriculture: Implications for agricultural extension and advisory services. *Journal of Food, Agriculture & Environment*, *9*(3–4), 486–491.

Balakrishnan, P., Shabbir, M. S., Siddiqi, A. F., & Wang, X. (2020). Current status and future prospects of renewable energy: A case study. *Energy Sources, Part A: Recovery, Utilization, and Environmental Effects*, *42*(21), 2698–2703.

Baldos, U. L. C., Fuglie, K. O., & Hertel, T. W. (2020). The research cost of adapting agriculture to climate change: A global analysis to 2050. *Agricultural Economics*, *51*(2), 207–220.

Barbosa, M., Pinheiro, E., Sokulski, C. C., Ramos Huarachi, D. A., & de Francisco, A. C. (2022). How to identify barriers to the adoption of sustainable agriculture? A study based on a multi-criteria model. *Sustainability*, *14*, 13277.

Bartke, S., & Schwarze, R. (2015). No perfect tools: Trade-offs of sustainability principles and user requirements in designing support tools for land-use decisions between greenfields and brownfields. *Journal of Environmental Management*, *153*, 11–24.

Baumann, M., Weil, M., Peters, J. F., Chibeles-Martins, N., & Moniz, A. B. (2019). A review of multi-criteria decision making approaches for evaluating energy storage systems for grid applications. *Renewable and Sustainable Energy Reviews*, *107*, 516–534.

Bohra, S. S., & Anvari-Moghaddam, A. (2022). A comprehensive review on applications of multicriteria decision-making methods in power and energy systems. *International Journal of Energy Research*, *46*(4), 4088–4118.

Cardoso, T. F., Watanabe, M. D. B., Souza, A., Chagas, M. F., Cavalett, O., Morais, E. R., Nogueira, L. A. H., Leal, M. R. L. V., Braunbeck, O. A.., Cortez, L. A. B., & Bonomi, A. (2018). Economic, environmental, and social impacts of different sugarcane production systems. *Biofuels Bioproducts & Biorefining-Biofpr*, *12*(1), 68–82.

Chang, Y. H., & Liang, Y. (2023). Intelligent risk assessment of ecological agriculture projects from a vision of low carbon. *Sustainability, 15*(7), 5765.

Charnes, A., Cooper, W. W., & Rhodes, E. (1978). Measuring the efficiency of decision making units. *European Journal of Operational Research, 2*(6), 429–444.

Cicciu, B., Schramm, F., & Schramm, V. B. (2022). Multi-criteria decision making/aid methods for assessing agricultural sustainability: A literature review. *Environmental Science and Policy, 138*, 85–96.

Coulibaly, T. P., Du, J., & Diakité, D. (2021). Sustainable agricultural practices adoption. *Agriculture (Poľnohospodárstvo), 67*(4), 166–176.

De Boeck, L., Van Asch, S., De Bruecker, P. & Audenaert, A. (2016). Comparison of support policies for residential photovoltaic systems in the major EU markets through investment profitability. *Renewable Energy, 87*, 42–53.

Deepa, N., Durai Raj, V. P. M., Senthil Kumar, N., Srinivasan, K., Chang, C.-Y., & Bashir, A. K. (2019). An efficient ensemble VTOPES multi-criteria decision-making model for sustainable sugarcane farms. *Sustainability, 11*, 4288.

Di Bene, C.; Gomez-Lopez, MD.; Francaviglia, R.; Farina, R.; Blasi, E.; Martinez-Granados, D., & Calatrava, J. (2022). Barriers and Opportunities for Sustainable Farming Practices and Crop Diversification Strategies in Mediterranean Cereal-Based Systems. *Frontiers in Environmental Science, 10*, 861225.

Dixit, A., Suvadarshini, P., & Pagare, D. V. (2022). Analysis of barriers to organic farming adoption in developing countries: A grey-DEMATEL and ISM approach. *Journal of Agribusiness in Developing and Emerging Economies.* https://doi.org/10.1108/JADEE-06-2022-0111

Ekinci, E., Mangla, S. K., Kazancoglu, Y., Sarma, P. R. S., Sezer, M. D., & Ozbiltekin-Pala, M. (2022). Resilience and complexity measurement for energy efficient global supply chains in disruptive events. *Technological Forecasting and Social Change, 179*, 121634.

Emenike, S. N., & Falcone, G. (2020). A review on energy supply chain resilience through optimization. *Renewable and Sustainable Energy Reviews, 134*, 110088.

European Commission. (2023). Agriculture and rural development. Available at: https://agriculture.ec.europa.eu/common-agricultural-policy/cap-overview/cap-glance_en

Eurostat. (2023). Available at: https://ec.europa.eu/eurostat/data/database

FAO, IFAD, UNICEF, WFP, & WHO. (2022). *The State of Food Security and Nutrition in the World 2022. Repurposing food and agricultural policies to make healthy diets more affordable*. Rome: FAO. https://doi.org/10.4060/cc0639en

Fishburn, P. C. (1967). Methods of estimating additive utilities. *Management Science, 13*, 435–453.

Fu, L. L., Mao, X. B., Mao, X. H., & Wang, J. (2022). Evaluation of agricultural sustainable development based on resource use efficiency: Empirical evidence from Zhejiang Province, China. *Frontiers in Environmental Science, 10*, 860481.

Fusco, G., Campobasso, F., Laureti, L., Frittelli, M., Valente, D., & Petrosillo, I. (2023). The environmental impact of agriculture: An instrument to support public policy. *Ecological Indicators, 14*, 109961.

Gabus, A., & Fontela, E. (1972). *World problems, an invitation to further thought within the framework of DEMATEL* (pp. 1–8). Geneva: Battelle Geneva Research Centre.

Ge, J. Q., Sutherland, L. A., Polhill, J. G., Matthews, K., & Miller, D., Wardell-Johnson, D. (2017). Exploring factors affecting on-farm renewable energy adoption in Scotland using large-scale microdata. *Energy Policy, 107*, 548–560.

Gerdessen, J. C., & Pascucci, S. (2013). Data envelopment analysis of sustainability indicators of European agricultural systems at regional level. *Agricultural Systems, 118*, 78–90.

Goldthau, A., & Hughes, L. (2020). Protect global supply chains for low-carbon technologies. *Nature, 585*(7823), 28–30.

Gorjian, S., Fakhraei, O., Gorjian, A., Sharafkhani, A., & Aziznejad, A. (2022). Sustainable food and agriculture: Employment of renewable energy technologies. *Current Robotics Reports, 3*, 153–163.

Gorjian, S., Ebadi, H., Trommsdorff, M., Sharon, H., Demant, M., & Schindele, S. (2021). The advent of modern solar-powered electric agricultural machinery: A solution for sustainable farm operations. *Journal of Cleaner Production, 292*, 126030.

Guitouni, A., & Martel, J.-M. (1998). Tentative guidelines to help choosing an appropriate MCDA method. *European Journal of Operational Research, 109*(2), 501–521.

Halla, P., & Binder, C. R. (2020). Sustainability assessment: Introduction and framework. In: C. R. Binder, R. Wyss & E. Massaro. *Sustainability assessment of urban systems* (pp. 7–29). Cambridge: Cambridge University Press. https://doi.org/10.1017/9781108574334

Hirschberg S., Bauer B., Burgherr P., Dones R., Schenler W., Bachmann T., Gallego-Carrera D. Final set of sustainability criteria and indicators for assessment of electricity supply options. NEEDS Deliverable D3.2 - RS2b. 2008, p. 36. Available at: https://www.psi.ch/sites/default/files/import/ta/NeedsEN/RS2bD3.2.pdf

Hitaj, C., & Loschel, A. (2019). The impact of a feed-in tariff on wind power development in Germany. *Resource and Energy Economics, 57*, 18–35.

Hwang, C. L., & Yoon, K. (1981). *Multiple attributes decision making methods and applications* (pp. 22–51). Berlin: Springer.

Hyland, J. J., Heanue, K., McKillop, J., & Micha, E. (2018). Factors underlying farmers' intentions to adopt best practices: The case of paddock based grazing systems. *Agricultural Systems, 162*, 97–106.

Indraningsih, K. S., Ashari, A., Syahyuti, S., Anugrah, I. S., Suharyono, S., Saptana, S., Iswariyadi, A., Agustian, A., Purwantini, T. B., Ariani, M., & Mardiharini, M. (2023). Factors influencing the role and performance of independent agricultural extension workers in supporting agricultural extension. *Open Agriculture, 8*(1), 20220164.

Isbasoiu, A., Jayet, P. A., & De Cara, S. (2021). Increasing food production and mitigating agricultural greenhouse gas emissions in the European Union: Impacts of carbon pricing and calorie production targeting. *Environmental Economics and Policy Studies, 23*, 409–440.

Jain, M., Soni, G., Verma, D., Baraiya, R., & Ramtiyal, B. (2023). Selection of technology acceptance model for adoption of Industry 4.0 technologies in agri-fresh supply chain. *Sustainability, 15*(6), 4821.

Jansson, T., Malmstrom, N., Johansson, H., & Choi, H. (2023). Carbon taxes and agriculture: the benefit of a multilateral agreement. *Climate Policy*. https://doi.org/10.1080/14693062.2023.2171355

Jeong, J. S. (2018). Design of spatial PGIS-MCDA-based land assessment planning for identifying sustainable land-use adaptation priorities for climate change impacts. *Agricultural Systems, 167*, 61–71.

Jiang, L. L., Huang, H. Q., He, S. R., Huang, H. Y., & Luo, Y. (2022). What motivates farmers to adopt low-carbon agricultural technologies? Empirical evidence from thousands of rice farmers in Hubei province, central China. *Frontiers in Psychology, 13*, 983597.

Kazemi, F., & Hosseinpour, N. (2022). GIS-based land-use suitability analysis for urban agriculture development based on pollution distributions. *Land Use Policy, 123*, 106426.

Kersuliene, V., Zavadskas, E. K., & Turskis, Z. (2010). Selection of rational dispute resolution method by applying new step-wise weight assessment ratio analysis (Swara). *Journal of Business Economics and Management, 11*(2), 243–258.

Keshavarz-Ghorabaee, M., Zavadskas, E. K., Olfat, L., & Turskis, Z. (2015). Multi-criteria inventory classification using a new method of evaluation based on distance from average solution (EDAS). *Informatica, 26*(3), 435–451.

Keshavarz-Ghorabaee, M., Amiri, M., Zavadskas, E. K., Turskis, Z., & Antuchevičienė, J. (2022). MCDM approaches for evaluating urban and public transportation systems: A short review of recent studies. *Transport, 37*(6), 411–425.

Khan, F., & Ali, Y. (2022). Moving towards a sustainable circular bio-economy in the agriculture sector of a developing country. *Ecological Economics, 196*, 107402.

Kumar, A., Mangla, S. K., & Kumar, P. (2022). Barriers for adoption of Industry 4.0 in sustainable food supply chain: A circular economy perspective. *International Journal of Productivity and Performance Management.* https://doi.org/10.1108/IJPPM-12-2020-0695.

Lang, T., Ammann, D., & Girod, B. (2016). Profitability in absence of subsidies: A techno-economic analysis of rooftop photovoltaic self-consumption in residential and commercial buildings. *Renewable Energy, 87*, 77–87.

Lehtonen, H., Huan-Niemi, E., & Niemi, J. (2022). The transition of agriculture to low carbon pathways with regional distributive impacts. *Environmental Innovation and Societal Transitions, 44*, 1–13.

Liu, F., Luo, M. C., Zhang, Y. Y., Zhou, S. L., Wu, X., Lin, A. Y., Guo, Y. X., & Liu, C. H. (2022). Data-driven evaluation of regional agricultural production efficiency for sustainable development. *Journal of Intelligent & Fuzzy Systems, 43*(6), 7765–7778.

Liu, Y., Ruiz-Menjivar, J., Zhang, L., Zhang, J. B., & Swisher, M. E. (2019). Technical training and rice farmers' adoption of low-carbon management practices: The case of soil testing and formulated fertilization technologies in Hubei, China. *Journal of Cleaner Production, 226*, 454–462.

Londo, M., Matton, R., Usmani, O., van Klaveren, M., Tigchelaar, C., & Brunsting, S. (2020). Alternatives for current net metering policy for solar PV in the Netherlands: A comparison of impacts on business case and purchasing behaviour of private homeowners, and on governmental costs. *Renewable Energy, 147*, 903–915.

MacCrimon, K. R. (1968). *Decision marking among multiple-attribute alternatives: a survey and consolidated approach.* RAND memorandum, RM-4823-ARPA (63 p). Santa Monica, CA: The Rand Corporation.

Majewski, S., Mentel, G., Dylewski, M., & Salahodjaev, R. (2023). Renewable energy, agriculture and CO2 emissions: Empirical evidence from the middle-income countries. *Frontiers in Energy Research, 10*, 921166.

Mangla, S. K., Luthra, S., Rich, N., Kumar, D., Rana, N. P., & Dwivedi, Y. K. (2018). Enablers to implement sustainable initiatives in agri-food supply chains. *International Journal of Production Economics, 203*, 379–393.

Mardones, C., & Lipski, M. (2020). A carbon tax on agriculture? A CGE analysis for Chile. A carbon tax on agriculture? *Economic Systems Research, 32*(2), 262–277.

Mareschal, B., & Brans, J. P. (1992). *PROMETHEE V: MCDM problems with segmentation constrains* (pp. 13–30). Brussels: Universite Libre de Brusells.

Maydana, G., Romagnoli, M., Cunha, M., & Portapila, M. (2020). Integrated valuation of alternative land use scenarios in the agricultural ecosystem of a watershed with limited available data, in the Pampas region of Argentina. *Science of the Total Environment, 714*, 136430.

Mendas, A., & Delali, A. (2012). Integration of multicriteria decision analysis in GIS to develop land suitability for agriculture: Application to durum wheat cultivation in the region of Mleta in Algeria. *Computers and Electronics in Agriculture, 83*, 117–126.

Mendas, A., Delali, A., Khalfallah, M., Likou, L., Gacemi, M. A., Boukrentach, H., Djilali, A., & Mahmoudi, R. (2014). Improvement of land suitability assessment for agriculture-application in Algeria. *Arabian Journal of Geosciences, 7*(2), 435–445.

Millar, R., Fuglestvedt, J., Friedlingstein, P., Rogelj, J., Grubb, M. J., Matthews, H. D., Skeie, R. B., Forster, P. M., Frame, D. J., & Allen, M. R. (2017). Emission budgets and pathways consistent with limiting warming to 1.5°C. *Nature Geoscience, 10*, 741–747.

Mng'ong'o, B. G., Rad, D., Koloseni, D., & Balas, V. E. (2022). Application of fuzzy TOPSIS in assessing performance of agro-processed organic crops: A case of Sustainable Agriculture Tanzania (SAT) in Morogoro. *Journal of Intelligent & Fuzzy Systems, 43*(2), 1811–1826.

Morkunas, M., & Volkov, A. (2023). The progress of the development of a climate-smart agriculture in Europe: Is there cohesion in the European Union? *Environmental Management, 71*, 1111–1127.

Moulogianni, C., & Bournaris, T. (2017). Biomass production from crops residues: Ranking of Agro-Energy regions. *Energies, 10*(7), 1061.

Mu, W., Kanellopoulos, A., van Middelaar, C. E., Stilmant, D., & Bloemhof, J. M. (2018). Assessing the impact of uncertainty on benchmarking the eco-efficiency of dairy farming using fuzzy data envelopment analysis. *Journal of Cleaner Production, 189*, 709–717.

Nelson, G., Bogard, J., Lividini, K., Arsenault, J., Riley, M., Sulser, T. B., & M. Rosegrant. (2018). Income growth and climate change effects on global nutrition security to mid-century. *Nature Sustainability, 1*(12), 773–781.

Olguín, P., De Kartzow, A., & Huenchuleo, C., (2019). Sustainability of quinoa in rainfed agricultural systems: a case study on the O'Higgins region Chile. *Ciencia E Investigacion Agraria, 46*, 197–207.

Olveira, M. W., Agostinho, F., Almeida, C. M. V. B., & Giannetti, B. F. (2016). Sustainable milk production: Application of the hierarchical analytical process towards a regional strategic planning. *Journal of Environmental Accounting and Management, 4*, 385–398.

Opricovic, S. (1998). *Multicriteria optimization of civil engineering systems*. PhD thesis, Faculty of Civil Engineering, Belgrade, 302 p.

Palma, J., Graves, A. R., Burgess, P. J., van der Werf, W., & Herzog, F. (2007). Integrating environmental and economic performance to assess modern silvoarable agroforestry in Europe. *Ecological Economics, 63*(4), 759–767.

Parra-López, C., Calatrava-Requena, J., & de-Haro-Giménez, T. (2008). A systemic comparative assessment of the multifunctional performance of alternative olive systems in Spain within an AHP-extended framework. *Ecological Economics, 64*, 820–834.

Pascaris, A. S. (2021). Examining existing policy to inform a comprehensive legal framework for agrivoltaics in the U.S. *Energy Policy, 159*, 112620.

Picazo-Tadeo, A.J., Gómez-Limón, J.A., Reig-Martínez, E. (2011). Assessing farming ecoefficiency: a data envelopment analysis approach. *Journal of Environmental Management, 92*, 1154–1164.

Pinter, L., Hardi, P., Martinuzzi, A., Hall, J. (2012). Bellagio STAMP: principles for sustainability assessment and measurement. *Ecological Indicators, 17*, 20–28.

Prol, J. L., & Steininger, K. W. (2020). Photovoltaic self-consumption is now profitable in Spain: Effects of the new regulation on prosumers' internal rate of return. *Energy Policy, 146*, 111793.

Rahman, M. M., Khan, I., Field, D. L., Techato, K., & Alameh, K. (2022). Powering agriculture: Present status, future potential, and challenges of renewable energy applications. *Renewable Energy, 188*, 731–749.

Rahut, D., Narmandakh, D., He, X., Yao, Y., Tian, S. (2023). Opportunities and challenges during low-carbon transition: A perspective from Asia's agriculture sector, 33 p. Available at: https://www.adb.org/sites/default/files/institutional-document/874256/adotr2023bp-low-carbon-asia-agriculture.pdf

Renwick, A., Dynes, R., Johnstone, P., King, W., Holt, L., & Penelope, J. (2019). Challenges and opportunities for land use transformation: Insights from the central plains water scheme in New Zealand. *Sustainability, 11* (18), 4912.

Rezaei-Moghaddam, K., & Karami, E. (2008). A multiple criteria evaluation of sustainable agricultural development models using AHP. *Environment, Development and Sustainability, 10*(4), 407–426.

Rickards, L., Alexandra, J., Jolley, C., & Frewer, T. (2018). *Final report: Review of agricultural extension* (53 p.). Canberra: Australian Centre for International Agricultural Research (ACIAR), ISBN 978-1-925747-64-5. Available at: https://www.aciar.gov.au/project/asem-2016-047 (Accessed 25 July 2023).

Rodrigues, S., Torabikalaki, R., Faria, F., Cafôfo, N., Chen, X., Ivaki, A. R., Mata-Lima, H., & Morgado-Dias, F. (2016). Economic feasibility analysis of small scale PV systems in different countries. *Solar Energy, 131*, 81–95.

Rodríguez Sousa, A. A., Parra-López, C., Sayadi-Gmada, S., Barandica, J. M., & Rescia, A. J. (2020). A multifunctional assessment of integrated and ecological farming in olive agroecosystems in southwestern Spain using the analytic hierarchy process. *Ecological Economics, 173*, 106658.

Rosegrant, M. W., Sulser, T. B., & Wiebe, K. (2022). Global investment gap in agricultural research and innovation to meet sustainable development goals for hunger and Paris Agreement climate change mitigation. *Frontiers in Sustainable Food Systems, 6*, 965767.

Roux, A., & Shanker, A. (2018). *Net metering and PV self-consumption in emerging countries.* IEA Photovoltaic Power System Program, International Energy Agency, Report IEA-PVPS T9-18, 56 p., ISBN: 978-3-906042-76-3.

Roy, B. (1978). ELECTRE III: un algorithme de classements fonde sur une representation floue des preference en presence de criteres multiples. *Cahiers de CERO, 20*, 3–24.

Saaty, T. L. (1980). *The analytic hierarchy process.* New York, NY: McGraw-Hill, pp. 11–29.

Sajadian, M., Khoshbakht, K., Liaghati, H., Veisi, H., & Mahdavi Damghani, A. (2017). Developing and quantifying indicators of organic farming using analytic hierarchy process. *Ecological Indicators, 83*, 103–111.

Saman, C. (2021). Does agricultural subsidies in the EU improved environmental outcomes? *Transformations in Business & Economics, 20*(3C(54C)), 642–652.

Santoso, I., & Darsono, S. (2019). Review of criteria on multi criteria decision making (MCDM) construction of dams. *International Journal of Geomate, 16*(55), 184–194.

Scarpellini, S., Gimeno, J. Á., Portillo-Tarragona, P., & Llera-Sastresa, E. (2021). Financial resources for the investments in renewable self-consumption in a circular economy framework. *Sustainability, 13*, 6838.

Shaw-Williams, D., & Susilawati, C. (2020). A techno-economic evaluation of virtual net metering for the Australian community housing sector. *Applied Energy, 261*, 114271.

Siksnelyte-Butkiene, I., Streimikiene, D., Balezentis, T., & Volkov, A. (2023). Enablers and barriers for energy prosumption: Conceptual review and an integrated analysis of business models. *Sustainable Energy Technologies and Assessments, 57*, 103163.

Siksnelyte-Butkiene, I., Streimikiene, D., Lekavicius, V., & Balezentis, T. (2021). Energy poverty indicators: A systematic literature review and comprehensive analysis of integrity. *Sustainable Cities and Society, 67*, 102756.

Siksnelyte-Butkiene, I., Zavadskas, E. K., & Streimikiene, D.; (2020). Multi-criteria decision-making (MCDM) for the assessment of renewable energy technologies in a household: A review. *Energies, 13*(4), 1164.

Siksnelyte, I., Zavadskas, E. K., Streimikiene, D., & Sharma, D. (2018). An overview of multi-criteria decision-making methods in dealing with sustainable energy development issues. *Energies, 11*(10), 2754.

Snow, S., Clerc, C., & Horrocks, N. (2021). Energy audits and eco-feedback: Exploring the barriers and facilitators of agricultural energy efficiency improvements on Australian farms. *Energy Research & Social Science, 80*, 102225.

Srinivasa Rao, C., Kareemulla, K., Krishnan, P., Murthy, G. R. K., Ramesh, P., Ananthan, P. S., & Joshi, P.K. (2019). Agro-ecosystem based sustainability indicators for climate resilient agriculture in India: a conceptual framework. *Ecological Indicators, 105*, 621–633.

Stepanyan, D., Heidecke, C., Osterburg, B., & Gocht, A. (2023). Impacts of national vs European carbon pricing on agriculture. *Environmental Research Letters, 18*(7), 074016.

Streimikiene, D., & Mikalauskiene, A. (2023). Assessment of agricultural sustainability: Case study of Baltic States. *Contemporary Economics, 17*(2), 128–141.

Streimikis, J., Shen, Z. Y., & Balezentis, T. (2023). Does the energy-related greenhouse gas emission abatement cost depend on the optimization direction: Shadow pricing based on the weak disposability technology in the European Union agriculture. *Central European Journal of Operations Research*. https://doi.org/10.1007/s10100-023-00866-0

Streimikis, J., Yu, Z. Q., Zhu, N., & Balezentis, T. (2022). Achievements of the European Union member states toward the development of sustainable agriculture: A contribution to the structural efficiency approach. *Technological Forecasting and Social Change, 178*, 121590.

Talukder, B., & Hipel, K. W. (2018). The PROMETHEE framework for comparing the sustainability of agricultural systems. *Resources, 7*(4), 74.

Talukder, B., Hipel, K. W., & van Loon, G. W. (2018). Using multi-criteria decision analysis for assessing sustainability of agricultural systems. *Sustainable Development, 26*, 781–799.

Talukder, B., Saifuzzaman, M., & Vanloon, G. W., (2016). Sustainability of agricultural systems in the coastal zone of Bangladesh. *Renewable Agriculture and Food Systems, 31*, 148–165.

Technology Executive Committee. (2017). Enhancing financing for the research, development and demonstration of climate technologies. Working Paper, 26 p. Available at: https://unfccc.int/ttclear/docs/TEC_RDD%20finance_FINAL.pdf

Thurman, M. (2022). Climate-smart agriculture certification: A call for federal action. *Columbia Law Review, 122* (3), 37–60.

Tran, D. D., van Halsema, G., Hellegers, P. J. G. J., Ludwig, F., & Seijger, C. (2018). Stakeholders' assessment of dike-protected and flood-based alternatives from a sustainable livelihood perspective in An Giang Province, Mekong Delta, Vietnam. *Agricultural Water Management, 206*, 187–199.

Troiano, S., Novelli, V., Geatti, P., Marangon, F., & Ceccon, L. (2019). Assessment of the sustainability of wild rocket (*Diplotaxis tenuifolia*) production: Application of a multi-criteria method to different farming systems in the province of Udine. *Ecological Indicators, 97*, 301–310.

Tuomisto, H. L, Angileri, V., De Camillis, C., Loudjani, P., Nisini, L., Pelletier, N., & Haastrup, P. (2013). *Final technical report: Certification of low carbon farming practices*, 36 p. Luxembourg: Publications Office of the European Union, ISBN 978-92-79-34849-5.

Tzouramani, I., Mantziaris, S., & Karanikolas, P. (2020). Assessing sustainability performance at the Farm level: Examples from greek agricultural systems. *Sustainability, 12*(7), 2929.

United Nations Food and Agriculture Organization. (2019). How to feed the world in 2050, 35 p. Available at: https://www.fao.org/fileadmin/templates/wsfs/docs/expert_paper/How_to_Feed_the_World_in_2050.pdf

Veisi, H., Liaghati, H., & Alipour, A., (2016). Developing an ethics-based approach to indicators of sustainable agriculture using analytic hierarchy process (AHP). *Ecological Indicators, 60*, 644–654.

Volkov, A., Morkunas, M., Balezentis, T., & Šapolaitė, V. (2020). Economic and environmental performance of the agricultural sectors of the selected EU countries. *Sustainability, 12*, 1210.

Wang, B., Song, J. N., Ren, J. Z., Li, K. X., Duan, H. Y., & Wang, X. E. (2019). Selecting sustainable energy conversion technologies for agricultural residues: A fuzzy AHP-VIKOR based prioritization from life cycle perspective. *Resources Conservation and Recycling, 142*, 78–87.

Wang, J. B. (2022). Drivers of the sustainable development of agro-industrial parks: Evidence from Jiangsu Province, China. *SAGE OPEN, 12*(4), 21582440221144415.

Wang, Z. G., Huang, L. F., Yin, L. S., Wang, Z. X., & Zheng, D. D. (2022). Evaluation of sustainable and analysis of influencing factors for agriculture sector: Evidence from Jiangsu Province, China. *Frontiers in Environmental Science, 10*, 836002.

Waseem, R., Mwalupaso, G. E., Waseem, F., Khan, H., Panhwar, G.M., Shi, Y. (2020). Adoption of Sustainable Agriculture Practices in Banana Farm Production: A Study from the Sindh Region of Pakistan. *International Journal of Environmental Research and Public Health, 17*, 3714.

Wen, Z., Liao, H., Zavadskas, E. K., & Antuchevičienė, J. (2021). Applications of fuzzy multiple criteria decision making methods in civil engineering: A state-of-the-art survey. *Journal of Civil Engineering and Management, 27*(6), 358–371.

Winter, S., & Schlesewsky, L. (2019). The German feed-in tariff revisited – An empirical investigation on its distributional effects. *Energy Policy, 132*, 344–356.

Xia, X. Q., & Ruan, J. H. (2020). Analyzing barriers for developing a sustainable circular economy in agriculture in China using Grey-DEMATEL approach. *Sustainability, 12*(16), 6358.

Xu, X., Sharma, P., Shu, S., Lin, T. S., Ciais, P., Tubiello, F. N., Smith, P., Campbell, N., & Jain, A. K. (2021). Global greenhouse gas emissions from animal-based foods are twice those of plant-based foods. *Nature Food, 2*, 724–732.

Yin, S., Wang, YX., Xu, JF. (2022). Developing a conceptual partner matching framework for Digital Green Innovation of agricultural high-end equipment manufacturing system toward Agriculture 5.0: A novel niche field model combined with fuzzy VIKOR. *Frontiers in Psychology, 13*, 924109.

Yucenur, G. N., Azakli, A. S., Bahadir, K., Tel, M. E., & Arabaci, SN. (2022). Prioritisation of Industry 4.0 implementations in agricultural sector with SWARA/EDAS. *International Journal of Sustainable Agricultural Management and Informatics, 8*(3), 326–344.

Zadeh, L. A. (1963). Optimality and non-scalar-valued performance criteria. *IEEE Transaction on Automatic Control, AC-8*, 59–60.

Zavadskas, E. K., Kaklauskas, A., & Sarka, V. (1994). The new method of multicriteria complex proportional assessment of projects. *Technological and Economic Development of Economy, 1*(3), 131–139.

Zekic, S., Kleut, Z., Matkovski, B., & Dokic, D. (2018). Determining agricultural impact on environment: Evidence for EU-28 and Serbia. *Outlook on Agriculture, 47*(2), 116–124.

Zhou, F., & Wen, C. H. (2023). Research on the level of agricultural green development, regional disparities, and dynamic distribution evolution in China from the perspective of sustainable development. *Agriculture, 13*(7), 1441.

Zhu, N., Streimikis, J., Yu, Z. Q., & Balezentis, T. (2023). Energy-sustainable agriculture in the European Union member states: Overall productivity growth and structural efficiency. *Socio-Economic Planning Sciences, 87*, 101520.

Index

agricultural input 66, 96
agricultural knowledge 33, 65, 197, 269
agricultural landscape 37
agricultural methods 117, 123, 195, 197, 222
agricultural organizations 26, 53, 205
agricultural productivity 16–17, 33, 38, 50–52, 63, 97–98, 115, 118, 124, 130, 175, 180, 197, 200, 224, 227–228, 232, 240
agricultural products 17, 20, 27–28, 76, 114, 121, 124, 126, 131–132, 135, 171, 181, 195, 203, 207, 219, 228, 235, 238–239, 281
agricultural subsidies 266, 309
agricultural sustainability indicators vi, 133, 135–136
agricultural waste x, 43, 160, 164–165, 171–173, 202, 249
agriculture schemes 228
agrochemicals vi, 108, 119, 123–125, 129, 149, 263, 277
agroecological vi, 19, 37–38, 40, 42, 60, 107, 114–115, 139, 221, 257, 260, 304
agroecosystems 19, 108, 114–117, 142, 144, 258, 309
agroforestry establishment 235
agroforestry methods 199
agroforestry practices 122–123, 258
agroforestry systems 36, 123, 188, 225, 254
agrotourism 228
AHP technique 107, 136, 140, 148, 278–281, 284–285, 303, 308, 310
air temperature 160, 162, 176
airflow speed 160
Algeria 282, 307
alignment efforts 132
alternative practices vi, 116
ammonia 134–135, 171, 177–180, 183, 191, 210, 293–296, 299–301, 303
amount 42, 107, 126, 156, 167, 232, 262, 270, 272, 287, 294, 299
analytic 136, 309–310
animal husbandry 98, 114, 117, 174
ash waste 169
Assessment Ratio Analysis 282, 306
atmosphere 3, 42, 44, 208–209
Australia 212, 220, 231, 250

Balezentis, T. iii–v, viii, 77, 141, 309–311
barriers v, 1, 6, 13, 16–17, 21–23, 25–27, 29, 59, 61–62, 65, 69, 74–75, 109–110, 113, 141, 143–144, 157–158, 162, 165, 205, 222, 235, 266, 273, 280–281, 304–305, 307, 309, 311

beverage waste 170, 187
biodiversity conservation 10, 19, 36–37, 39, 42, 116, 122–123, 131, 140, 143, 172, 188, 193–195, 197, 201, 203, 221, 238, 243, 254, 258–259, 277, 291
biodiversity enhancement 38
biodiversity protection 37–39
biodynamic 282
bioenergy ix, 151, 154, 158, 163, 165–167, 171–172, 183, 186–187, 190–192, 200, 209, 229, 260, 268, 281, 304
bioethanol 159, 166, 192
biological activity 121
biopower 155
biotechnology 60, 73, 165, 184–185, 188, 191–192, 304
biotic factors 19
Brazil 147, 166, 169, 190, 223, 231, 250, 256–257, 260
buildings ix, 5, 7–8, 30, 43, 155–156, 160, 201, 209, 307
business activities 99
business development 52, 69
business expansion models 164
business management 61, 73, 174
business model 147, 184, 203, 254
business models i, ix, 128, 168, 173–174, 190, 202–205, 245, 249, 258, 269
business opportunities x, 4, 112, 154, 170, 181, 192

capable 216, 220
capacity building 50–51, 54, 57, 145, 153, 225–227, 235, 292
capital 14, 21, 23–24, 26, 39, 47, 105–106, 111, 129, 134, 157, 189, 202, 251, 255–256, 269
carbon capture ix, 7, 14, 44, 75, 178, 180, 208–209, 247–248, 251, 258
carbon footprint 16, 43, 126–127, 143, 160, 167, 171, 175, 177, 184, 193, 197, 200–201, 208–209, 211, 263, 269
carbon pricing 8–9, 11–12, 26, 28, 44, 56, 62, 64–65, 68, 71–72, 75, 268–269, 301–302, 306, 309
carbon sequestration x, 8, 10, 36–37, 39, 44, 54, 59, 73, 115, 119, 126, 140, 191, 201, 204, 221, 224, 229, 238, 254, 268, 277, 295
carbon tax 269, 303–304, 307

313

Index

case study i, vii, x–2, 64, 70, 74, 146, 150, 171, 181–185, 192, 247, 250–251, 256, 260, 283–284, 292–293, 303–304, 308–309
cashew industry 241, 246
CCS 7–9, 44–45, 60, 180, 208–209, 216, 249, 251
cereal productivity 175
certification vi, 27, 119, 127, 131–133, 202, 204, 206–207, 230, 235, 269, 294, 310
CH_4 ix, 126, 171, 208
challenges of agricultural sustainability vi, 110
changing conditions 197, 226, 229, 234–235
chemical inputs 19, 34, 132, 197, 220–221, 224, 231, 238
chemical pollution 17, 30
Chile 269, 307–308
China viii, 13–14, 43, 62, 68–69, 72, 75–76, 115–116, 145, 148, 150, 165, 167, 183, 185, 192, 205–206, 210, 222, 232, 249–251, 253, 255–257, 259–260, 271, 280–281, 284, 305–307, 310–311
Chinese 13, 69, 75, 165, 167, 206, 222, 232, 249, 256, 280
circular business 173–174, 184, 190, 245, 249
circular economy vi, ix, 7, 14, 30–31, 44, 59, 80, 99, 101, 138, 141, 149, 156, 158, 163–174, 180–192, 194, 201–202, 246, 252, 281, 304, 307, 309
civil society 6–7, 11, 42, 57, 70, 212
clean energy infrastructure 13, 264
clean energy transition 30–31, 262
climate change adaptation 24, 52, 71, 74, 117, 139, 141, 147, 206–207
climate change impacts 15, 19, 23, 40–41, 48, 58–61, 70, 115, 141, 144, 152, 174, 306
climate change mitigation v–vi, ix, 1, 3–11, 13–14, 16–18, 44–45, 54, 58, 70–71, 74, 78, 116, 127, 154, 157, 163, 174–177, 181, 197, 203, 219, 224, 236, 238, 243, 250, 252, 262, 266, 309
climate mitigation 5, 7, 10, 68, 218, 221
climate resilience 9–10, 15, 37, 51–52, 260
CNG 15
CO_2, 7–8, 14, 18, 43–44, 71, 77–78, 85–86, 89–91, 116, 126, 171, 175, 177–180, 190, 208–209, 212, 247, 256, 266, 268–269, 303, 307
cognitive development 100
collaboration 6, 8, 17, 29, 34, 42, 44–48, 52–59, 61–63, 65–66, 68–69, 72–74, 76, 146, 151, 186, 202, 209, 211, 216, 222, 246
collaborative efforts 48, 52, 221
combustion ix, 15, 94, 153–154, 158, 163, 176, 211, 248
commitment 6, 10, 29, 32, 42, 44, 53–57, 102, 124–126, 151, 193, 206, 213–214, 219, 222, 239, 242–245
common barriers 21

communication 7, 46–47, 53, 55, 62, 68, 72–73, 110, 112–113, 142, 228, 242, 271, 288–289, 291–292
communication technology 62, 112
community engagement 129, 239–240
community resilience 37, 145, 225–226
comparison vii, 10, 17, 71, 79, 272, 284–285, 287, 298–299, 305
compatibility 209, 215, 282
competitive ix, 27, 94–95, 100, 131, 165, 172, 202, 210, 212–213, 215, 255–256, 258
competitive advantage 27, 131, 172, 202, 213, 255–256
competitive prices ix
complementary expertise 53
Complex Proportional Assessment 282
comprehensive evaluation 280, 285, 291compressed natural gas 15
concept and agroecological characteristics vi, 115
conclusion 229
conservation efforts 122–123, 204
consumer demand 26–27, 109, 131, 198
continuous improvement 12, 47, 56, 131, 202
conventional energy 158, 209
conventional greenhouse 160
conventional heating systems 160
cooperation v, 1, 4, 6, 10–12, 31–32, 42, 45–48, 52, 54–60, 75–76, 113, 129, 188, 228
cooperative efforts 56
cooperative opportunities 164
cooperatives 66, 140, 164, 166, 169–171, 173, 175, 228
coordination 12, 47, 53, 55, 57, 74
course correction 6, 47
cover cropping 37, 39, 114, 117, 124, 126, 201, 220–221, 224, 229–231, 277
crafts 228
Crete 159–160, 192
criteria vi–viii, 2, 34, 119, 131, 133, 136–137, 140, 144, 146–147, 149, 175, 200, 241–242, 246, 273–281, 283–288, 298, 304–306, 308–311
critical barriers 21
critical materials 169, 215
crop health 39, 230, 232–233
crop production 61, 121, 179, 189, 221, 224–225, 258
crop productivity 108, 114, 159, 247
crop varieties 37–38, 52, 59, 116–117, 119, 123, 202, 222, 224, 226, 229, 234–235, 238
crop yield 121, 131, 159, 260
CSA 112, 114–115, 118, 203, 234
CSR vii, 193, 214, 236–246, 248, 254–255, 260
CSR Commitment 243–245
cultivation costs 17
cultural diversity 16
cultural factors 20

Index

customer 55–56, 109, 168, 173, 213, 242, 260
cybersecurity 215–216
cybersecurity threats 216

damage 4, 6, 15, 39, 98
data analytics 42, 201–203, 219, 232–233, 242
data collection 108, 120, 234, 287
Data Envelopment Analysis (DEA) 137, 283–285, 287–288, 305, 308
data sharing 53, 60decarbonize 178, 209, 295
decisions 11, 17, 23–24, 44, 52, 56, 100, 109, 127–128, 133, 149, 152, 193, 198, 214, 216–217, 219, 231–232, 265, 277, 287, 304
definition vii, 20, 101–104, 119, 164, 207, 220, 223, 236, 257, 289
deforestation 18, 44–45, 140, 196–197, 223, 246
DEMATEL 137, 280–281, 284–286, 304–305, 311
dependency 92, 154, 175, 181, 196
deployment 11, 13, 52, 153, 157–158, 162, 209, 211, 215, 268
detrimental 33, 103
Developing Countries 6, 8, 24, 35, 61, 74, 97, 114–115, 139, 141, 145, 175, 246, 255, 264–265, 281, 305
DFS 16
diesel 15, 153, 170, 176–177
digital infrastructure 216
dimensions of sustainability 291
disaster risk 21, 52, 146
distribution i–311
diversified farming systems 16, 61
domain 137
downtime 216
drip irrigation 36, 177, 200, 224–225, 232, 255

earth elements 215
ecological balance 103, 132, 194
ecological focus areas 34, 38
ecological footprint 45, 194, 219
ecological health 125, 144, 219
economic benefits 5, 14, 17, 72, 109, 163, 198, 232, 250, 255
economic challenges 21, 262
economic considerations 40–41, 131
economic development 69, 100–103, 115, 164, 194–195, 199, 226–228, 256, 270, 311
economic efficiency 133, 201, 232–233
economic growth 7–8, 14, 18, 29–30, 45, 101, 149, 194, 198, 200, 209, 218, 221, 226–228, 252–253, 258
economic impact 14, 17, 60, 250
economic productivity 199
economic resource 15
economic risks 216

economic sustainability 5, 37, 105–106, 119, 238, 240
economic viability vi, 26, 28, 97, 105–106, 119, 127–128, 136, 215–216, 226, 229, 239–240, 273, 277
economics i, viii, 4, 62, 68–74, 140–145, 147–151, 166, 190, 217, 246–248, 250–252, 254–255, 257–259, 304, 306–309
ecosystem functioning 37
ecosystem health 40–41, 226, 229, 259
ecosystem resilience 36, 196–197
ecosystem services 20, 34, 38, 61, 104, 108, 115–116, 120, 122, 130–131, 142, 149, 194, 196, 201, 250–251
ecosystems 4, 9, 34, 37, 39, 45, 75, 77, 96, 102, 119, 122, 124, 132, 141, 144, 151, 194–197, 199, 204, 219, 223, 231, 234, 237, 239–240
effective leadership 53
efficient water management techniques 225
ELECTRE 136, 147, 280, 282, 284–285, 287, 309
electric 5, 15, 64, 138, 157, 162–163, 170, 175–177, 185, 201, 216, 306
electricity grid 153
electrification 7, 9, 13–15, 43, 66, 76, 82, 88, 159, 163, 175–176, 178, 180, 191
emerging opportunities 53, 56
emerging technology uncertainty 215
emerging trends 75, 216, 251
emphasis v, 1, 16, 29, 33, 42, 48, 98, 100, 104, 115, 118–119, 129, 165, 220, 222–223, 242, 250
employment 16, 77, 99, 104, 106, 113, 128, 134–135, 157, 185, 218, 243, 259–260, 306
employment opportunities 128, 157, 218, 243
empower 20, 27, 50, 59, 133, 228, 272
energy consumption v, ix, 1, 17, 43, 81–84, 88, 90, 92, 94–95, 119, 125, 134–135, 162, 172, 175–179, 209–210, 213, 219, 270, 294, 296
energy crops 151, 154, 157–158, 170, 260
energy efficiency vi, 5, 7–8, 14, 18, 30, 43, 45, 54, 82, 95–96, 125–126, 160, 162–163, 171, 173, 176, 178, 194, 209–210, 213, 246, 249–252, 270, 309
energy grids 215
energy infrastructure 8, 13, 180, 263–264
energy input 87, 92, 95, 177
energy policy viii, 70, 184, 250, 260, 264, 274, 305, 308, 311
energy production vi, ix–1, 43, 153, 158, 190–191, 211, 215
energy security 8, 63, 157, 175, 211, 217–218
energy services 16, 153
energy storage 8, 30, 159, 166, 215–217, 219, 260, 268, 304

energy systems 16, 23, 63, 67, 153, 159, 215–216, 219, 248, 253–254, 260, 265, 304
energy technologies 61, 75, 125, 153–154, 156, 161, 174, 178, 185, 212, 215–217, 219, 252–253, 256, 261–263, 265–267, 269, 271–272, 306, 309
energy technology 252
energy use v, 25, 77–87, 89–91, 93–95, 119, 125–126, 135, 154, 175–176, 180, 215, 221, 270
energy waste 125, 219
enhancement 15, 38, 98, 101, 115, 162–163, 167, 222
enterprise 24, 100, 203, 258
entrepreneurship 137, 202, 205, 227–228, 256
environmental benefits 15, 27, 66–67, 167, 207, 221
environmental challenges 11, 30, 205, 223
environmental criteria 175, 279
environmental degradation 9, 16, 44, 98, 114, 193–195, 211, 219
environmental footprint i, ix, 51, 141
environmental impacts ix, 20, 35–36, 96, 98, 104, 107, 126, 130, 136, 138, 164, 172, 176, 183, 199, 201, 203–204, 217, 222–223, 234, 243, 254, 277
environmental improvement 15, 109
environmental protection 4, 17–18, 45, 101, 121, 141, 187, 195, 200, 222, 231, 243
environmental quality 17, 19, 40, 68, 97, 101, 115, 117, 219, 222, 304
environmental reasons 4
environmental stewardship 44–45, 127, 132, 198, 210, 214, 228, 233, 237–238, 240, 243
environmental sustainability 18, 29, 96, 128, 131, 140–141, 147, 161, 183, 186, 194, 198, 205–206, 252, 255–256
environmentally conscious 29, 44, 52, 116, 198, 201, 213, 220–221, 245–246
environmentally favorable strategy 160
environmentally friendly 31, 39, 42, 44, 51, 56, 97, 128, 150, 168, 198, 200, 202–203, 205, 212, 219–221, 227, 229, 231, 234, 239, 270, 294
environmentally friendly business 202
environmentally friendly industry 42
environmentally friendly practices 31, 51, 128, 219, 229
equity 12, 15–16, 20, 37, 62, 104–106, 110, 129–130, 132, 134, 146, 166, 195, 226, 243, 247, 282
ethanol 154, 161, 170, 176
ethics 16, 141, 187–188, 249, 253, 279, 310
EU i, v, vii, ix–2, 11, 13, 29–33, 64–65, 67–68, 70–71, 73–74, 77–87, 89–96, 109, 133, 138, 142, 146, 148, 151, 163, 168, 174, 183, 186, 192, 220–221, 249–250, 254–256, 258, 266, 280–283, 292–297, 299, 301–303, 305, 309–311

EU farmers 295
Europe 16, 29–32, 63, 73, 75, 94, 116, 140–142, 145–146, 150, 167, 181, 191, 243, 251, 254, 257–258, 283, 308
European Commission 29–31, 133, 135, 142, 200, 221, 249, 295, 305
European Union i, iii, ix–2, 29–30, 38, 63–64, 66, 78, 96, 99, 144, 167–168, 221–222, 243, 249, 306, 308, 310–311
EV infrastructure 14, 30
evaluation method 136
evaluation of energy efficiency 125
expertise 24, 46–48, 50–51, 53, 58–59, 62, 117, 150, 236, 285, 287, 292
export opportunities 214

fair trade 119, 131, 199, 204, 243
FAO 115, 184, 224, 250, 262, 305, 310
farm industry 101
farming inputs 127
farming system 107, 146, 191
farming systems 16, 20–21, 29, 37, 52, 61, 63, 118–119, 122, 126, 139, 146, 202, 204, 220, 229, 234, 249, 255, 282, 310
farmland 24, 60, 206, 221
feasibility 160–161, 166, 215, 247, 271, 309
feedback 47, 53–54, 56–57, 69, 309
fertilizer application 37, 112, 120, 206fertilizer application behavior 206
fertilizer consumption 135
fertilizer industry 178
fertilizer management 124
fertilizer production 170, 177–180
fertilizer supply chain 180
fertilizer use 124, 126, 224, 237
fiber productivity 103
financial barriers 235
financial challenges 216
financial constraints 23, 29
financial viability 105, 238
finite resources 19, 219
fiscal 14
flexibility 12, 53, 55–57, 233, 287
floor space 16
Food and Agriculture Organization 115, 250, 252, 261, 310
food chain x, 153, 222
food processing 98, 153, 166, 174, 228
Food Procurement Program 223
food sector ix–x, 2, 167, 248, 251, 259, 294
food security and nutrition 20, 51–52, 129, 305
food security and resilience 224–226
food security concerns 223, 229
food security level 100
food supply chain 33–34, 75, 141, 143, 167, 277, 281, 307
food supply instability 4
food supply system 175

Index

food waste x, 33–34, 155, 165–168, 170, 183–184, 186, 190, 201
footprint of carbon vi, 119, 126
foreign policy 29, 31–32
forest resources 268
forestry ix, 5, 78, 135, 174, 194, 223, 248–249, 254, 260, 268, 293, 296
fossil energy 82, 92
fossil fuel consumers 180
fossil fuel extraction 45
fossil fuel subsidies 9, 14, 68, 70
fossil resource 169
foster 8, 28–29, 41, 44, 46, 48, 54, 56, 58–59, 113, 122–123, 173, 193–194, 202, 209, 211, 213, 221–222, 226, 246, 292
France 69, 83–89, 91, 93, 95, 220, 282, 300–302
fully renewable energy 209
future generations 45, 60, 99, 102, 104, 195, 198, 200, 217, 291

GAHE 160
GAI 193, 219, 229
gasoline 160–161, 170, 176
genetic diversity 38, 123, 225–226, 234
genetic resources 38, 122–123, 226, 234
geopolitical tensions 45, 219
geothermal energy 75, 155, 208
GHG emission factor 179
GHG emission growth 179
GHG emission intensity 178
GHG emission lifecycle 179
GHG emission reduction 2, 115, 138, 157, 160, 176, 178, 181
GHG emissions ix–2, 4–8, 11, 16–18, 26, 29–30, 32, 37, 44, 46, 58, 77–81, 96, 108, 114–116, 118, 124–127, 136, 157, 163, 166, 168, 170–172, 175–181, 257, 262, 265–266, 268–270, 277, 283, 293, 295, 301, 303
GHG reduction 75, 171, 265–266
GHGs 4, 10, 16, 44, 114, 126
global efforts 31, 208, 211, 246
global food security 98, 117, 138, 140, 233–234
global priority 16, 100
global trend 15
global trends 19, 43
global warming 3–4, 44, 74, 113–116, 140, 156, 180, 208, 211, 218, 265
GlobalGAP 132
goals and objectives 53, 55, 57
Good Agricultural Practices 124, 132
government 12, 20–21, 23–26, 28, 41, 48, 50, 52, 100–101, 134, 137, 146, 167, 170, 202, 205, 217, 221, 231–232, 247, 249, 252, 263, 266–270, 281, 293–294, 297, 300–301, 303
government agencies 26, 50, 202, 231

government expenditure 134
government policy guidance 217
government support 21, 24, 137, 269, 294, 300–301
governmental funds 135
Green Agricultural Initiatives vii, 219, 223, 227
Green Deal v, ix, 1, 29–35, 37–40, 42, 59, 61, 63–68, 71, 73–75, 78, 109, 248–249, 253, 295–296
green entrepreneurship 202, 205, 256
green investment 9, 13, 72, 250
Green Revolution 111, 114, 189, 222
greenhouse cultivation 159
greenhouse gas emissions 65, 68, 96, 112, 119, 134–135, 151, 167, 171–172, 174, 177, 179–180, 184, 192–194, 196–197, 199, 201, 206–208, 211, 218–219, 223–224, 238, 243, 250, 306, 311
greenhouse heating 159, 186
greenhouse thermal model 160
groundwater 17, 103, 108, 121, 134, 149, 176, 186, 282, 304

harmful pesticides 38
harmonization 56–58, 64, 132
heat capacity 160
heating 82, 125, 153–157, 159–160, 162, 175, 180, 186, 201, 208, 210, 261
higher productivity 228
Himalayas 115
holistic approach 20, 193, 217, 230, 234, 259
human activities 42, 45, 106, 113, 208
human health 3–4, 16, 33, 39–40, 45, 67, 74, 107, 114–115, 119, 121, 124, 218, 228, 277
hydropower 155, 184, 208

implementation cost 271
implementation of agricultural strategies 266
implementation plan 30–31
implementation tools 30
implemented strategies 266
implementing policies 15, 211
implementing supportive policies 29, 59
implementing technologies 36, 194
improvement opportunities 125
improvement policies 101
inclusivity 20, 57, 194–195, 198–199, 243
India 71, 115–116, 147, 162, 222–223, 231, 250, 255–256, 279, 309
indigenous 20, 33, 60, 114–118, 123, 139, 148, 197
indigenous technology 115
individualized livestock care 113
industrial agriculture 116
industrial development 100
industrial ecology 163–166, 182, 184–185, 188–189, 192
industrial facilities 44, 209

industrial policies 30
industrial revolution 112
industrial symbiosis 163–164, 166, 182–183, 185, 189–192
industrialization 114, 217
industrialized countries 113
Industry 4.0 281, 306–307, 311
industry players 26
infrastructure and supply chain 216
infrastructure development 211, 227, 236, 240, 250
infrastructure limitations 24
innovative technologies 43, 181, 203
input-output analysis 169institution 167
institutional barriers 13, 109
Instrument 65, 137, 268, 274, 277, 305
insufficiency 99
integrated assessment 282
integrated pest management 17, 34, 36, 65, 72, 119, 124, 200, 230, 255, 277
intellectual property 53
intensification practices 113
intensive monocultures 19
intercropping 37, 114, 117, 130, 225
interest rates 14, 166, 170, 207
intermittency 15, 215–216
intermittency and reliability 216
intermittent energy supply 264, 303
international climate policy 10
international cooperation 4, 6, 10, 12, 31, 45, 57
introduction v, ix, 1, 19, 62, 77, 80, 99, 101, 114, 140, 151, 179, 258, 306
investing x, 14, 31, 41, 207, 212–213, 215, 219, 235, 246
investment costs 128, 156, 180
investment decisions 23, 128, 214, 216–217
investor 214
Iran 137, 205, 279
irrigation water 177
Italy 84–86, 88–89, 91, 93, 95, 146, 166–167, 183–184, 188, 220, 257, 280, 284, 300–302

job creation 5, 32, 45, 157–158, 209, 218, 238
job opportunities 20, 219

Knowledge Sharing 48, 54, 57, 59, 207, 225–226

labour 252, 260
land productivity 159, 179
land's capacity 226
landscape 11, 37–38, 108, 122–123, 132, 191, 193, 211–212, 282, 295
landscape diversity 108
Latin America 19, 60, 116, 141, 147
leadership 45–46, 53, 57, 74, 202, 213, 218
learning curve 28, 216

legal framework 113, 308
livelihood 20, 139, 266, 279, 310
livestock buildings 155
livestock density 135
livestock emissions 18
livestock farmers 164
livestock husbandry 179
livestock integration 130, 146
livestock output 87
livestock production 18, 114, 116, 121, 179, 256, 277
livestock propagation 98
livestock rearing 228
livestock sector 201
livestock strains 123
local farmers 33, 234, 280
low-carbon practices 27
low-carbon products 27

Macedonia 282
macroeconomic performance 14
main commodities 113
maintenance staff 216
major challenges 103, 264
manufacturing ix, 32, 44, 63, 163, 174, 210, 215, 218, 247, 281, 311
market access 26–27, 51, 56, 100, 119, 127, 131, 198, 206–207, 214, 235, 238
market entry 27, 222
market fluctuations 56–57, 128, 197, 227
market opportunities 27, 59, 201
market uncertainties 216
marketing infrastructure 50
mathematical model 75
MAVT 282
MCDM vi–vii, 2, 133, 136–138, 142, 152, 273–278, 280, 283–287, 292, 298, 306–307, 309
microbial biomass 120–121
mitigation actions ix, 174
mitigation benefits 15
mitigation efforts 5, 14, 17, 127, 209
mitigation goals 5, 17, 175
Mitigation in Traditional Agriculture vi, 116
mitigation instruments 265
mitigation measures 5, 16, 96
mitigation policies 68, 174, 266
mitigation strategies 5, 54, 66, 117, 126–127, 142, 146, 187, 212, 214, 266
mitigation targets 17, 174
mitigation technologies 266
mobility ix, 30–31
Multiple Attribute Value Theory 282
multipurpose productivity 283
municipal waste 181

N_2O ix, 126, 171, 191, 208
natural environment 3–4, 99, 138, 194

Index

natural fertilizer 168
natural resource 103, 227, 246, 248
natural resources 20, 35, 72, 97–99, 113, 116, 141, 144–145, 149, 164, 168, 194, 198–199, 219, 222, 226, 228, 240, 246, 251, 295
negative consequences 195–197
negative effect 266
negative emission technologies 42
negative environmental impacts 35, 98, 130, 138, 172, 199, 222–223, 234
negative growth 83, 86
negative health factors 5
negative ideal solution 285
negative impact 3–4, 177, 237, 261–262, 270, 273
negative indicator 296
negative outcomes 28
net metering 266, 307, 309
New Zealand 69, 220, 268, 308
nitrogen 24, 70–71, 114, 116, 120, 135, 167–169, 171, 177–180, 183–185, 190, 192, 218
nitrous oxide ix, 44, 114, 119, 126, 177, 196–197, 208
NMSA 222
Nordic 23, 67, 83
North America 16, 64, 109, 258
nuclear 8, 153, 175, 208, 211, 216, 248, 256
nutrient cycling 37–38, 116, 120–122, 124, 231, 258

objectives vi, x, 11, 30, 32–33, 53, 55, 57, 69, 78, 80, 94, 98, 102, 105–108, 115, 118, 120, 136, 138, 171, 173, 195, 198, 221–222, 240, 249, 253, 274, 284, 291, 294
obsolete 215
oil 14, 44, 81–82, 89–91, 95, 132, 135, 150, 153–155, 187, 293, 295, 299
optimize 19, 36–37, 41–42, 121–122, 126, 200–204, 221, 225, 227, 232
organic adopters 109
organic agriculture 19, 75, 96, 98, 101, 111, 146, 148–150, 152, 176, 225, 229, 247, 251, 254, 256–257, 259
organic certification 119, 132, 235
organic commodity 110
organic cultivation 111, 222
organic decomposition 114, 117
organic detritus 154
organic feed 234
organic food 17, 19, 70, 109, 151
organic land usage 221
organic material 170
organic matter 36, 108, 119–121, 124, 130, 167, 171, 231
organic pioneers 110
organic producers 111
organic production 109

organic products 17, 227, 279
organic systems 279
organic wastes 35, 165, 171, 183
outperform 162, 205
output 87–88, 90, 105–106, 169, 189, 191, 222, 266, 283, 287–288
overview vii, 62, 138, 150, 249, 256, 294–295, 305, 309

Pakistan 24, 60, 66, 205, 280, 311
palm oil 132
pathway 181, 186, 195, 205, 223, 246
payback period 155, 160, 170, 262
penetration of renewables 16, 87–88, 138
performance evaluation 55–56, 62, 76, 255
performance reliability 215
pesticide application 96, 232
pesticide inputs 176
pesticide toxicity 124
pesticide usage 17, 180
pesticide use 65, 98, 138, 180, 235
phasing out 9, 14, 43, 70
phosphorus 120, 135, 168–169, 171, 177–178, 180, 183, 185, 187, 189, 191
photothermal 160
policy alignment 56–58
policy and regulatory 25–26, 28, 217
policy barriers 25
policy challenges 29
policy changes 8, 56
policy coherence 57, 62, 68, 70
policy coordination 57
policy development 6, 202, 231
policy dialogue 51–52
policy directions 28
policy domains 29–30, 70
policy environment 28, 57–58, 113, 143
policy evaluation 25, 57policy evaluators 25
policy execution 220
policy failures 21
policy formulation 222
policy framework 11, 29, 63, 148
policy implementation 57, 133
policy incentives 205, 260
policy interventions 21
policy measures 21, 29, 59, 78, 183, 207, 254, 265, 297, 300–301
policy objectives 11
policy package 29
policy reform 106
policy reforms 21
policy scenarios 149, 268
policy space 30
policy support 6, 8, 59, 74, 158, 211
policy tools 206, 266
policy uncertainty 110

policymakers i, 11, 19, 26, 29, 40, 48, 54, 57–58, 103, 121–123, 127–128, 198, 205, 207, 264, 292, 298
policymaking 12, 74
political support v, 1, 3, 6–9, 11, 15, 58
pollutants 121, 158, 181, 218–219, 258
population 16, 20, 77, 97–98, 101–102, 118, 134, 138, 178, 198, 219, 221, 224, 234, 261–262, 294
potential ventures 175
poverty level 265
poverty measurement 288
poverty reduction 128, 198
power generation 14, 43, 153, 155–156, 158, 175–176, 209, 218, 256
power grid 8, 15, 154, 166
power plant 15
PPP 48, 50
practical obstacles 177
precision agriculture 19, 23, 34–35, 42, 52, 59, 62, 70, 73, 96, 112, 125, 185, 189, 197, 200–203, 221, 232–234, 238, 251, 253
precision technology 36
preservation vi, 37–38, 41, 122–123, 132, 154, 163, 197, 199, 221, 224, 229–230, 243
preservation of biodiversity vi, 41, 122, 224
preventive measures 231
primary objective 115
processed foods 228
producer reputation 131
production processes 94, 127, 178, 238
productivity analysis viii
productivity decline 117
productivity indicator 105–106
productivity measures 105–106, 121
profitability 19–20, 23–24, 28, 39, 56, 105–106, 109, 112, 119, 127, 136, 194, 198, 205, 212, 227–228, 232, 237–238, 240, 262, 272, 305, 307
promotion 38–39, 50, 138, 222, 230, 243, 264
public awareness 12, 67, 229, 236, 264, 273
public policy 7, 48, 73–74, 254, 305
public relations 214

quality of life 97, 101–103, 129, 199, 210, 226, 273
quantitative framework 94
quotas 221

radioactive waste 208
rapid changes 216
realization 186, 221
recognition 9–10, 15, 116
reduce emissions 13–14, 16, 44, 48, 65, 119, 126, 210–211
reduced dependence on imports 225–226
reduced input costs 226–227
reduced input waste 232–233
reducing carbon emissions 43, 208, 217
reducing demand 7
reducing emissions 7, 11–13, 59, 61, 118, 178, 303
reducing energy costs 5
reducing environmental impact 19, 108
reducing food waste 33
reducing GHG emissions 115, 126, 257, 265, 295
reducing greenhouse 65, 172, 180, 199, 206, 211
reducing pesticides 33
reducing waste 20, 44, 54, 163–164, 182, 198, 232
reduction 2, 7, 10–11, 14–15, 18, 25, 30–31, 34, 39, 42, 44, 46, 52, 56, 66, 75, 78, 81, 95–96, 115, 118, 123–125, 127–128, 138, 154, 157, 160, 162–163, 167, 171, 176, 178, 181, 189, 192, 198, 208, 212, 219, 261, 265–266, 268–270, 282, 302–303
reference 105–106
regional disparities 223, 253, 311
regular evaluation 54, 57
regulatory changes 28, 214
regulatory compliance 132, 213, 243
regulatory framework 272
renewable energy policies 9, 13–14, 61
renewable energy production vi, ix–1, 43, 153
renewable energy sources vi, 4–5, 7–8, 13–14, 16, 18, 43, 45, 71, 82, 88, 92, 119, 125, 153, 158, 193–194, 197, 200, 207–209, 215–219, 229, 236, 264, 266
renewable energy subsidies 266
renewable energy usage i, vi, 157, 170renewables in agriculture vi, 153, 155, 157, 159, 161–163, 165, 167, 169, 171, 173, 175, 177, 179, 181, 183, 185, 187, 189, 191
replace 160, 176
replacing ix, 157–158, 210, 218
researchers i, 6, 8, 19, 29, 33, 57, 59, 66, 98, 103, 121, 123, 131, 140, 163–164, 220, 292
resilience building 15
resilience to climate change impacts 15, 19
resilient 15, 20–22, 29, 38–39, 42, 44, 52, 58–60, 109, 116, 119, 123, 131, 139, 195, 197–198, 200, 202, 220, 223–226, 229–230, 234, 236, 238, 243, 245–247, 253, 279, 309
resilient agricultural 22, 197, 220, 234
resistance issues 113
resource allocation 53, 112, 137
resource application 232
resource commitment 53
resource conflicts 122
resource conservation 56, 67, 103
resource consumption 30, 194
resource depletion 194, 198, 219, 223, 229
resource efficiency 42, 119, 128, 164, 189, 195, 197–200, 203, 220, 222

Index

resource endowment 280
resource enhancement 15
resource extraction 44
resource inputs 112
resource management 19–20, 62, 103, 126, 129, 197, 224, 227, 238, 277
resource optimization 213
resource scarcity 195, 214
resource sharing 46–47
resource supply 170
resource usage 280
resource use 36, 42, 102, 108, 198, 200–202, 204, 227, 232, 234, 277, 305
resource utilization 136, 221–222, 225
responsibilities 46–47, 53, 55, 100, 237, 240, 245
return on investment 216
rice farmers 138, 279, 306–307
risk mitigation 214
risk reduction 15, 52
river 72
road infrastructure 153
robust 12, 14, 34, 132, 141, 165, 216–217, 226, 285
rural communities 18, 20, 22, 51, 59, 69, 103–104, 128–129, 153, 157, 166, 198, 219, 223, 226, 228, 234, 238, 243, 245, 294
rural development viii, 58, 144, 174, 189, 198, 200, 218, 238, 243, 249, 252, 259, 305
rural revitalization 222

satellite data 232
SAW 78, 82, 85, 92–93, 99, 280, 282
scalable 216
scientific advances 132
scientific literature 101, 165, 173, 181, 268, 272–274, 278, 284, 288, 301
seasonal trends 270
sensitivity analysis 137, 284, 286, 288
service providers 264
shared resources 50–51
shared vision 46–48, 53, 57
Siksnelyte-Butkiene, I. iii–iv, vii–viii, 261, 273–274, 288, 309
Simple Additive Weighting 282
skills 24, 109, 113, 217, 226, 228, 243–244, 266, 269, 292
slow down 14
smooth transition 21
social and community 217
social barriers 6
social benefits 16–17, 146, 217, 244, 280
social capital 21, 24, 129, 269
social disparities 129
social equity 12, 20, 37, 110, 130, 132, 166, 226, 243
social indicator 106
social prosperity vi, 119, 128
soil degradation 158, 195–196, 222–224, 228, 247
soil fertility 36, 38–39, 108, 120–124, 188, 195, 222, 224–227, 229, 231, 240, 277
soil health 19, 34, 36, 39, 73, 119–121, 123–124, 130–131, 136, 140, 171, 182, 199–201, 203–204, 220–222, 224, 229–231, 291
soil nutrients 225
soil quality vi, 19, 103, 106, 108, 120, 142, 152, 155, 182, 246–247
soil temperature 160, 184
solar greenhouses 156, 159
solar power 153, 156
Spain 84–91, 93, 95, 161, 167–168, 184–185, 254, 278–279, 283, 300–302, 308–309
steelmaking 7
stochastic trends 83
Streimikiene, D. iii–viii, 3, 97, 133, 135–136, 141, 150, 153, 193, 283, 309
subsidies 9, 13–14, 25–27, 44, 66, 68, 70, 101, 137, 158, 174, 205–207, 217, 221, 235, 251, 259, 266, 269, 307, 309
substitution 16, 32, 105–106, 149
successful transition process 217
succession 104, 108
sugar cane 170
summarized indicators framework 133
supply capacity 222
supply chain actors 238
supply chain criteria 241, 246 supply chain disruptions 216
supply chain efficacy 180
supply chain infrastructure 27
supply chain initiatives 242
supply chain management 16, 67, 73, 149, 199–200, 240, 258, 277
supply chain responsibility 240, 242, 245
supply chain risks 216
supply chain standards 137
supply chain vulnerabilities 215
susceptible 19, 103, 106, 114
sustainability approach 290
sustainability assessment vi–viii, x–1, 133, 136, 138–139, 141–142, 147, 166, 183, 273–274, 279, 282–283, 288, 291–292, 303–304, 306
sustainability challenges 34, 255
sustainability criteria 34, 136, 200, 306
sustainability dimensions 37, 291
sustainability efforts 10
sustainability goals 21, 198, 221, 243, 277
sustainability investments 13
sustainability issues 278, 289, 291
sustainability matters 16
sustainability measurement 274
sustainability objectives 120
sustainability of agriculture x–1, 141, 144, 303
sustainability performance 137, 279, 310
sustainability principles 109, 195, 288, 304

Index

sustainable agriculture concepts x
sustainable agriculture decision making vii, 273, 276–278, 280, 284, 291
sustainable agriculture development v–vi, x–2, 98, 100–101, 108, 133, 138, 303
sustainable agriculture dimensions v, 101
sustainable agriculture goals 291
sustainable agriculture indicators 133
sustainable agriculture issues 277, 282, 284
sustainable agriculture practices 31, 35, 37, 67, 260, 271, 292, 311
sustainable agriculture research 291
sustainable agriculture studies 280, 287, 291–292
sustainable agriculture supply chains 242
sustainable agriculture theory 110
sustainable certifications 131
sustainable farming methods 25, 27–28, 42
sustainable farming practices 20, 25, 27–28, 34–35, 42, 48, 52, 56, 58, 115, 141, 199–200, 203–204, 220–224, 226, 228, 238, 240, 263, 270–271, 280, 293–294, 305
sustainable farming systems 122
sustainable food production 34, 39–40, 68, 114–115, 168, 197–198, 203, 221
sustainable livestock management 36, 197, 201, 234
sustainable resource management 19–20, 62, 126, 197
SWARA 282–283, 306, 311
synthetic chemicals 200, 203, 230
synthetic inputs 38, 227
synthetic nitrogen 116, 179–180
synthetic pesticides 123, 132, 180, 224, 228–230

tax 25, 206, 221, 235, 257, 259, 268–269, 303–304, 307
tax incentives 206, 235, 259
tax policy 206
technological advancement 99
technological advancements 8, 44–45, 51, 57, 195, 215
technological risks 216
technology adoption 75, 145, 271
technology development 5–6, 66, 118
technology risks 212, 214–216
technology transfer 73, 207
terrestrial carbon sequestration 268
test 282, 291
TFP 105–106
tidal energy 161, 187
TOPSIS 136, 138, 146, 280–281, 284–286, 298, 307
total factor productivity 105
toxic pesticides 124
trade certification 132, 204
traditional agrarian knowledge 114

traditional agriculture vi, 37, 115–117, 139, 150, 195–197, 261
traditional agroecosystems 115–116
traditional business 118
traditional crops 117, 228
traditional diffusion frameworks 110
traditional energy sources 14
traditional expertise 117
traditional farmers 116
traditional farming 294
traditional heating systems 160
traditional knowledge 115, 117, 196
traditional parametric methods 287
traditional practices 114
traditional producers 110
traditional sustainability approach 290
transformation 8, 29, 43, 78, 87, 118, 139, 144, 146–147, 152, 223, 246, 308
Transitioning Workforce 216
transparency 5–6, 12, 34, 47, 54, 56, 69, 72, 113, 127, 132, 238–239, 243, 246, 289, 291–292
transparent 11, 34, 53, 55–56, 72, 113, 131, 160, 192, 291–292
transportation ix, 5, 7, 9, 13–14, 18, 24, 27, 30, 43, 65–66, 119, 125, 127, 153, 155, 163, 171–172, 177–179, 203, 210–211, 228, 306
travel 16
trust and communication 53

UK iv, 6, 16, 60, 63, 69, 151, 222, 253
undermine 14, 57, 73–74
United States 75, 140, 144, 151–152, 187, 220–221, 250, 254–255, 265, 268, 270, 273
united territorial communities 220–221
upfront costs 23, 216
Uzbekistan 50, 64, 139

value addition 173, 228
value chains 51, 148, 188
variability 42, 60, 69, 103, 105, 197, 215, 233, 264
variable rate application 232–233
VIKOR 137, 146, 280–281, 284–286, 310–311
vulnerabilities 130, 197, 215
vulnerability 8–9, 15, 24, 45, 67–68, 117, 130–131, 139, 195, 228, 238, 304

waste assessment 8
waste collection 173
waste disposal 208
waste generation 164, 169, 174, 194, 201, 211
waste management 18, 44, 132, 146, 164, 186, 190, 192, 201, 234
waste products 176
waste recycling 166, 168–171

Index

waste reduction 30, 34, 192, 265
waste valorization 43
wasteful expenditures 127
wastewater x, 134, 167, 188
water conservation vi, 36, 41, 70, 119, 121–122, 143, 231, 235, 254, 277
water footprint 121
water management 37, 40–41, 62–64, 69, 72, 74, 119, 121–122, 130–131, 177, 200, 221–222, 225, 228, 232, 257–259, 291, 310
water productivity 121, 159, 181, 260
water resources 19, 40–41, 60, 62–63, 72, 121–122, 137, 143, 152, 180, 196, 199, 230–233, 237–238
water source 119, 155
water stress 117, 121–122, 158, 176
whole process vii, 273, 288–289
willingness 17, 23, 28, 143, 216, 265, 272, 303
workforce 52, 103, 216–217
workshop 64, 210, 249
worldwide 13–14, 97, 101, 112, 118, 189, 246, 262

Taylor & Francis eBooks

www.taylorfrancis.com

A single destination for eBooks from Taylor & Francis with increased functionality and an improved user experience to meet the needs of our customers.

90,000+ eBooks of award-winning academic content in Humanities, Social Science, Science, Technology, Engineering, and Medical written by a global network of editors and authors.

TAYLOR & FRANCIS EBOOKS OFFERS:

- A streamlined experience for our library customers
- A single point of discovery for all of our eBook content
- Improved search and discovery of content at both book and chapter level

REQUEST A FREE TRIAL
support@taylorfrancis.com